Gene Regulation by Steroid Hormones III

Gene Regulation by Steroid Hormones III

Edited by
A.K. Roy and J.H. Clark

With 138 Figures

Springer-Verlag
New York Berlin Heidelberg
London Paris Tokyo

ARUN K. ROY
Department of Biological Sciences, Oakland University, Rochester, Michigan 48063, USA.

JAMES H. CLARK
Department of Cell Biology, Baylor College of Medicine, Houston, Texas 77030, USA.

Cover Picture: Electronmicrograph of a hybrid molecule between ovalbumin gene and ovalbumin mRNA. The intervening sequences are shown as loops. Courtesy of Dr. Eugene Lai, Baylor College of Medicine.

Library of Congress Cataloging in Publication Data
Gene regulation by steroid hormones III.
Proceedings of the 3rd Meadow Brook Conference on Hormones, held at Meadow Brook Mansion, Oakland University, Rochester, Mich., Sept. 15–17, 1985.
Includes bibliographies and index.
1. Steroid hormones—Physiological effect— Congresses. 2. Steroid hormones—Receptors—Congresses. 3. Gene expression. 4. Genetic regulation. I. Roy, A.K. (Arun K.) II. Clark, James H. (James Henry), 1932– . III. Meadow Brook Conference on Hormones (3rd : 1985 : Oakland University) [DNLM: 1. Gene Expression Regulation—drug effects—congresses. 2. Hormones—pharmacodynamics—congresses. 3. Steroids—pharmacodynamics—congresses.
WK 150 G3261 1985]
QP572.S7G462 1987 574.87′328 86-17807

Typeset by David Seham Associates, Metuchen, New Jersey.
Printed and bound by Arcata Graphics/Halliday, West Hanover, Massachusetts.
Printed in the United States of America.

9 8 7 6 5 4 3 2 1

ISBN 0-387-96436-3 Springer-Verlag New York Berlin Heidelberg
ISBN 3-540-96436-3 Springer-Verlag Berlin Heidelberg New York

Preface

The field of steroid hormone action has continued to expand into the realm of molecular biology at a pace even faster than most of us ever imagined. The techniques of molecular biology have made it possible to clone hormone-regulated genes and to examine steroid-receptor interactions with these genes. Nucleotide sequences of these genes, which show preferential binding of steroid receptors, have been identified. These results are complemented by the identification of chromatin acceptor proteins, which also show preferential binding for steroid-receptor complexes. Thus, one can envision the day when cloned genes, purified acceptor proteins, and receptor-steroid complexes will be recombined in vitro to form a functional unit. Cellular localization of steroid receptors has undergone recent revision, and it now appears that receptors are localized primarily in the nuclear compartment. These findings, although controversial, will lead to a reassessment of many of the previous concepts of steroid-receptor interactions and regulation. The way in which these observations at the molecular and cellular levels fit into the overall scheme of physiology, development, and evolution are continuing to progress, and the future promises some very exciting syntheses of understanding at all levels of biological organization. The third Meadow Brook Conference on hormones was held in order to bring together investigators who will undoubtedly contribute heavily to this future synthesis and to permit a free exchange of ideas and concepts as they relate to the current state-of-the-art in molecular endocrinology.

Spring 1986

ARUN K. ROY
JAMES H. CLARK

Contents

Contributors

C.W. BARDIN, The Population Council, The Rockefeller University, New York, NY 10021, USA

K. BHATTACHARYYA, McArdle Laboratory for Cancer Research, University of Wisconsin, Madison, WI 53706, USA

A.B. CHAPMAN, Department of Pharmacology, Stanford University School of Medicine, Stanford, CA 94305, USA

J.H. CLARK, Department of Cell Biology, Baylor College of Medicine, Houston, TX 77030, USA

E.M. CORMIER, Departments of Human Oncology and Biochemistry, University of Wisconsin, Madison, WI 53706, USA

M. DANIELSEN, Department of Pharmacology, Stanford University School of Medicine, Stanford, CA 94305, USA

W.F. DEMYAN, Department of Biological Sciences, Oakland University, Rochester, MI 48063, USA

C.A. DONALDSON, Department of Biochemistry, The University of Arizona, College of Medicine, Tucson, AZ 85724, USA

M.D. FRANCIS, Department of Medicine and Pathology, University of Colorado Health Science Center, Denver, CO 80262, USA

H. FREAKE, Division of Endocrinology and Metabolism, Department of Medicine, University of Minnesota, Minneapolis, MN 55455, USA

A. GOLDBERGER, Department of Cell Biology, Mayo Medical School, Rochester, MN 55905, USA

J. GORSKI, Department of Biochemistry, University of Wisconsin, Madison, WI 53706, USA

G.L. GREENE, The Ben May Laboratory, University of Chicago, Chicago, IL 60637, USA

G.L. GUNSALUS, The Population Council, The Rockefeller University, New York, NY 10021, USA

J.A. GUSTAFSSON, Department of Medical Nutrition, Huddinge University Hospital, F69, S-141 86 Huddinge, Sweden

G.L. HAMMOND, Department of Obstetrics and Gynecology, University of Western Ontario, London, Ontario, N6A 4G5, Canada

M.R. HAUSSLER, Department of Biochemistry, The University of Arizona, College of Medicine, Tucson, AZ 85724, USA

N.J. HICKOK, The Population Council, The Rockefeller University, New York, NY 10021, USA

J. HORA, Department of Cell Biology, Mayo Medical School, Rochester, MN 55905, USA

M. HORTON, Department of Cell Biology, Mayo Medical School, Rochester, MN, 55905, USA

K.B. HORWITZ, Department of Medicine and Pathology, University of Colorado Health Science Center, Denver, CO 80262, USA

O.A. JÄNNE, The Population Council, The Rockefeller University, New York, NY 10021, USA

V.C. JORDAN, Department of Biochemistry, University of Wisconsin, Madison, WI 53706, USA

W.B. KINLAW, Division of Endocrinology and Metabolism, Department of Medicine, University of Minnesota, Minneapolis, MN 55455, USA

E.S. KLEIN, Department of Pharmacology, Stanford University School of Medicine, Stanford, CA 94305, USA

D.M. KNIGHT, Department of Pharmacology, Stanford University School of Medicine, Stanford, CA 94305, USA

K.K. KONTULA, The Population Council, The Rockefeller University, New York, NY 10021, USA

M. LE CUNFF, Department of Biochemical Endocrinology and Reproduction, University of Paris, 94270, Bicetre, France

S. LIAO, Department of Biochemistry, The Ben May Laboratories, University of Chicago, Chicago, IL 60627, USA

B. LITTLEFIELD, Department of Cell Biology, Mayo Medical School, Rochester, MN 55905, USA

F. LOGEAT, Department of Biochemical Endocrinology and Reproduction, University of Paris, 94270 Bicetre, France

C. MACGEOCH, Department of Medical Nutrition, Huddinge University Hospital, F69, S-141 86, Huddinge, Sweden

D. MAJUMDAR, Department of Biological Sciences, Oakland University, Rochester, MI 48063, USA

D.J. MANGELSDORF, Department of Biochemistry, University of Arizona, College of Medicine, Tucson, AZ 85724, USA

C.N. MARIASH, Division of Endocrinology and Metabolism, Department of Medicine, University of Minnesota, Minneapolis, MN 55455, USA

S.L. MARION, Department of Biochemistry, University of Arizona, College of Medicine, Tucson, AZ 85724, USA

B.M. MARKAVERICH, Department of Cell Biology, Baylor College of Medicine, Houston, TX 77131, USA

E. MELANITOU, The Population Council, The Rockefeller University, New York, NY 10021, USA

E. MILGROM, Department of Biochemical Endocrinology and Reproduction, University of Paris, 94270 Bicetre, France

E.T. MORGAN, Department of Medical Nutrition, Huddinge University Hospital, F69, S-141 86 Huddinge, Sweden

G.C. MUELLER, McArdle Laboratory for Cancer Research, University of Wisconsin, Madison, WI 53706, USA

C.V.R. MURTY, Department of Biological Sciences, Oakland University, Rochester, MI 48063, USA

N.A. MUSTO, The Population Council, The Rockefeller University, New York, NY 10021, USA

W.C. NG, National Institute for Medical Research, The Ridgeway, Mill Hill, London, NW7 1AA, England

J.P. NORTHROP, Department of Pharmacology, Stanford University School of Medicine, Stanford, CA 94305, USA

M.R. OLSEN, The Ben May Laboratory, University of Chicago, Chicago, IL 60637, USA

J.H. OPPENHEIMER, Division of Endocrinology and Metabolism, Department of Medicine, University of Minnesota, Minneapolis, MN 55455, USA

R. PAMPHILE, Department of Biochemical Endocrinology and Reproduction, University of Paris, 94270, Bicetre, France

A.J. PERLMAN, National Institute for Medical Research, The Ridgeway, Mill Hill, London, NW7 1AA, England

J.W. PIKE, Department of Biochemistry, University of Arizona, College of Medicine, Tucson, AZ 85724, USA

J. Reventos, Department of Obstetrics and Gynecology, University of Western Ontario, London, Ontario, N6A 4G5 Canada

G.M. Ringold, Department of Pharmacology, Stanford University School of Medicine, Stanford, CA 94305, USA

A.K. Roy, Department of Biological Sciences, Oakland University, Rochester, MI 48063, USA

F.H. Sarkar, Department of Biological Sciences, Oakland University, Rochester, MI 48063, USA

T.J. Schuh, The Ben May Laboratory, University of Chicago, Chicago, IL 60637, USA

H.L. Schwartz, Division of Endocrinology and Metabolism, Department of Medicine, University of Minnesota, Minneapolis, MN 55455, USA

P.J. Seppänen, The Population Council, The Rockefeller University, New York, NY 10021, USA

N.M. Sleator, Department of Biochemistry, University of Arizona, College of Medicine, Tucson, AZ 85724, USA

T. Spelsberg, Department of Cell Biology, Mayo Medical School, Rochester, MN 55905, USA

J.R. Tata, National Institute for Medical Research, The Ridgeway, Mill Hill, London, NW7 1AA, England

J.L. Vannice, Department of Pharamacology, Stanford University School of Medicine, Stanford, CA 94305, USA

L.L. Wei, Department of Medicine and Pathology, University of Colorado Health Science Center, Denver, CO 80262, USA

W.V. Welshons, Department of Human Oncology, University of Wisconsin Hospital and Clinics, Madison, WI 53792, USA

D. Witte, Department of Biochemistry, The Ben May Laboratories, University of Chicago, Chicago IL 60637, USA

A.P. Wolffe, National Institute for Medical Research, The Ridgeway, Mill Hill, London, NW7 1AA, England

Chapter 1

Biochemical Evidence for the Exclusive Nuclear Localization of the Estrogen Receptor

W.V. WELSHONS, E.M. CORMIER, V.C. JORDAN, AND J. GORSKI

Introduction

In this chapter we describe data from cell enucleation experiments that indicate that the unoccupied estrogen receptor is a nuclear rather than a cytoplasmic protein. In the revised model for estrogen action (Fig. 1), the estrogen receptor is assumed to be a nuclear protein both before and after hormone binding. We also discuss other published work that is related to the proposition that the unoccupied estrogen receptor is a nuclear protein (reviewed in Welshons and Gorski, 1986) and that suggest a reevaluation of the translocation model of steroid action.

Evidence for the translocation model of estrogen action was published independently by Gorski and Jensen in 1968 (Gorski et al., 1968; Jensen et al., 1968). When the immature rat uterus is homogenized, the unoccupied estrogen receptors are found predominantly in the cytosol. However, when the uterus is exposed to labeled estradiol before homogenization, most of the now occupied estrogen receptors are found in the particulate fraction, with the nuclei. In a time course of hormone binding, the receptor that apparently translocates from the cytosol can be extracted from the nuclear pellet (Fig. 2). This process can be reproduced in vitro, using cytosol containing unoccupied receptor and isolated nuclei, which initially contain no receptor. The translocation process requires first that the hormone binds to the receptor and second that the receptor is transformed or activated by brief warming, before the receptor-hormone complex acquires the ability to bind to nuclei.

The results obtained with subcellular fractionation of homogenized tissue were apparently confirmed by autoradiography (Jensen et al., 1968) using techniques for diffusable substances worked out by Stumpf (Stumpf and Roth, 1966; Stumpf, 1971). These workers found that when immature rat uteri were incubated with [^{3}H]estradiol at 37°C and then processed for autoradiography, silver grains corresponding to the location of radioactive estradiol were found predominantly over the nuclei of the cells. If the labeled tissue was instead homogenized, [^{3}H]estradiol was recovered bound to receptor in the nuclear fraction of the tissue.

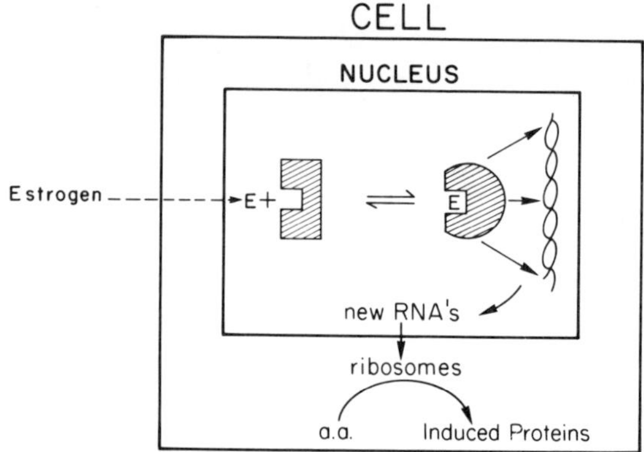

Fig. 1. Model of nuclear estrogen receptor. E = estrogen.

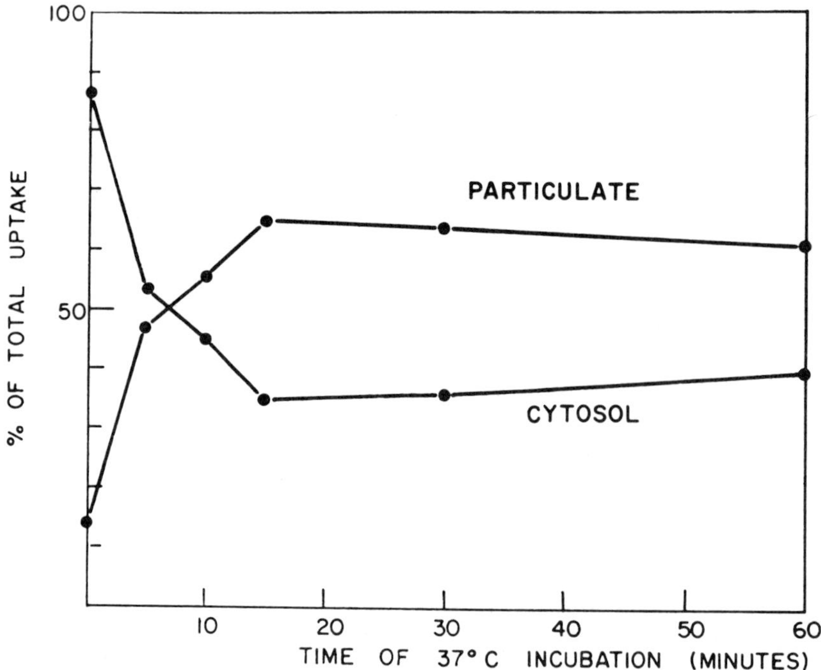

Fig. 2. The release of radioactive estradiol from cytosol into particulate fraction upon incubation at 37°. In each group two uteri were incubated at 0° for 1 min in 2 ml of Eagle's medium containing 0.02 μg of ^{3}H-estradiol-17β. The uteri were then transferred to fresh medium without any estradiol and incubated at 37° for various times as indicated (Shyamala and Gorski, 1969).

When the tissue was incubated briefly at 2°C, conditions under which little nuclear translocation is observed in the fractionated tissue, grains were found generally distributed over all parts of the cells throughout the tissue. This was assumed to represent hormone bound to nontransformed receptor in the cytoplasm, because after homogenization, most of the radioactivity of the tissue was recovered bound to nontransformed receptor, in the cytosol extract. Quantitatively, the amount of cytosolic receptor recovered was close to the expected receptor content of the uterus, not some small nonrepresentative fraction of the receptor. Therefore, the results obtained using autoradiography seemed to be in quantitative agreement with the subcellular fractionation data.

The evidence for the translocation model has been very convincing, because of its great reproducibility, quantitative concurrence with a morphological technique (autoradiography), and the absence of other localization techniques with contradictory results.

More recently, however, studies conducted by Sheridan (Sheridan et al., 1979, 1981) have suggested that autoradiography does not show cytoplasmic binding, but rather free hormone that is trapped in the tissue after the low temperature incubation, and which then binds to receptor during the homogenization after the cells are broken open. These results indicate that the best evidence for the translocation model is derived from the fractionation studies with homogenized cells, and may be subject to extraction artifacts, as noted in the original papers (Gorski et al., 1968; Jensen et al., 1968).

Estrogen Receptor Distribution Using Cell Enucleation

To avoid potential extraction artifacts, we applied cell enucleation techniques to separate cytoplasm from nuclei without breaking open the cell (Fig. 3). After treatment of the cells with the drug cytochalasin B, which stimulates or facilitates enucleation of cells (Carter, 1967; Poste, 1972; Prescott and Kirkpatrick, 1973), equilibrium density centrifugation is used to pull the dense nucleus away from the less dense cytoplasm of the cell.

Enucleation of freshly dispersed rat uterine or pituitary cells, or of primary cultures derived from both organs, was not satisfactory because of cell damage or incomplete separation of the fractions. To avoid these problems we selected GH_3 rat pituitary cells to analyze receptor distribution. The GH_3 cell line is derived from a rat pituitary tumor (Tashjian et al., 1970; Tashjian and Hoyt, 1972) and contains estrogen receptor that is "translocated" from the cytosol to the nucleus in homogenized cell studies (Haug et al., 1978). The cells are estrogen responsive and show increased prolactin synthesis in the presence of physiological concentrations of estradiol (Haug and Gautvik, 1976; Tate et al., 1984).

The GH_3 cells were enucleated (Fig. 4) to yield an almost homogeneous

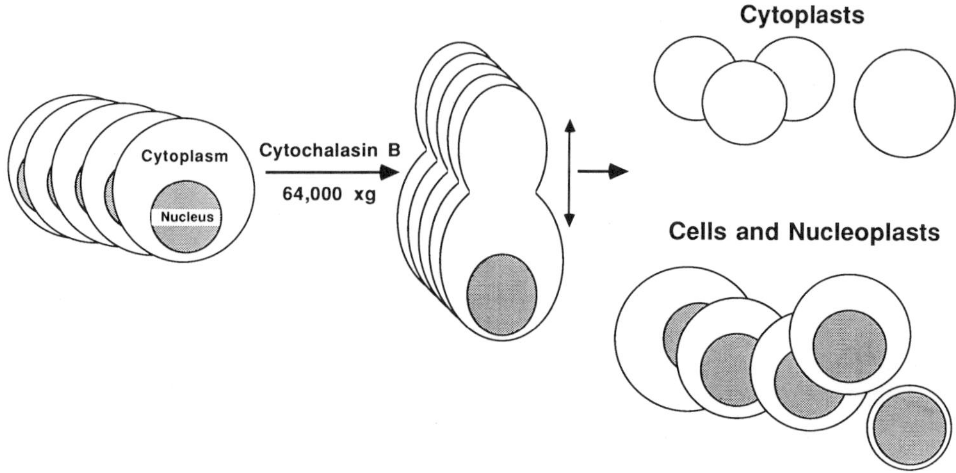

Fig. 3. Enucleation of GH_3 cells (Welshons and Gorski, 1986).

fraction of cytoplasts (enucleated cytoplasm), but a more heterogeneous fraction containing cells from which varying amounts of cytoplasm had been removed (cells plus nucleoplasts). When the estrogen receptor concentration per protein was measured in the cells or in the enucleation fractions, it was found that the cytoplasts contained consistently low concentrations of receptors, which were recovered in the nucleus-containing cell plus nucleoplast fraction (Fig. 5). Receptor per DNA was similar in all fractions, suggesting that most of the receptors were nuclear. The distribution of the cytosolic enzyme marker lactate dehydrogenase followed the pattern expected for a cytoplasmic protein; the concentration of enzyme per protein in the cytoplasts was similar to or greater than that in the whole cells (Fig. 6). The subcellular distributions of estrogen receptors and lactate dehydrogenase were essentially opposite.

It was found that the GH_3 cells could be enucleated with centrifugation alone, without using the drug cytochalasin B. The distribution of estrogen receptors was the same as was observed previously using the drug (Table 1). By the DNA:protein ratio, the cytoplast fraction showed a tenfold reduction of DNA compared with the whole cells. The concentration of estrogen receptor per protein in the cytoplasts was only 5% of the concentration in the whole cells. The similar values per DNA in all fractions suggest that most of the residual cytoplast estrogen receptor may have been contained in the contaminating whole cells rather than in the cytoplasts themselves.

The cytoplasts formed from GH_3 cells were an average of 25%–35% of the cell (Welshons et al., 1985). Fractionation of the cells and nucleoplasts by density to isolate nucleoplasts from which more of the cytoplasm had been removed indicated that for removal of at least half of the cytoplasm,

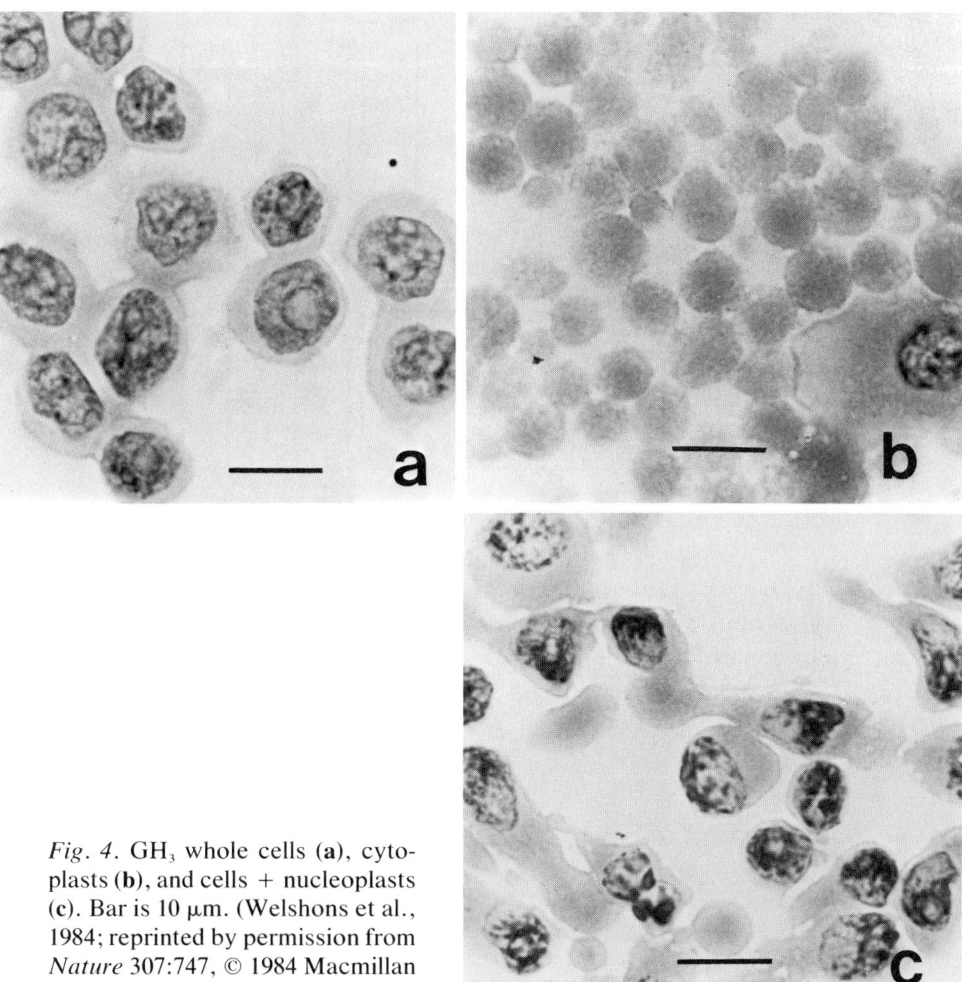

Fig. 4. GH_3 whole cells (**a**), cytoplasts (**b**), and cells + nucleoplasts (**c**). Bar is 10 μm. (Welshons et al., 1984; reprinted by permission from *Nature* 307:747, © 1984 Macmillan Journals Limited)

there was no loss of receptor from the nucleoplast (Welshons et al., 1984). However, it was considered that the receptor might still be cytoplasmic in the cell, near the nucleus in the perinuclear cytoplasm. Although GH_3 cells were not enucleated to remove all of the cytoplasm from very many cells, the human breast cancer cell lines MCF-7 or T47D could be so enucleated. In nucleoplasts from which 75%–80% of the cell protein had been removed (and nearly all the cytoplasm), all the unoccupied estrogen receptors of the cells remained with the nucleoplasts (W. Welshons, E. Cormier, and V. Jordan, written communication, August 1985). This indicates that the receptors are not in the perinuclear cytoplasm, unless confined entirely within a very narrow zone around the nucleus.

In the enucleation experiments with GH_3 cells, cytoplasts were formed

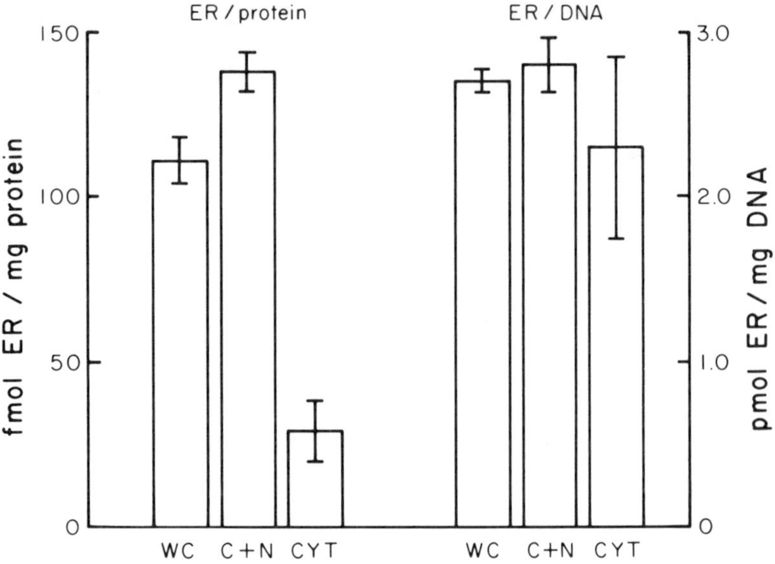

Fig. 5. Intracellular distribution of estrogen receptor (ER). GH_3 cells were enucleated using cytochalasin B and centrifugation. Estrogen receptor, protein, and DNA were measured in untreated whole cells (WC), in the cells + nucleoplasts (C + N), and in the cytoplasts (Cyt). (Welshons, Lieberman, and Gorski, 1984; reprinted by permission from *Nature* 307:747; © 1984 Macmillan Journals Limited)

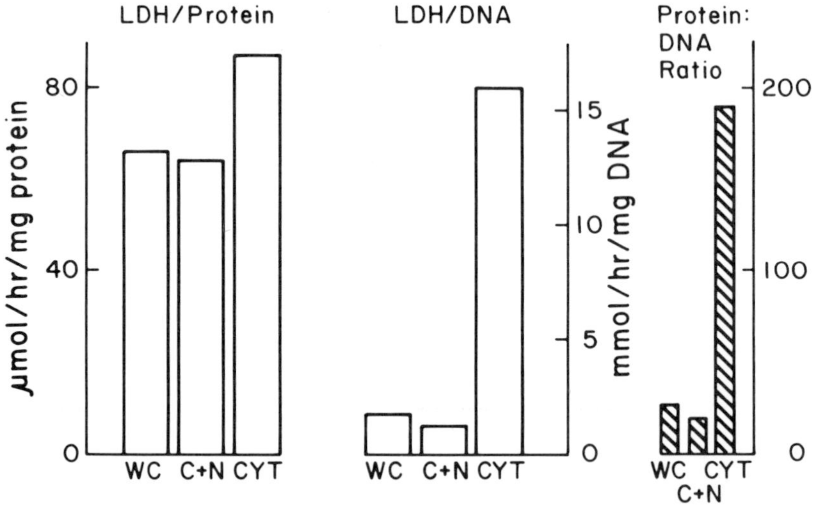

Fig. 6. Lactate dehydrogenase distribution in enucleated GH_3 cells. GH_3 cells were enucleated using centrifugation alone. Protein, DNA, and the cytosolic marker lactate dehydrogenase (LDH) were measured in the whole cells (WC), in the cells + nucleoplasts (C + N), and in the cytoplasts (Cyt). (Welshons, Krummel, and Gorski, 1985; © 1985, The Endocrine Society)

Table 1. Distribution of Estrogen Receptors in GH_3 Cells Enucleated Without Cytochalasin

Fraction	fmol/mg protein	pmol/mg DNA	Molecules per cell or equivalent	DNA/protein μg/mg
Whole cells	130 ± 3	3.5 ± 0.1	32,000	37.
Cells + nucleoplasts	207 ± 9	3.5 ± 0.1	30,000	60.
Cytoplasts	7.1 ± 1	2.2 ± 0.3	1,500[a]	3.2

Note: Estrogen receptor was measured in whole cells and in cells plus nucleoplasts and cytoplasts after enucleation using equilibrium density centrifugation without the use of cytochalasin B. Values are shown per protein, per DNA, or per number of intact cells.
Source: Welshons, Krummel, and Gorski, 1985; © 1985, The Endocrine Society.
[a]Since cytoplasts were one fifth the size of whole cells, the receptors per whole cell equivalent tabulated here were calculated as five times the actual number of receptors per cytoplast.

from approximately 90% of the cells (Welshons et al., 1985), and full recovery of protein, DNA, and estrogen receptor was obtained after the enucleation (Welshons et al., 1984; Shull et al., 1985). This indicates that receptor was not selectively lost from the cytoplasts, for example, and that most of the cells were sampled, not some small fraction that might not contain estrogen receptor. The cytoplasts and cells plus nucleoplasts were intact after the procedure. Dye exclusion was high, the fractions synthesized protein generally and prolactin specifically (Shull et al., 1985) and were observed to attach to culture dishes at 37°C.

When the cell plus nucleoplast fraction was homogenized and receptors measured in subcellular fractions by the usual techniques, the receptors for estrogen, progesterone, and glucocorticoids were found in the cytosol (Table 2). Comparison of the hormone-binding rates and the exchange rates in the cells indicated that the receptors were indeed unoccupied before assay (Welshons et al., 1985). Nuclear localization of the unoccupied estrogen receptor using cell enucleation has also been reported independently in an endometrial cell line (Gravanis and Gurpide, 1986).

Table 2. Distribution of Steroid Receptors in *Homogenized* Cells

Steroid receptor	Fraction cytosolic (%)	Receptor recovery compared with whole cell uptake (%)
Glucocorticoid	96	71
Estogen	85	69
Progesterone	93	81

Note: The cell plus nucleoplast fraction was homogenized immediately after enucleation, and the "cytosolic" and "nuclear" receptors (extracted from the pellet with 0.4 *M* KC1) were measured using hydroxylapatite. When homogenized, most of the receptor in these cells was recovered in the cytosol. The receptor recovered in the homogenized cells plus nucleoplasts is compared with the receptor in a sample of the intact cells plus nucleoplasts measured by whole-cell uptake.
Source: Welshons, Krummel, and Gorski, 1985; © 1985, The Endocrine Society.

Table 3. Distribution of Progesterone Receptors in GH_3 Cells Enucleated Without Cytochalasin

Fraction	fmol/mg protein	pmol/mg DNA	Molecules per cell or equivalent	DNA/protein μg/mg
Whole cells	246 ± 3.	7.4 ± 0.1	63,000	34.
Cells + nucleoplasts	172 ± 9.	3.8 ± 0.1	33,000	45.
Cytoplasts	8 ± 0.5	3.8 ± 0.3	2,000[a]	2.1

Note: Progesterone receptor was measured in whole cells, cells plus nucleoplasts, and cytoplasts after enucleation using equilibrium density centrifugation without the use of cytochalasin B. Values are shown per protein, per DNA, or per number of intact cells.
Source: Welshons, Krummel, and Gorski, 1985; © 1985, The Endocrine Society.
[a]Since cytoplasts were one fifth the size of whole cells, the receptors per whole cell equivalent tabulated here were calculated as five times the actual number of receptors per cytoplast.

Little of the receptors for progesterone or glucocorticoids were found in cytoplasts of GH_3 cells (Table 3, Fig. 7), indicating that these steroid receptors may also be located in the nucleus (Welshons et al., 1985). A complicating observation, however, was that the receptors for both hormones were not fully recovered after the enucleations. Losses of 25%–50% were usually seen. This did not seem to affect the distribution of the receptors that we observed, for the following reasons:

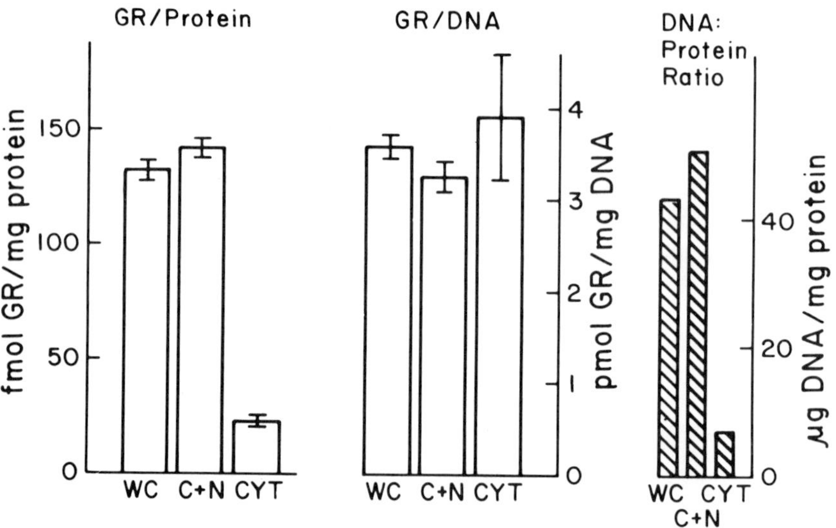

Fig. 7. Distribution of glucocorticoid receptor in enucleated GH_3 cells. GH_3 cells were enucleated using centrifugation alone. Glucocorticoid receptor, protein, and DNA were measured in the whole cells (WC), in the cells + nucleoplasts (C + N), and in the cytoplasts (Cyt). (Welshons, Krummel, and Gorski, 1985; © 1985, The Endocrine Society)

1. The same nuclear distribution of glucocorticoid receptors was observed in an experiment where little receptor loss was observed (Fig. 7).
2. When the cells were "mock enucleated" without removal of cytoplasm, the same loss of receptor was observed.
3. When cells were enucleated in the presence of dexamethasone, to make the glucocorticoid receptor unambiguously nuclear, again the same receptor loss was observed (Welshons et al., 1985).

Therefore, it would not appear that the receptor loss could be due to selective loss of receptor in cytoplasts after enucleation. Data on the synthesis and degradation rates of proteins in cytoplasts indicate that there is either no change or sometimes reduced protein turnover in cytoplasts (Fan et al., 1977; Gopalakrishnan and Thompson, 1977; Freikopf-Cassel and Kulka, 1981; Knecht et al., 1982).

Data That Support Nuclear Location of Unoccupied Steroid Receptors

The nuclear localization of the unoccupied estrogen receptor obtained by enucleation is not an isolated observation. Recent immunocytochemistry of the estrogen receptor (King and Greene, 1984; McClellan et al, 1984) and of the progesterone receptor (Gasc et al., 1984; Perrot-Applanat et al., 1985) has shown an exclusively nuclear location for those receptors, whether occupied by hormone or not. Many earlier reports have described a cytoplasmic location (Morel et al., 1981; Raam et al., 1982, 1983; Greene and Jensen, 1982; Gasc et al., 1982). However, the more recent techniques using shorter fixation, and differentiating an antigenic nonreceptor protein that binds to the receptor but does not itself bind hormone (Gasc et al., 1984), seem to show consistently a nuclear localization for the two receptors.

Immunocytochemistry of the glucocorticoid receptor so far has shown cytoplasmic or cytoplasmic and nuclear immunoreactivity, with nuclear translocation after hormone binding (Papamichail et al., 1981; Govindan, 1980; Gustafsson et al., 1983; Bernard and Joh, 1984; Antakly and Eisen, 1984). Whether an exclusively nuclear localization of the glucocorticoid receptor will be observed as for the estrogen and progesterone receptors, or whether the glucocorticoid receptor is in fact cytoplasmic before hormone binding, remains to be seen. However, as noted above, cytoplasts formed from GH_3 cells do not contain the glucocorticoid receptor (Welshons et al., 1985).

Sheridan et al. (1979, 1981) provided important evidence that most of the unoccupied estrogen and progesterone receptors were nuclear, using autoradiography of hormone bound under nontranslocating conditions.

Immature rat uteri were incubated briefly at 4°C with [^{3}H]estradiol or [^{3}H]R-5020 (progestin), then thoroughly rinsed of free hormone and processed for autoradiography of diffusable compounds. Only slight penetration of the hormone into the tissue was observed after 5 min at 4°C, but where present at the periphery, bound hormone was found mostly over nuclei of the cells (Fig. 8). If the labeled tissue was homogenized instead of used for autoradiography, most of this bound hormone was extracted, especially for the bound [^{3}H]R-5020 (Sheridan et al., 1981), and was found on sucrose gradients to migrate as the 8S nontransformed receptor.

The receptor was not transformed, according to the sucrose gradient analysis, and it was not activated because it was extracted to the cytosol when the tissue was homogenized, yet by autoradiography the receptor was nuclear, not cytoplasmic. Since a clearly nuclear form of the receptor was extracted during homogenization, and since the nontransformed, occupied receptor is found with the unoccupied receptor in the cytosol, there was little reason to assume that most of the unoccupied receptor was not also nuclear but extractable. Sheridan et al. postulated that the unoccupied receptors were in equilibrium between the cytoplasm and the nucleus, which was shifted in favor of the cytosol when the cell was homogenized and its contents diluted (Sheridan et al., 1979, 1981; Martin and Sheridan,

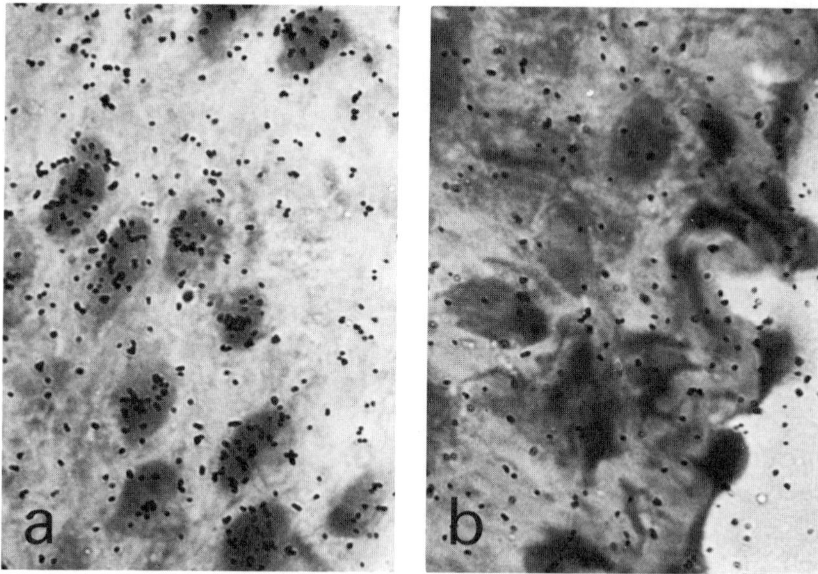

Fig. 8. Autoradiograms of estrogen-primed rat uteri from the periphery of sections incubated for 5 min at 4° with [^{3}H]R5020 (5 n*M*; **a**) or with [^{3}H]R5020 (5 n*M*) plus unlabeled R5020 (500 n*M*; **b**). (Sheridan et al., 1981; © 1981, The Endocrine Society)

1982). However, whether a significant amount of receptor might actually be cytoplasmic in the intact cell was not shown, and the data are also consistent with an essentially nuclear receptor.

Related findings demonstrating that a large fraction of nontransformed (4S) estrogen receptor could be found in the nucleus of the cell were described by Siiteri and Linkie (Siiteri et al., 1973; Linkie, 1977; Linkie and Siiteri, 1978). The time course of the nuclear receptor content suggested that the nontransformed 4S receptor-hormone complex was acting as the precursor of the transformed 5S receptor-hormone complex, within the nucleus. These observations would be hard to reconcile with the model of transformation-driven translocation from the cytoplasm to the nucleus, but are consistent with a receptor that is nuclear before and after hormone binding.

Jordan et al. (1985) described a number of low-affinity estrogens and antiestrogens that show full or partial estrogenic effects in the rat uterus, yet no nuclear translocation is observed. This has been explained as extraction of the receptor from the nuclear compartment when the ligand dissociates from the receptor during homogenization, and the extraction model has been extended to the unoccupied receptor and to other antihormones that had earlier been thought to paralyze the receptor in the cytosol extract.

The compounds are derivatives of tamoxifen, the principal antiestrogen used in the treatment of hormone-dependent breast cancer (Furr and Jordan 1984). The parent structure is a triphenylethylene, with an ethoxyaminomethyl side chain in the *para* position of the middle phenyl ring (Fig. 9). The side chain on the middle ring confers antiestrogenic activity on the molecule. Replacing it with a much smaller hydroxyl group as in ICI 77949, or shifting the side chain to an end phenyl ring as in ICI 47699 (Fig. 9), converts the molecule to an estrogen. The affinity of the compounds for the estrogen receptor, through which their action is mediated, is determined by position 4 on the first ring (the X position in Fig. 9). A compound with a hydrophobic group at position 4 has low affinity for the estrogen receptor and exhibits low potency. A hydroxyl group, however, confers high receptor affinity on the compound, and high potency. Affinity of the compound for the receptor is separate from the estrogenicity; 4-hydroxy-tamoxifen has an affinity for the estrogen receptor that is slightly higher than that of estradiol, yet is antiestrogenic because of the ethoxyaminomethyl side chain (Jordan et al., 1977).

The compounds ICI 77949 and ICI 47699 are fully as estrogenic as estradiol, if given in sufficient quantity, in stimulating either uterine weight gain (Fig. 10) or uterine progesterone receptor in immature rats (Jordan et al., 1985). All three of the antiestrogens tamoxifen, 4-methyl-tamoxifen and 4-hydroxytamoxifen (monohydroxytamoxifen) are partially estrogenic in the rat uterus to the same degree (Fig. 10).

However, when tissue from the immature rats was homogenized and

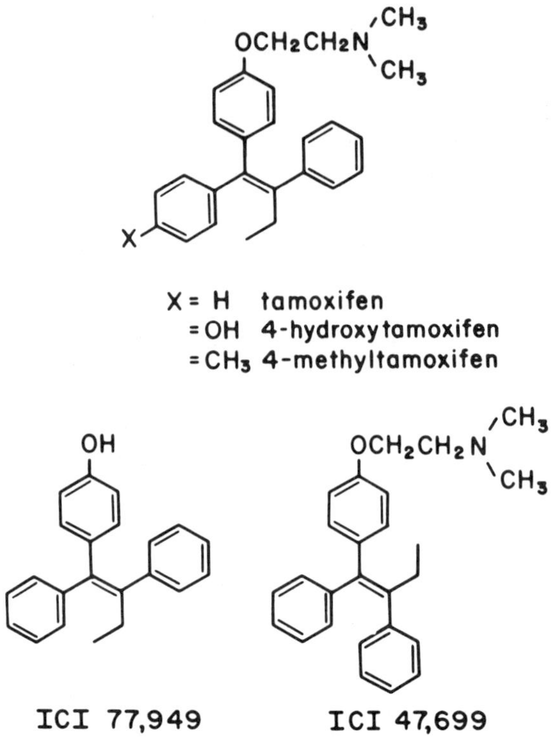

Fig. 9. Tamoxifen and derivatives described here. (Jordan et al., 1985; © 1985, The Endocrine Society)

the receptor content of nuclear and cytosolic fractions measured by exchange assay, neither one of the ICI compounds translocated estrogen receptor to the nucleus, or depleted receptor from the cytosol in tissue from animals treated with the compounds (Fig. 11). Nuclear translocation, as assessed in the homogenized tissue, can be demonstrated for only the high-receptor-affinity compounds such as estradiol or 4-hydroxytamoxifen, even though all the compounds showed similar biological activity within the estrogen or antiestrogen groups.

Whether or not the receptor can be detected in the nuclear compartment following tissue homogenization, the estrogen receptor is occupied in vivo, since [^{3}H]estradiol binding to receptor is competitively reduced by the prior injection of compounds. The receptor occupied by the low-affinity estrogens appears to be nuclear in vivo because the full estrogenic effects are produced by the compounds, and these effects include direct regulation of gene expression in the nucleus.

If nuclear receptor is extracted into the cytosol when the tissue is homogenized, when a low-affinity ligand binds to the receptor, then it is

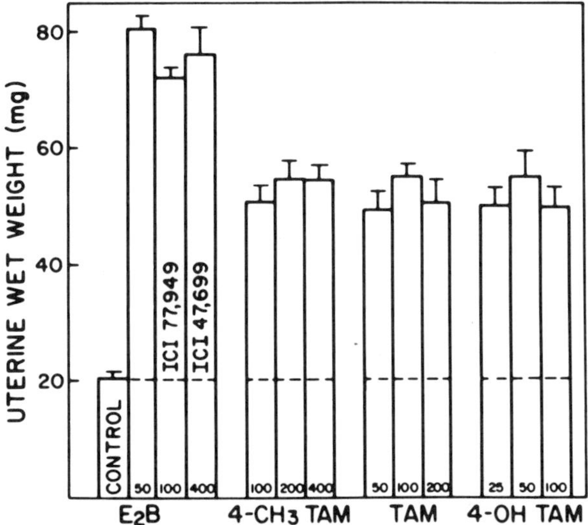

Fig. 10. Immature rat uterine weights 48 h after administration of estrogens or antiestrogens. Immature female rats (18–20 days old) were injected subcutaneously with the indicated doses of compounds (μg) in 0.1 ml peanut oil. The animals were killed 48 h after injection, and the uteri were removed, blotted dry, and weighed. The antiestrogens were tested at several doses to show that the agonist activity was partial, not the result of being in the middle of the dose-response curve. (Jordan et al., 1985; © 1985, The Endocrine Society)

suggested that the ligand dissociates during homogenization and unoccupied receptor is extracted into the cytosol. Since receptors believed to be in the nucleus can be extracted from the nucleus under the standard homogenization conditions, why then should initially occupied receptor behave differently? There seems little reason to postulate that the unoccupied receptor is cytoplasmic, and a single compartment model for the receptor and its interactions with the estrogens and antiestrogens has been suggested (Jordan et al., 1985).

A number of steroid antagonists have been described that do not demonstrate "nuclear translocation" in homogenized tissue, but like the tamoxifen derivatives, these compounds all have low affinities for their respective receptors. It is suggested that, in fact, the compounds bind to the appropriate steroid receptor in the nucleus, but without producing the necessary conformational change for hormone action. The apparent lack of translocation would be from dissociation of the ligand from the receptor in vitro and return of the receptor to a readily extractable form (Jordan et al., 1985). The absence of demonstrable nuclear translocation in these hormone antagonists is suggested to be from their low affinity for the receptor, and not the cause of their antagonistic activity. Antagonistic

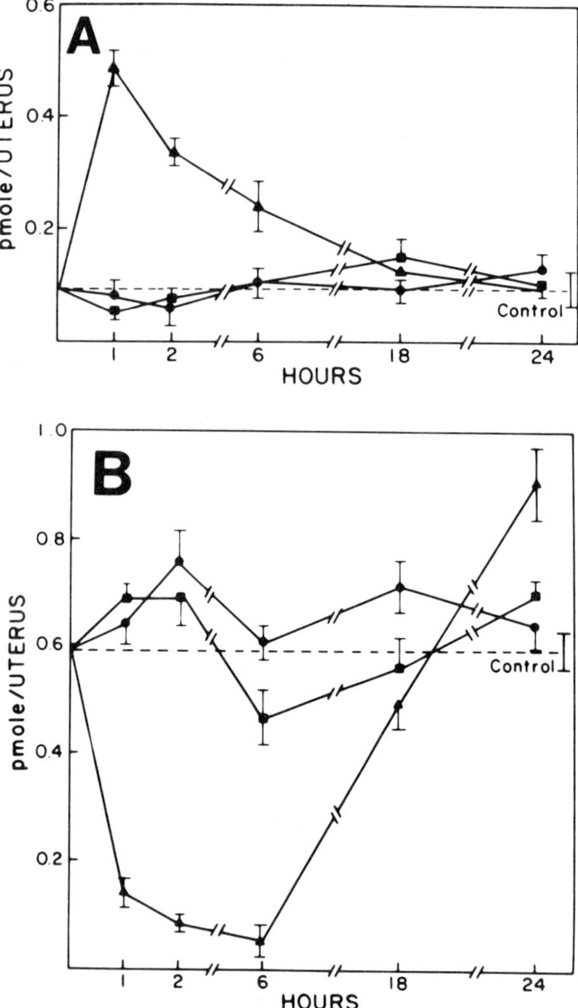

Fig. 11. The determination of nuclear and cytosolic ER levels in uteri by exchange assay after administration of E_2, 4-CH_3-TAM, or ICI 77,949. After administration of a single subcutaneous injection of E_2 (10 μg: ▲), 4-CH_3-TAM (200 μg; ●), or ICI 77,949 (100 μg; ■), uteri were removed and nuclear (**A**) and cytosol (**B**) exchange assays were performed. (Jordan et al., 1985; © 1985, The Endocrine Society)

activity is related to the inappropriate conformation that is produced by the drug-receptor complex within the nucleus.

Equilibrium binding data described by Sakai and Gorski (1984) suggest that hormone binding and receptor transformation in whole cells are more like those of immobilized receptor than of soluble receptor in solution. Estrogen receptor in cytosol extracts displays cooperative estradiol binding, as described by Notides et al. (1981), at receptor concentrations of 0.5 nM or higher. Aggregated 8S receptor (low salt extract) bound to hydroxylapatite, and "monomeric" 4S receptor in solution in 0.4 M KCl both showed the cooperative hormone binding. But when the "monomeric" 4S receptor was bound to the hydroxylapatite, the cooperative binding was lost, only straight-line Scatchard plots were observed (Fig. 12).

Transformation or activation of the receptor is accompanied by a transition to tighter binding of the hormone by the receptor, which could be measured in receptor bound to hydroxylapatite by the dissociation rate of bound hormone. Receptor in solution could be fully converted to the higher affinity form, but "monomeric" receptor immobilized on hydroxylapatite could be only partially converted, to about 80% (Sakai and Gorski, 1984). The receptor was capable of 100% transformation when it was eluted from the hydroxylapatite.

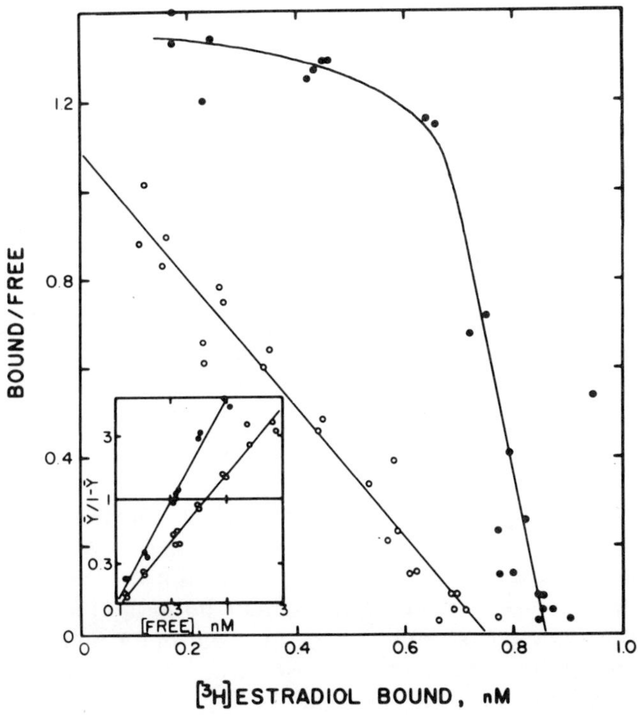

Fig. 12. Scatchard plot of equilibrium binding at 0°C of [^{3}H]estradiol to native and monomeric estrogen receptors. *Inset:* Hill plots. (●) Binding to receptor in solution; receptor concentration = 0.85 nM, half-maximal binding = 0.32 nM E_2, n_H = 1.54. (○) Binding to monomeric receptor generated by treatment with 0.4 M KCl and adsorption onto HAP; receptor concentration = 0.73 nM, half maximal binding = 0.64 nM E_2, n_H = 1.01. (From Sakai and Gorski, 1984; reprinted with permission from *Biochemistry*, Vol 23, pp 3541–3547; © 1984, American Chemical Society)

Both noncooperative hormone binding and incomplete receptor transformation are what is observed in the intact cell (Williams and Gorski, 1972, 1974). Therefore, estrogen receptor in the intact cell seems to behave as though it is immobilized within the cell, not soluble. The experiments also imply that there is no soluble movement of the receptor from one site to another after hormone binding (Sakai and Gorski, 1984). Consequently, this evidence suggests that the unoccupied receptor is in the nucleus already bound at or near the site of hormone action before the hormone is bound.

Function of an Exclusively Nuclear Estrogen Receptor

An exclusively nuclear receptor raises some questions about receptor action which are different from those related to a translocating receptor mechanism:

1. Does the receptor relocate within the nucleus after binding hormone, or does the unoccupied receptor already reside at the site of hormone action in the nucleus before the hormone binds? Perhaps the importance of the unoccupied receptor (rather than occupied forms) in finding the specific sites of action has been overlooked.
2. How is the increased affinity of the transformed receptor for the nucleus related to changes in gene expression? Is the tight binding a part of the hormone response, does it occur only at the sites of hormone action in the nucleus, or is it nonspecific? Does the receptor regulate gene expression positively or negatively? Since the unoccupied receptor is nuclear, it theoretically could work negatively by blocking gene expression until hormone is bound.
3. How is the number of estrogen receptors regulated? Since unoccupied receptor is nuclear, does it regulate its own expression at the receptor gene, or is regulation mediated through other factors?
4. Why does the unoccupied receptor accumulate in the nucleus? Does the receptor protein contain a specific nuclear transport signal, or does the receptor accumulate in the nucleus because of an affinity for nuclear elements, or is exclusion from the cytoplasm involved?
5. Since the receptor is nuclear throughout hormone binding and early action, it may be important to have a mechanism for removing receptors after action to limit the hormonal response. Phosphorylation of the receptor may be involved in this function.

Summary

Cell enucleation (as well as recent immunocytochemistry) has provided direct evidence that the unoccupied estrogen receptors are located in the nucleus of the responsive cell, not in the cytoplasm as has been thought for more than 15 years. Biochemical evidence from a number of other experiments supports these observations. A receptor located exclusively in the nucleus throughout hormone binding and early hormone action may function quite differently from a receptor that must translocate to the nucleus after hormone is bound.

Acknowledgments. We thank Mary Pankratz for help preparing the manuscript. Supported in part by ACS IN-35 and 5-F32-HD-06008 (WVW), by HD-08192, CA-18110, and 1-T32-HD-07259 (JG) and by P30-CA-14520, P01-CA-20432, CA-32713, and grants from ICI Pharmaceuticals (VCJ).

References

Antakly T, Eisen HJ (1984) Endocrinology 115: 1984–1989
Bernard P, Joh TH (1984) Arch Biochem Biophys 229: 466
Carter SB (1967) Nature 213:261–264
Fan WWJ-W, Ivarie RD, Levinson BB (1977) J Biol Chem 252:7834–7841
Freikopf-Cassel A, Kulka RG (1981) FEBS Lett 124: 27–30
Furr BJA, Jordan VC (1984) Pharmacol Ther 25: 127–205
Gasc J-M, Renoir J-M, Radanyi C, Joab I, Baulieu E-E (1982) CR Acad Sc Paris 295: 707–713
Gasc J-M, Renoir J-M, Radanyi C, Joab I, Tuohimaa P, Baulieu E-E (1984) J Cell Biol 99: 1193–1201
Gopalakrishnan TV, Thompson EB (1977) J Cell Physiol 93: 69–80
Gorski J, Toft D, Shyamala G, Smith D, Notides A (1968) Recent Prog Horm Res 24: 45–80
Govindan MV (1980) Exp Cell Res 127: 293–297
Gravanis A, Gurpide E (1986) J Ster Biochem 24: 469–474
Greene GL, Jensen EV (1982) J Ster Biochem 16: 353–359
Gustafsson JA, Okret S, Wikstrom AC, Andersson B, Radojcic M, Wrange O, Sachs W, Doupe AJ, Patterson PH, Cordell B, Fuxe K (1983) In: Eriksson H, Gustafsson JA (eds) Nobel symposium on steroid hormone receptors: Structure and function. Elsevier, Amsterdam, p 355
Haug E, Gautvik KM (1976) Endocrinology 99:1482–1489
Haug E, Naess O, Gautvik KM (1978) Mol Cell Endocrinol 12: 81–95
Jensen EV, Suzuki T, Kawashima T, Stumpf WE, Jungblut PW, DeSombre E (1968) Proc Natl Acad Sci USA 59: 632–638
Jordan VC, Collins MM, Rowsby L, Prestwich G (1977) J Endocrinol 75: 305–316
Jordan VC, Tate AC, Lyman SD, Gosden B, Wolf MF, Bain RR, Welshons WV (1985) Endocrinology 116: 1845–1857
King WJ, Greene GL (1984) Nature 307:745–747
Knecht E, Hernandez-Yago J, Grisolia S (1982) FEBS Lett 150: 473–476
Linkie DM (1977) Endocrinology 101: 1862–1870
Linkie DM, Siiteri PK (1978) J Ster Biochem 9: 1071–1078
Martin PM, Sheridan PJ (1982) J Ster Biochem 16: 215–229
McClellan MC, West NB, Tacha DE, Greene GL, Brenner RM (1984) Endocrinology 114: 2002–2014

Morel G, Dubois P, Benassayag C, Nunez E, Radanyi C, Redeuilh G, Richard-Foy H, Baulieu E-E (1981) Exp Cell Res 132: 249–257
Notides AC, Lerner N, Hamilton DE (1981) Proc Natl Acad Sci USA 78: 4926–4930
Papamichail M, Ioannidis C, Tsawdaroglou N, Sekeris CE (1981) Exp Cell Res 133: 461–465
Perrot-Applanat M, Logeat F, Groyer-Picard MT, Milgrom E (1985) Endocrinology 116: 1473–1484
Poste G (1972) Exp Cell Res 73: 273–286
Prescott DM, Kirkpatrick JB (1973) Methods Cell Biol 7: 189–202
Raam S, Nemeth E, Tamura H, O'Briain DS, Cohen JL (1982) Eur J Cancer Clin Oncol 18: 1–12
Raam S, Richardson GS, Bradley F, MacLaughlin D, Sun L, Frankel F, Cohen JL (1983) Breast Cancer Res Treat 3: 179–199
Sakai D, Gorski J (1984) Biochemistry 23: 3541–3547
Sheridan PJ, Buchanan JM, Anselmo VC, Martin PM (1979) Nature 282: 579–582
Sheridan PJ, Buchanan JM, Anselmo VC, Martin PM (1981) Endocrinology 108: 1533–1537
Shull JD, Welshons WV, Lieberman ME, Gorski J (1985) In: Moudgil V (ed) Molecular Mechanisms of Hormone Action. Walter De Gruyter Publishing, Berlin, pp 539–562
Shyamala G, Gorski J (1969) J Biol Chem 244: 1097–1103
Siiteri PK, Schwarz BE, Moriyama I, Ashby R, Linkie D, MacDonald PC (1973) Adv Exp Med Biol 36: 97–112
Stumpf WE (1971) Acta Endocrinol [Suppl] Copenh), 153: 205–221
Stumpf WE, Roth LJ (1966) J Histochem Cytochem 14: 274–287
Tashjian AH Jr, Hoyt RF Jr (1972) In: Sussman M (ed) Molecular genetics and developmental biology. Prentice-Hall, Englewood Cliffs, pp 353–387
Tashjian AH Jr, Bancroft FC, Levine L (1970) J Cell Biol 47: 61–70
Tate AC, Lieberman ME, Jordan VC (1984) J Ster Biochem 20: 391–395
Welshons WV, Gorski J (1986) In: Conn PM (ed) The Receptors Vol. IV. Academic Press (in press)
Welshons WV, Krummel BM, Gorski J (1985) Endocrinology 117: 2140–2147
Welshons WV, Lieberman ME, Gorski J (1984) Nature 307: 747–749
Williams D, Gorski J (1972) Proc Natl Acad Sci USA 69: 3464–3468
Williams D, Gorski J (1974) Biochemistry 13: 5537–5542

Discussion of the Paper Presented by W. Welshons

SCHRADER: When you make the nuclei by this cytochalasin treatment, you end up with a pretty nice clean population of nuclei. I understand that there is no hormone present, but you're claiming that the receptors are still in that fraction?
WELSHONS: Yes, the unoccupied estrogen receptors are in that fraction.
SCHRADER: If the receptors are not held tightly enough in the normal cell, they leak out during homogenization. Why don't the receptors leak out of these nuclei when you prepare them? Why don't the receptors leak out of the nuclei when you get to the end of the experiment? Why do you think they are held there?

WELSHONS: These are not isolated nuclei; they are nucleoplasts that still have the plasma membrane around them and a thin rim of cytoplasm. In essence they are still live cells, but with very little cytoplasm. There has been no dilution of the nuclear contents.
SCHRADER: Alright then let me propose an experiment. Suppose you take those nucleoplasts and you pretend that they are tissue culture cells and you go through a standard hypotonic shock, or whatever you normally do to make cell fractionation studies by the classical means. If you disrupt them in that way, do the nucleoplasts then dump their unoccupied receptor into the soluble fraction very readily?
WELSHONS: For the GH_3 cells clearly. The receptor distribution data that were shown earlier (Table 2) were for cells that had already been enucleated; 85% to 95% of the receptors were extractable into low-salt cytosol.
SCHRADER: So, as with the vitamin D case, you then would favor the idea that the steroid receptors are weakly held nuclear nonhistone chromosomal proteins? Would that be a fair assesment of what you are trying to say?
WELSHONS: Yes.
MILGROM: Just one point of clarification. When you showed the progesterone and glucocorticoid receptors, what cell types were they?
WELSHONS: Those were in the GH_3 cell type.
MILGROM: An Italian group reported that if you homogenize tissues at, not zero degrees, not low temperature, then even after homogenization the receptor in the absence of the hormone remains attached to the nucleus. Did you try this method?
WELSHONS: No we didn't, but there are actually a number of reports in which the distribution is not the expected one. For example, McCormick and Glasser (Endocrinology 106:1634–1649, 1980) dispersed and separated the immature rat uterine cell types. When the uterus itself was homogenized, they found that 90% of the estrogen receptor was extractable in the cytosol as usual, but when they homogenized the dispersed cells from that same tissue, 90% of that unoccupied receptor remained in the nuclei. They couldn't explain the phenomenon but certainly it suggests that where the receptor is found, the extent of extraction from nucleus can depend on details of the homogenizing conditions.
MILGROM: One question about the turnover of receptor in the nucleoplasts. Did you look at this problem?
WELSHONS: We're beginning to. A problem we have at this point in the MCF-7 cells is that when the cells are replated even without enucleation, the receptor content drops initially for the first hour or two. This confounds initial rate measurements, and we are trying to work around it by using other cell types or other media. At any rate, the nucleoplasts at 37°C lose estrogen receptor after the enucleation faster than do whole cells (or the various cytoplast fractions) replated at the same time, and in fact that difference in the degradation rate has a half-time of around 3 to 5 hours.
MILGROM: What happens if you give hormone to the nucleoplast?
WELSHONS: We haven't done that, although we are interested in what regulates both degradation and synthesis in this model, where the compartments for synthesis and degradation have presumably been separated.
MUELLER: Since you have such a weak association of the receptor apparently with the nucleus, have you been able to establish any conditions that would retain this in your homogenization state?
WELSHONS: We have not actually looked in that direction. Peter Sheridan, though,

in papers in *Endocrinology* and *Nature,* examined different extraction volumes, and he found that the less the dilution of the tissues when he homogenized, the larger the fraction of unoccupied receptor that was retained by the nuclei, up to 40% of the total unoccupied sites.

MUELLER: It seems to me that what one is talking about is some type of phase distribution and that basically to talk about the receptors as being concentrated with nucleus represents a status quo situation for phase distribution. If one looked at a dynamic state, receptors may be floating in and out at a fantastic rate and the representation one sees may not have a real significance.

CLARK: There was actually one paper published by Jack Gorski, and I can't remember the other author, in which they homogenized tissues in a very peculiar thing.

WELSHONS: Sucrose?

CLARK: No, it was worse than sucrose; maybe methanol? Very peculiar things and they found nuclear localization.

WELSHONS: To address the earlier comment, a dynamic phase distribution of the receptor between nuclear and cytoplasmic compartments has been suggested by several groups. However, we think that data on cooperative hormone binding (Fig. 12) and extent of receptor transformation suggest that in intact cells (rather than in homogenized, cell-free systems), the receptors behave as though they are bound in the cell rather than soluble, and this would not be consistent with a highly dynamic state. These data would be more consistent with the unoccupied receptor as an integral chromatin protein, at or close to the sites of action before hormone binding. But of course the physical properties of the unoccupied receptor, why it appears to be so loosely associated with the nucleus, and whether it is bound or soluble in the nucleus are currently unresolved questions.

Discussants: J. CLARK, E. MILGROM, G. MUELLER, W. SCHRADER, and W. WELSHONS

Chapter 2

Structure, Dynamics, and Cloning of the Estrogen Receptor

G.L. GREENE

Introduction

Recent biochemical (Welshons et al., 1984) and immunocytochemical data (King and Greene, 1984; Press and Greene, 1984; Press et al., 1985) suggest that the majority of functional estrogen receptor may reside in the nucleus, regardless of hormone status, and that binding of hormone to receptor leads to a tighter association of steroid-receptor complex with nuclear components. The nature of this association is not known, although a number of nuclear acceptor sites have been proposed, including specific DNA sequences (Cato et al., 1984; Jost et al., 1985; Compton et al., 1983; Payvar et al., 1983), ribonucleoprotein (Liang and Liao, 1974), basic nonhistone proteins (Puca et al., 1974), the nuclear matrix (Barrack and Coffey, 1980; Barrack, 1983), and acidic nonhistone protein/DNA complexes (Spelsberg et al., 1983). The biological significance of these results has not been established, and, as yet, no one has been able to reconstitute all of the cellular components required for steroid hormone response in any cell-free system.

The availability of specific polyclonal and monoclonal antibodies to various steroid receptors has led to new approaches to the study of the structure, composition, and dynamics of steroid receptors. As independent probes for the receptor molecule, these antibodies are being used to detect, measure, and purify receptor in tissue extracts, to determine the distribution and intracellular location of receptors in various responsive tissues, as well as to map the hormone- and DNA-binding domains of the receptor and to study the structural changes that accompany the binding of steroids to their receptors. This chapter summarizes the results of our efforts to purify and characterize calf and human estrogen receptors as well as to prepare specific monoclonal antibodies to these proteins. As a consequence of their high specificity and affinity for receptor, the resulting antibodies have been particularly useful for purifying and comparing estrogen receptor (ER) from various sources, mapping functional domains on the receptor molecule, and for localizing ER in reproductive tissues and cancers in the presence and absence of hormone or antagonist. Our monoclonal antibodies have also proved invaluable for isolating cDNA clones corresponding to the human ER gene (Walter et al., 1985).

Purification of Estrogen Receptor

Early attempts to purify ER from calf uterus relied on the use of multiple-step protocols that were laborious, required large amounts of tissue, and afforded very low yields of highly purified receptor. In addition, receptor was frequently degraded during purification. However, it was possible to obtain enough partially purified nuclear [^{3}H]estradiol-receptor complex (E^*R_n) from calf uteri to generate the first polyclonal (Greene et al., 1977; Greene et al., 1979) and monoclonal (Greene et al., 1980a) antibodies to the steroid-binding protein. Purification of this complex was achieved by a sequence of extraction of E^*R_n from nuclear pellets with 400 m*M* KCl, followed by ammonium sulfate precipitation, gel filtration, and polyacrylamide gel electrophoresis (Greene et al., 1979). A 12,000-fold purification of receptor afforded a 1% yield of E^*R_n that was essentially pure and that contained one molecule of [^{3}H]estradiol per protein molecule of M_r 68,000. Immunizations were carried out with receptor that was about 20% pure.

As part of our more recent efforts to isolate ER in a pure form for detailed analysis of amino acid composition and sequence as well as physicochemical properties, we have developed a two-step affinity chromatography procedure for the purification of unoccupied cytosolic ER (Greene, 1984). The use of steroid affinity chromatography for the purification of receptors has generally been limited by the resistance of the bound receptor to elution under conditions compatible with its stability as well as by cleavage of ester and amide groups in the spacer arms that link steroids to the supporting matrix. For ERs, the elution problem was solved by including chaotropic salts such as sodium thiocyanate (Greene et al., 1980b) with dimethylformamide (Musto et al., 1977) in the eluting medium with estradiol to facilitate release of the receptor protein. A stable steroid affinity adsorbent was prepared by linking estradiol to Sepharose 6B via a thioether bridge in the 17α position of the steroid (Greene et al., 1980b). As a result, we have established a purification scheme that is simple, reproducible, and that affords a good yield of highly purified receptor as the steroid-receptor complex (E^*R_c) (Greene, 1984; Greene et al., 1980b). Following the examples of Puca (Molinari et al., 1977) and Bresciani (Sica and Bresciani, 1979), who used heparin-Sepharose to improve their purification of calf uterine ER, we included heparin-Sepharose in our protocol.

The cytosolic forms of ER from calf uterus and from MCF-7 human breast cancer cells are now routinely purified to virtual homogeneity by the above protocol. For MCF-7 receptor, the overall recovery of receptor as E^*R_c is 30%–45% and the purity ranges from 60% to greater than 90% of the specific radioactivity expected for one molecule of [^{3}H]estradiol bound to a 4S monomer of M_r 65,000. We have isolated as much as 5 nmol (315 μg) of receptor in a single experiment. Recent modifications of the protocol include omission of the gel filtration step prior to heparin-Sepharose chromatography and replacement of the original di-n-propyl thioether bridge in the steroid resin with a 1,4-bis(2,3-epoxypropoxy)butyl thioether spacer. The latter estradiol resin affords more highly purified receptor (300- to 1000-fold), indicating that less nonspecific adsorption of proteins to the adsorbent occurs when the steroid is linked through the longer diglycidyl ether bridge.

The highly purified MCF-7 E^*R_c has properties that are similar to, if not the same as, the activated steroid-receptor complex found in high-salt nuclear extracts of MCF-7 cells. Thus, the purified cytosol E*R binds DNA and sediments as a 5.3S species in sucrose gradients containing 400 m*M* KCl. In addition, an apparent molecular weight of 140,000, calculated from a Stokes radius of 5.74 A and the 5.3S sedimentation coefficient, is consistent with the formation of a homodimer of two 65K (4S) monomers. Chemical cross-linking experiments and dense amino acid labeling of unpurified nuclear MCF-7 ER also indicate that activated E*R is a homodimer (Miller et al., 1985). Interestingly, the purified E*R has lost its ability to form an 8–9S complex in low-salt gradients and sediments as a 5.9S species in 10 m*M* KCl, indicating that the factors, or factor, responsible for the formation of these larger complexes in cytosols are removed during purification. In fact, if purified receptor is added to receptor-depleted MCF-7 cytosol, a 7–8S complex is observed in 10 m*M* KCl. When highly purified receptor is analyzed by sodium dodecyl sulfate (SDS)-gel electrophoresis under reducing conditions, one major silver-stained band, M_r 65,000, is seen. The same band can be visualized by autoradiography if E* is exchanged with [^{3}H]tamoxifen aziridine, a specific covalent tag for estrogen receptor (Katzenellenbogen et al., 1983). When incubated with ^{32}P-adenosine triphosphate (ATP) in the presence of a purified cytosolic calcium/calmodulin-dependent protein kinase isolated from rat brain by Howard Schulman of Stanford, both calf and human 65K E*R proteins were efficiently labeled, indicating that these receptors can serve as substrates for this enzyme in vitro. It is not clear whether ER is phosphorylated in vivo, although in vitro or in vivo phosphorylation of calf (Migliaccio et al., 1984) and chicken (Raymoure et al., 1985) ER, mouse glucocorticoid receptor (Housley and Pratt, 1983), and avian (Weigel et al., 1981; Dougherty et al., 1984) progesterone receptor have been reported. It is possible that two closely spaced protein bands observed by us in stained and immunoblotted SDS gels of purified calf uterine ER (Greene, 1984) represent different states of phosphorylation of the receptor.

Monoclonal Antibodies to Estrogen Receptor

The first monoclonal antibodies to mammalian ERs were prepared in our laboratory by polyethylene glycol-mediated fusion of splenic lymphocytes from male Lewis rats, which were immunized with partially purified nuclear E*R from calf uterus, with mouse myeloma cells (P3-X63-Ag8, P3-NSI/1-Ag4-1, and Sp2/0-Ag14) (Greene et al., 1979). Rat antibodies were detected in hybridoma culture medium by double antibody precipitation of crude nuclear E*R from calf uterus. Both IgM- and IgG2a-secreting hybridomas were obtained. Like the polyclonal rat antiserum, all of the monoclonal antibodies recognized 4S cytosol E*R and 5S nuclear E*R from calf uterus. However, in contrast to monoclonal IgG, which showed comparable affinity for the cytosol and nuclear forms of calf E*R, IgM reacted preferentially with the nuclear form. The reasons for this distinction are not clear. All these monoclonal antibodies were specific for calf ER, as were the polyclonal antibodies present in the serum of the immunized rat. These antibodies recognized occupied as well as unoccupied receptors and did not interfere with the binding of steroid to receptor. All the data accumulated thus far indicate that the 10 monoclonal antibodies (IgM and IgG) recognize either the same epitope or mutually exclusive epitopes on the calf receptor.

Although the monoclonal antibodies prepared against calf ER have proved useful for the characterization and purification of cytosol and nuclear forms of calf ER, they are limited by their specificity and recognition of only one region of the receptor molecule. Because of our interest in being able to study ERs in other species, particularly in human reproductive tract and breast cancer, all subsequent efforts have been directed toward preparing monoclonal antibodies against human ER. With the successful partial purification of cytosol ER obtained from MCF-7 human breast cancer cells, we began, in 1979, to immunize male Lewis rats with E^*R_c, eluted from the estradiol affinity resin, which was about 5%–10% pure. Fusion of splenic lymphocytes from immunized animals with two different mouse myeloma lines (P3-X63-Ag8 and Sp2/0-Ag14) yielded three cloned hybridomas (D58, D75, D547) (Greene et al., 1980b), each of which secretes a unique idiotype of antibody that recognizes a distinct region of the ER molecule. Subsequent fusions, carried out both in our own laboratory (Greene et al., 1984) and at Abbott Laboratories (Miller et al., 1982), have produced a total of 13 monoclonal antibodies, all of which (with one possible exception) recognize distinct regions of the receptor molecule. These antibodies have high affinity ($K_d = 10^{-9} - 10^{-10} M$) for both steroid-occupied and unoccupied ER and recognize nuclear as well as cytosol forms of the receptor molecule. Although they vary in their cross-reactivity with ERs from various animal species, each antibody appears to be completely specific for the 65,000-dalton steroid-binding subunit of the ER complex, as judged by extensive sucrose gradient and immunoblot analyses of cytosol and nuclear extracts from a variety of tissues and cell lines. Cross-reactivity patterns (Greene et al., 1984) indicate both sequence homology and heterogeneity among mammalian and nonmammalian estrophilins. Some determinants (e.g., H222 and H226) are common to all tested ERs, including those from hen oviduct, whereas others (e.g., D547 and D58) are present only in mammalian receptors, and one (D75) appears to be restricted to primate estrophilin.

Immunochemical Analysis of Receptor Structure and Dynamics

Two of the more important uses of the monoclonal ER antibodies have been the immunochemical purification of ER from different sources and the mapping of functional domains of the receptor. MCF-7 cytosol ER has been purified to near homogeneity in a single step by chromatography on an immunoadsorbent consisting of D547 IgG conjugated to Sepharose 4B (Greene, 1984). Intact estradiol-receptor complex (E*R) can be eluted in good yield with a buffer containing sodium thiocyanate and dimethylformamide, similar to the conditions used for the elution of receptor from the estradiol affinity adsorbent. The E*R obtained by this procedure appears to be identical to receptor purified by affinity chromatography. The D75, D547, H222, and H226 antibodies have also been used to screen a λgt11 cDNA library for expressed MCF-7 ER peptide fragments (Walter et al., 1985), to immunopurify ^{35}S-labeled ER in mRNA translation mixtures, and to identify ER in liver nuclear matrix preparations from estrogen-treated rats (Alexander et al., 1986, submitted for publication).

In ongoing studies to map the location of various determinants in relation to each other and to the steroid- and DNA-binding domains, the relative positions of nine unique determinants have been determined by density gradient analysis of antibody-E*R interaction after limited proteolysis of MCF-7 cytosol E*R with trypsin, chymotrypsin, or papain (Greene et al., 1984). As shown in Fig. 1, determinants for three of the monoclonal rat antibodies (D75, D547, H226) are susceptible to selective cleavage by one or more of the enzymes tested. Six other antibodies are capable of binding

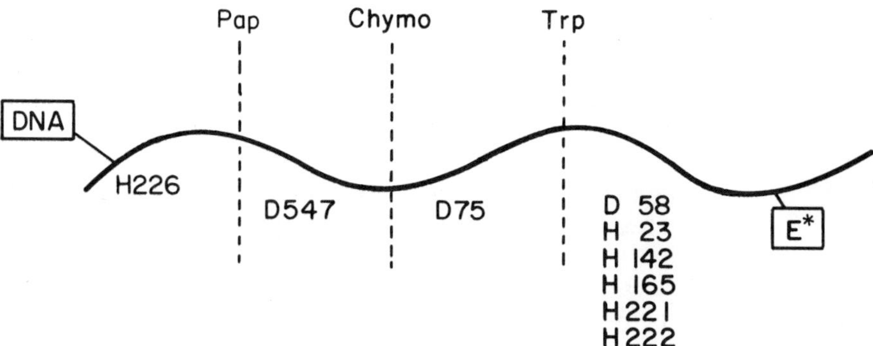

Fig. 1. Map of antigenic determinants in relation to steroid-binding and DNA-binding domains of the MCF-7 cytosol estrogen receptor. Dashed lines indicate sites of cleavage for listed enzymes. Postulated recognition sites for nine monoclonal antibodies are shown. E* = estrogen-binding region; DNA = DNA-binding region. (Reproduced with permission from Greene et al., 1984, *J Steroid Biochem* 20:51–56)

the smallest (2.6S) steroid-binding fragment remaining after cleavage with trypsin. When tested for their ability to associate with φX174 double-stranded DNA in sucrose density gradients, none of the E*R fragments was able to bind DNA, whereas the intact 5S nuclear receptor cosedimented with the DNA. Although the H226 and H222 determinants appear to be well separated from each other on the polypeptide chain, the corresponding antibodies can not bind simultaneously, suggesting that these regions of the receptor are spatially proximal to each other. Interestingly, the determinants that are most well conserved across all tested species are located either near the steroid-binding domain (H23, H142, H165, H221, H222) or near the DNA-binding domain (H226).

The major disadvantage of all receptor assays performed on tissue extracts is their inability to provide information about inter- and intracellular receptor distribution. To overcome this problem, five monoclonal antibodies (D547, D58, D75, H222, H226) have been used individually to localize estrophilin by an indirect immunoperoxidase technique in frozen, fixed sections of human breast tumors (King et al., 1985), human uterus (Press and Greene, 1984), rabbit uterus (King and Greene, 1984), and in other mammalian reproductive tissues, as well as in fixed MCF-7 cell cultures (King and Greene, 1984) and in paraffin-embedded sections of breast tumors (Poulsen et al., 1985) and human endometrium. Specific immunoperoxidase staining for receptor in estrogen-sensitive tissues is confined to the nucleus of all stained cells, regardless of hormone status. Staining is absent in nontarget tissues, such as colon epithelium, and in receptor-negative breast cancers; in addition, it can be abolished by the addition of highly purified receptor to primary antibody. Heterogeneous staining has been observed in MCF-7 cells as well as in receptor-poor and receptor-rich breast cancers (King et al., 1985), possibly reflecting either the variations in cell cycle or the presence of estrogen-sensitive and insensitive cells. Little or no cytoplasmic staining for ER has been observed in any of the tissues or tumor cells examined thus far, including those deprived of exogenous estrogens. Treatment of cells or tissues in vivo or in vitro with estradiol alters the intensity but not the distribution of specific staining for ER. Thus, when immature rabbit uteri were incubated in vitro with 20 n*M* estradiol for 60 min at 37°C, a slight decrease in the intensity of specific nuclear staining was observed. A similar result was obtained with MCF-7 cells that been cultured on charcoal-stripped serum for 4 days. However, when postmenopausal human uterus was treated in the same manner, an increase in nuclear staining intensity was observed similar to the results reported by McClellan et al. (1984) for uterus and oviduct obtained from estrogen-treated monkeys. In all cases, no specific cytoplasmic staining was observed by light microscopy.

In view of the exclusively nuclear localization of specific immunoperoxidase staining for receptor in all estrogen-sensitive tissues and cells studied thus far, it appears that both cytosol and nuclear forms of the receptor reside in the nuclear compartment in the presence and absence of steroid. These observations are consistent with the hypothesis that unoccupied estrophilin recovered in the low-salt cytosol fraction of a tissue homogenate represents receptor that is loosely associated with nuclear components and that binding of estradiol to receptor leads to a tighter association, a phenomenon previously interpreted as indicating translocation of the receptor from the cytoplasm to the nucleus (Jensen et al., 1968; Gorski et al., 1968). This revised interpretation of steroid hormone action is shown schematically in Fig. 2 (DeSombre et al., 1984). According to this model the steroid passes through the cell to the nucleus, either unaided or perhaps bound loosely to low affinity sites in the cytoplasm, where it interacts with unoccupied receptor in the nucleus, resulting in the formation of the activated (dimeric?) steroid-receptor complex. Several lines of evidence support this hypothesis, including the observation by Welshons et al. (1984; Welshons et al., 1985) that cytochalasin-induced enucleation of rat pituitary (GH_3) cells leads to partitioning of unoccupied ER almost exclusively into the nucleoplast fraction. Also, progesterone receptor has been localized to the nuclei of hormone-responsive cells in chick oviduct (Gasc et al., 1984) and in rabbit uterus (Perrot-Applanat et

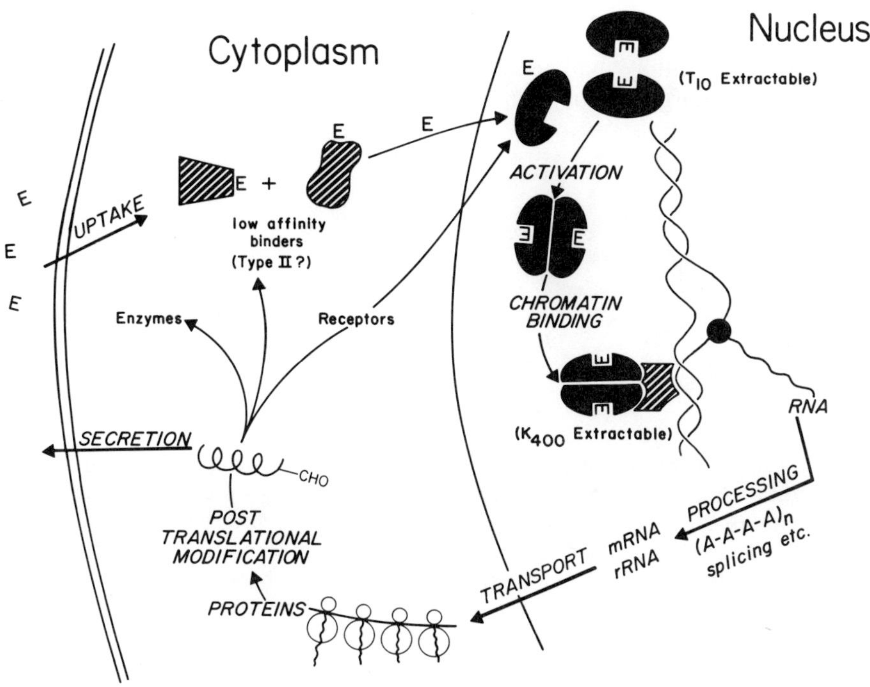

Fig. 2. Revised schematic diagram of estrogen action in a target cell. (Reproduced with permission from DeSombre et al, 1984, in: *Hormones and Cancer,* vol. 142. Alan R. Less, Inc., New York, pp 1–21)

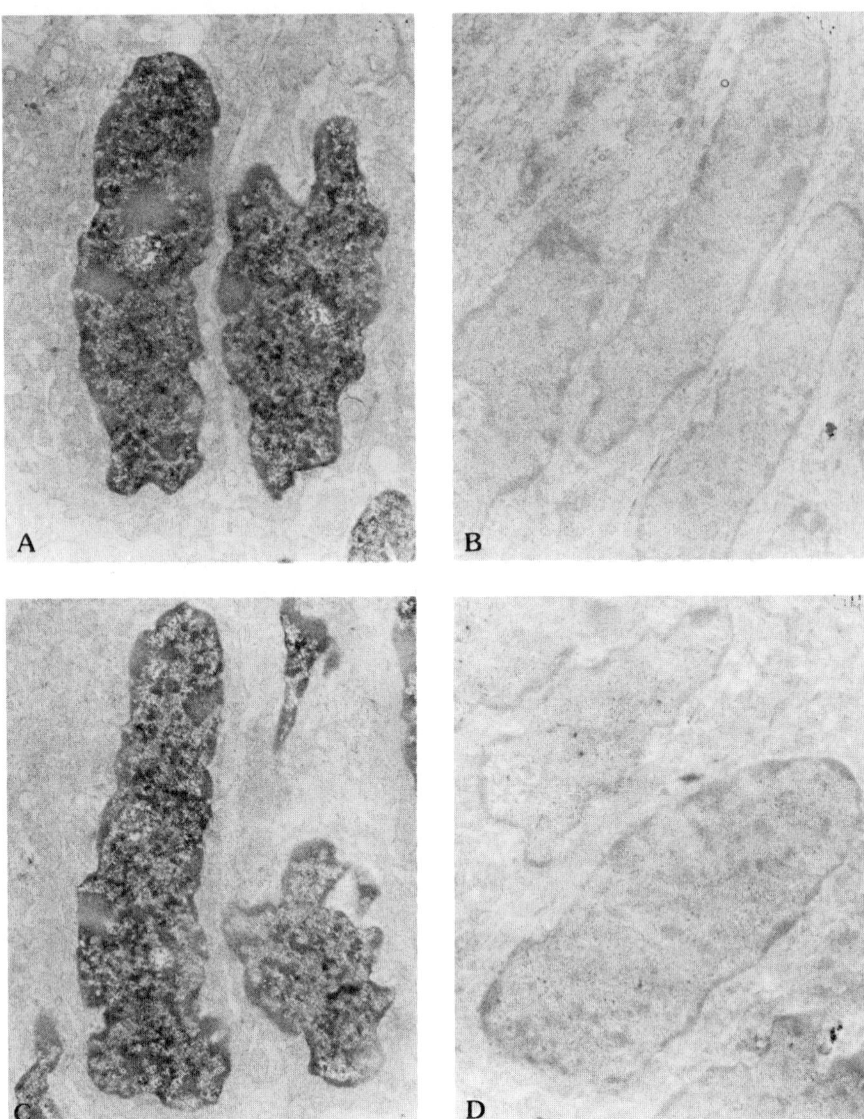

Fig. 3. Immunoelectron microscopic localization of estrogen receptor in human proliferative phase endometrium. Fixed, frozen sections were treated identically except for the first incubation, which varied as follows: **A**: H226 ER antibody alone (3.3 pmol/ml); intranuclear localization of diaminobenzidine reaction product is observed. **B**: H226 (3.3 pmol/ml) in MCF-7 human breast cancer cytosol (20 pmol ER/ml); the formation of immunoprecipitate is inhibited. **C**: H226 (3.3 pmol/ml) in receptor-depleted MCF-7 cytosol (0.2 pmol ER/ml; ER was depleted by estradiol affinity chromatography); intranuclear reaction product is observed. **D**: H226 (3.3 pmol/ml) in ER-depleted cytosol to which highly purified MCF-7 ER (50 pmol ER/ml) was added; the formation of immunoprecipitate is inhibited. All magnifications are 10,000×. No uranyl acetate or lead citrate was used. (Reproduced with permission from Press et al. 1985, *J Histochem Cytochem* 33:915–924)

al., 1985) with polyclonal and monoclonal antibodies, respectively. In contrast, immunoreactive glucocorticoid receptor has been observed in both the cytoplasm and the nucleus of target cells (Fuxe et al., 1985a; Fuxe et al., 1985b). The reasons for this discrepancy are not clear at this time. However, thus far no one has been able to demonstrate conclusively the hormone-induced translocation of any steroid receptor from the cytoplasm to the nucleus of a target cell.

At least two forms of steroid-receptor complex appear to be tightly bound to chromatin. One of these forms can be extracted with salt (0.4 *M* KCl) and one is resistant to salt extraction and may be associated with transcriptionally active DNA on the nuclear matrix (Alexander et al., 1986, submitted for publication). Both forms are recognized by one or more of our monoclonal antibodies and both migrate as M_r 65,000 proteins in reducing SDS gels. We are currently trying to identify specific chromatin-binding sites for ER as a function of occupancy by estrogens and estrogen antagonists by immunoelectron microscopy. Preliminary results (Press et al., 1985) indicate that unoccupied ER is localized in the dispersed euchromatin, but not in the condensed heterochromatin or nucleoli of epithelial and stromal nuclei of human endometrium (Fig. 3A,C) and MCF-7 cells. Competition studies (Fig. 3B,D) indicate that only ER inhibits specific nuclear staining. The absence of appreciable specific cytoplasmic localization at the electron microscopic level is consistent with light microscopic results. However, one would predict that detectable ER should be associated with ribosomes at some stage of the cell cycle, especially in view of the relatively rapid turnover of receptor ($t_{1/2}$ = 4 h) (Miller et al., 1985). It is possible that ribosome-associated ER can be detected by immunoelectron microscopy with colloidal gold (Christensen et al., 1985). In addition, this technique should allow us to determine whether chromatin structure and/or receptor sites are altered when cells are exposed to estrogens and estrogen antagonists.

Cloning of Human Estrogen Receptor cDNA

The isolation of cDNA clones corresponding to part or all of the translated sequence of ER mRNA from MCF-7 human breast cancer cells has now been achieved (Walter et al., 1985). Estrogen receptor sequences were identified in randomly primed λgt10 and λgt11 MCF-7 cDNA libraries by screening either with monoclonal ER antibodies or with synthetic oligonucleotides corresponding to two peptide sequences obtained from purified MCF-7 human ER. Among the cDNA clones isolated by oligonucleotide hybridization was a 2.1 kb cDNA (OR8), which cross-hybridized with all other cDNAs and which contained the expected sequences for the two ER peptides. In addition, this cDNA hybridized selectively to a 6.2 kb poly(A)$^+$RNA, which, when translated in vitro in the presence of [^{35}S]methionine, coded for the synthesis of immunoreactive 65 kDa ER as well as a smaller amount of an immunoreactive 46 kDa protein (Walter et al., 1985). The molecular weight of the major product is in good agreement with published values of 65 to 70 kDa for ER from several sources (Redeuilh et al., 1980; Lubahn et al., 1985; Katzenellenbogen et al., 1983; Sakai and Gorski, 1984; Van Oosbree et al., 1984). The identity of the smaller peptide is not known, but it may represent an in vitro degradation product of the 65 kDa ER. From the size of the mRNA for ER, it is clear that a large portion should be untranslated. A 65 kDa protein would require about 1.8 kb of coding sequence, leaving more than 4 kb untranslated. In addition, the translated region is likely to be at the 5′ end of the gene, similar to the structure of several other receptor mRNAs, including the recently cloned human glucocorticoid receptor mRNA, which also contains a long 3′-untranslated region (Hollenberg et al., 1985). Studies are in progress to determine the complete amino acid sequence of human ER as well as to look for any sequence homologies among steroid receptors, especially in regions comprising putative DNA- and steroid-binding domains.

Acknowledgments. These investigations were supported by grants from the National Institutes of Health (CA 02897), the American Cancer Society (BC 86), Abbott Laboratories, and the Women's Board of the University of Chicago Cancer Research Foundation.

References

Alexander RB, Greene GL, Barrack ER (1986) Endocrinology (unpublished)

Barrack ER (1983) Endocrinology 113: 430–432

Barrack ER, Coffey DS (1980) J Biol Chem 255: 7265–7275

Cato ACB, Geisse S, Wenz M, Westphal HM, Beato M (1984) EMBO J 3: 2771–2778

Christensen AK, Komorowski TE, Wilson B, Ma SF, Stevens RW III (1985) Endocrinology 116: 1983–1996

Compton JG, Schrader WT, O'Malley BW (1983) Proc Natl Acad Sci USA 82: 16–20

DeSombre ER, Greene GL, King WJ, Jensen EV (1984) In: Hormones and cancer Vol 142. Alan R Liss Inc, New York, pp 1–21

Dougherty JJ, Puri RK, Toft DO (1984) J Biol Chem 259: 8004–8009

Fuxe K, Harfstrand A, Agnati LF, Yu ZY, Wikstrom AC, Okret S, Cantoni E, Gustafsson JA (1985a) Neurosci Lett 60: 1–6

Fuxe K, Wikstrom AC, Okret S, Agnati LF, Harfstrand A, Yu ZY, Granholm L, Gustafsson JA (1985b) Endocrinology 117: 1803–1812

Gasc JM, Renoir J, Radanyi C, Joab I, Tuohimaa P, Baulieu EE (1984) J Cell Biol 99: 1193–1201

Gorski J, Toft D, Shymala S, Smith D, Notides A (1968) Recent Prog Horm Res 24: 45–80

Greene GL, Closs LE, Fleming H, DeSombre ER, Jensen EV (1977) Proc Natl Acad Sci USA 74: 3681–3685

Greene GL, Closs LE, DeSombre ER, Jensen EV (1979) J Steroid Biochem 11: 333–341
Greene GL, Fitch FW, Jensen EV (1980a) Proc Natl Acad Sci USA 77: 157–161
Greene GL, Nolan C, Engler P, Jensen EV (1980b) Proc Natl Acad Sci USA 77: 5115–5119
Greene GL (1984) In: Biochemical Action of Hormones Vol XI. Academic Press, New York, pp 207–239
Greene GL, Sobel NB, King WJ, Jensen EV (1984) J Steroid Biochem 20: 51–56
Hollenberg SM, Weinberger C, Ong ES, Cerelli G, Oro A, Lebo R, Thompson EB, Rosenfeld MG, Evans RM (1985) Nature 318: 635
Housley PR, Pratt WB (1983) J Biol Chem 258: 4630–4635
Jensen EV, Suzuki T, Kawashima T, Stumpf WE, Jungblut PW, DeSombre ER (1968) Proc Natl Acad Sci USA 59: 632–638
Jost JP, Geiser M, Seldran M (1985) Proc Natl Acad Sci USA 82: 988–991
Katzenellenbogen JA, Carlson KE, Heiman DF, Robertson DW, Weill LL, Katzenellenbogen BS (1983) J Biol Chem 258: 3487–3495
King WJ, Greene GL (1984) Nature 307: 745–747
King WJ, DeSombre ER, Jensen EV, Greene GL (1985) Cancer Res 45: 293–304
Liang T, Liao S (1974) J Biol Chem 249: 4671–4678
Lubahn DB, McCarty Jr. KS, McCarty Sr. KS (1985) J Biol Chem 260: 2215
McClellan MC, West NB, Tacha DE, Greene GL, Brenner RM (1984) Endocrinology 114: 2002–2014
Migliaccio A, Rotondi A, Auricchio F (1984) Proc Natl Acad Sci USA 81: 5921–5925
Miller LS, Tribby IIE, Miles MR, Tomita JT, Nolan C (1982) Fed Proc 41: 520
Miller MA, Mullick A, Greene GL, Katzenellenbogen BS (1985) Endocrinology 117: 515–522
Molinari AM, Medici N, Moncharmont B, Puca GA (1977) Proc Natl Acad Sci USA 74: 4886–4890
Musto NA, Gunsalus GL, Miljkovic M, Bardin CW (1977) Endocr Res Commun 4: 147–157
Payvar F, DeFranco D, Firestone GL, Edgar B, Wrange O, Okret S, Gustafsson JA, Yamamoto KR (1983) Cell 35: 381–392
Perrot-Applanat M, Logest F, Groyer-Pickard MT, Milgrom E (1985) Endocrinology 116: 1473–1484
Poulsen HS, Ozzello L, King WJ, Greene GL (1985) J Histochem Cytochem 33: 87–92
Press MF, Greene GL (1984) Lab Invest 50: 480–486
Press MF, Nousek-Goebl NA, Greene GL (1985) J Histochem Cytochem 33: 915–924
Puca GA, Sica V, Nola E (1974) Proc Natl Acad Sci USA 71: 979–983
Raymoure WJ, McNaught RW, Smith RG (1985) Nature 314: 745–747
Redeuilh G, Richard-Foy R, Secco C, Torelli V, Bucourt R, Baulieu EE, Richard-Foy A (1980) Eur J Biochem 106: 481
Sakai D, Gorski J (1984) Endocrinology 115: 2379
Sica V, Bresciani F (1979) Biochemistry 18: 2369–2378
Spelsberg TC, Littlefield BA, Seelke R, Martin-Dani G, Toyoda H, Boyd-Leinen P, Thrall C, Kon OL (1983) Recent Prog Horm Res 39: 463–517
Van Oosbree TR, Kim UH, Mueller GC (1984) Anal Biochem 136:321
Walter P, Green S, Greene G, Krust A, Bornert JM, Jeltsch JM, Staub A, Jensen

E, Scrace G, Waterfield M, Chambon P (1985) Proc Natl Acad Sci USA 82: 7889–7893
Weigel NL, Tash JS, Means AR, Schrader WT, O'Malley BW (1981) Biochem Biophys Res Commun 102: 513–519
Welshons WV, Lieberman ME, Gorski J (1984) Nature 307: 747–749
Welshons WV, Krummel BM, Gorski J (1985) Endocrinology 117: 2140–2147

Discussion of the Paper Presented by G. Greene

ROY: Is there any sequence homology between the estrogen receptor cDNA and the glucocorticoid receptor cDNA?
GREENE: We don't know.
SPELSBERG: Your localization in the euchromatin, or dispersed chromatin, is that with, or without estrogen?
GREENE: That was without estrogen.
SPELSBERG: Have you done it with estrogen?
GREENE: I don't have any immunoelectron micrographs of treated cells.
SPELSBERG: Is the only time you see matrix-associated receptors after estrogen treatment?
GREENE: That is correct.
SPELSBERG: In your northern blot you showed a very large mRNA and implied that the 2.1 kilobase (kb) cDNA (OR8) contained the start site for translation. Is that true?
GREENE: Yes, that is true as far as I know.
SPELSBERG: So that cDNA contains a whole coding sequence?
GREENE: Yes, although there is still some ambiguity with that cDNA at the moment. There is definitely a 1.5 kb open reading frame in that segment, which would correspond to the immunoreactive peptide that we have observed in transformed *Escherichia coli*. However, we believe that OR8 contains the entire coding sequence for ER with the correct 5′ start site.
SPELSBERG: In your map of all the cDNAs, you said that the steroid-binding site was on one end. On which end is the steroid-binding site?
GREENE: The 3′ end of the cDNA, which corresponds to the carboxy terminus of the protein.
OLSEN: A recent paper in the *Journal of Biological Chemistry* reported on a putative molibdate stabilized receptor in calf uterus of 89,000 molecular weight. I notice that your antibodies don't show any protein of this size. First, how do you reconcile this, and, second, is there any evidence from your cloning that this possibility exists?
GREENE: I'm glad you brought that up. Obviously I can't answer that question directly, although Tom Ratajczak and I have communicated on this issue, and he is sending me some purified 89K protein for analysis. My suspicion is that this 89K protein may be similar to the 90K protein that Dave Toft isolated in association with progesterone receptor. There is no direct evidence that Ratajczak was looking at the calf receptor on SDS gels. Since tamoxifen aziridine is available commercially now, I am surprised that he didn't label the 89K protein that was on the gel and show that it indeed was the right molecule. He only showed silver-stain data on the purified protein, so my feeling is that this protein will not turn out to be a

steroid-binding component. We have never seen anything larger than 65 KDa for ER with any of our antibodies, or with tamoxifen aziridine or in translation products.

SPELSBERG: One last question. Concerning your 46K protein, do you have any evidence that this could be a regulated transcription of that gene?

GREENE: No, there isn't any good data on that question right at the moment. The only thing one can say is that there seems to be a variable amount of the 46K protein at different times, which would be more consistent with degradation of the 65K species, but I really can't answer the question.

O'MALLEY: It seems that in most presentations, these smaller molecular forms are always written off as degradation. I would say I don't see any evidence that the 46K protein is degradation. In fact, it does not look like degradation to me. You're getting two different products from m an authentic ER clone, one of the expected molecular weight, and one of 46K. I think many of us have synthesized proteins in the reticulocyte system and protease is not a problem with protein products, so you could say this is exquisitely sensitive. It gives only one clip; you get only a single clear band and 46K out of it. You can resolve this very simply by doing a time-course incubation in the reticulocyte mixture. You just let it go another 30 minutes and if that is degradation, it will be in small molecular weight form. Then you will know. But I would say if you write it off to degradation, you're likely to be missing something important, which means another related gene, a truncated message from aberent processing, or perhaps a molecule not binding hormone, which is very important to the complex. In fact it may well be the other member of the dimeric complex for nuclear activation to the 5.3S form, so I would concentrate on that smaller molecular form at this point because with the in vitro translation, I think you have a mechanism to show whether or not it is degradation and in enough form to look at complex interactions between the 56 and 46.

GREENE: I think it is something that needs to be explored and we certainly can't say for certain that it is a degradation product.

MILGROM: Do you have any kinase activity associated with the purified receptor?

GREENE: No, we haven't observed such activity.

MILGROM: When you purify the receptor, is it still native; does it still bind steroid? Can it bind DNAs and so on?

GREENE: It will bind to DNA and it still binds steroid.

TATA: I have a question and a comment. The question is: It wasn't clear from your southern blot whether you have one large gene or in fact multiple genes?

GREENE: The available data support the idea that there is one gene.

TATA: I thought I would mention some additional evidence for nuclear localization, or at least the ability of estrogen receptor to come into the nucleus, from some unpublished experiments done in my lab by Sharon Oxenberg, over three years ago. The experiment consisted of homogenizing Xenopus liver and extracting all the receptor and then partially purifying it on Geoff's estradiol affinity matrix; the estradiol-eluted receptor was then injected into Xenopus ocytes. You may have seen that Nolan has recently reported similar experiments. Injections were carried out in different locations outside of the nucleus, and the nuclei were then manually decepted, just popped out, without homogenization. The cells are living and the experiment involved determining the kinetics of the transfer of the receptor to the nucleus; at the start, the receptor was entirely in the cytoplasmic fraction. To our surprise ER disappeared very rapidly from the cytoplasm and there was a corresponding accumulation of ER in the nucleus. The time course was about 30 minutes and, talking to Ron Lasky, who studied this problem with a large number of nuclear proteins, we concluded that it is one of the most rapidly moving proteins in the nucleus. Considering the viscosity of occyte cytoplasm, one can only conclude that what one is injecting is truly a nuclear protein, which is retained in the nucleus in an intact cell that has not been homogenized or insulted in any other way.

Discussants: G. Greene, E. Milgrom, M. Olsen, B. O'Malley, A.K. Roy, T. Spelsberg, and J.R. Tata

Chapter 3

Physical and Functional Parameters of Isolated Estrogen Receptors

G.C. MUELLER, M.R. OLSEN, K. BHATTACHARYYA, AND T.J. SCHUH

The current view of estrogen receptor action is that these molecules, on binding estradiol at 37°C, undergo an activation that facilitates their association with specific DNA or DNA/chromatin complexes. Little is known at this time, however, as to the molecular nature of such interactions or, more specifically, the manner in which they modulate the expression of specific genes. Accordingly, the goal of this chapter is to draw attention to some little-studied properties of estrogen receptors—properties arising from their interactions with other common molecules which may play a role in their interactions with chromatin and are likely to influence the cascading genetic events.

Our initial objective in this chapter is to demonstrate that estrogen receptors are dynamic molecules—highly active conformationally and capable of exhibiting a surprising degree of functional heterogeneity. The evidence for this view of receptors comes out of experiments in which the simple agents—*p*-secondary amylphenol, tetracaine, heparin, and arachidonic acid—have been found to dramatically influence the affinity of estrogen receptors for estradiol (Fig. 1).

In a series of experiments several years back (Mueller et al., 1984), our laboratory discovered that both the binding of estradiol and its release were influenced greatly by the presence of simple phenols in the binding assay. For example, 2-tetra hydronaphthol, an analogue of the A and B rings of estradiol (Fig. 1), prevented the forward binding of estradiol by cytosolic estrogen receptors, but did not displace the estradiol once it was bound (Mueller et al., 1984). In contrast, *p*-secondary amylphenol (*p*SAP), which possesses the same number of carbons but has a flexible alkyl side chain in place of the rigid B-ring of tetrahydronaphthol, both prevented the forward binding of estradiol and caused a striking concentration-dependent release of prebound estradiol at 0–4°C. These observations prompted us to conclude that in the formation of a complex with the receptor, estradiol very likely interacts first through its A-ring but its high-affinity binding is stabilized through secondary interactions and folding of the receptor molecule around the C/D rings of estradiol. In fact, it would

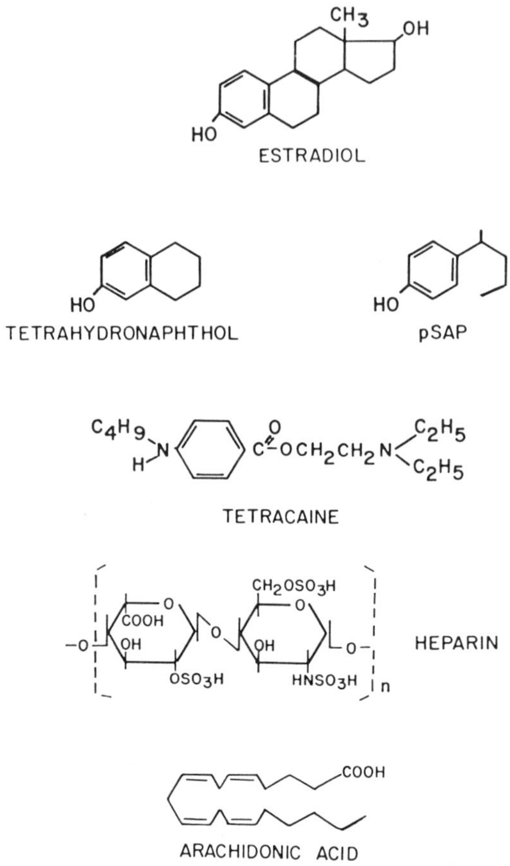

Fig. 1. Structures of some compounds that modify estrogen receptor structure and function.

appear that once the estradiol has been bound securely, the region of the receptor molecule that is approached by the A-ring of estradiol is very likely open for other approaching phenolic entities. It is proposed that when such a site is addressed by *p*SAP, the flexible side chain induces a conformational change in the receptor molecule, which relaxes the hydrophobic binding of the receptor to the C/D-rings of estradiol. In contrast, it would appear that 2-tetrahydronaphthol with its rigid B-ring may, like a second molecule of estradiol, actually reinforce the binding of the estradiol molecule that is already bound.

As is discussed later, this unusual effect of *p*SAP on estradiol binding of the receptor has proved very useful in the purification of receptors. In addition, this agent has also provided the initial evidence for functional heterogeneity of estrogen receptors. Experimentally, estrogen receptors, occupied by [^{3}H]estradiol, were adsorbed on heparin-agarose and then submitted to a linear gradient of *p*SAP. In agreement with the binding studies in cytosol, *p*SAP effected a release of the prebound estradiol;

however, the estradiol eluted in peaks characteristic of subsets of receptors (i.e., approximately five subsets) (Fig. 2). In this process, tightly bound estradiol is displaced by the low-affinity ligand, *p*SAP, while the receptor itself is retained on the heparin-agarose in a state that can rebind estradiol.

A second agent that can influence estradiol binding is the local anesthetic tetracaine (Fig. 1). As shown in an earlier study, this compound dramat-

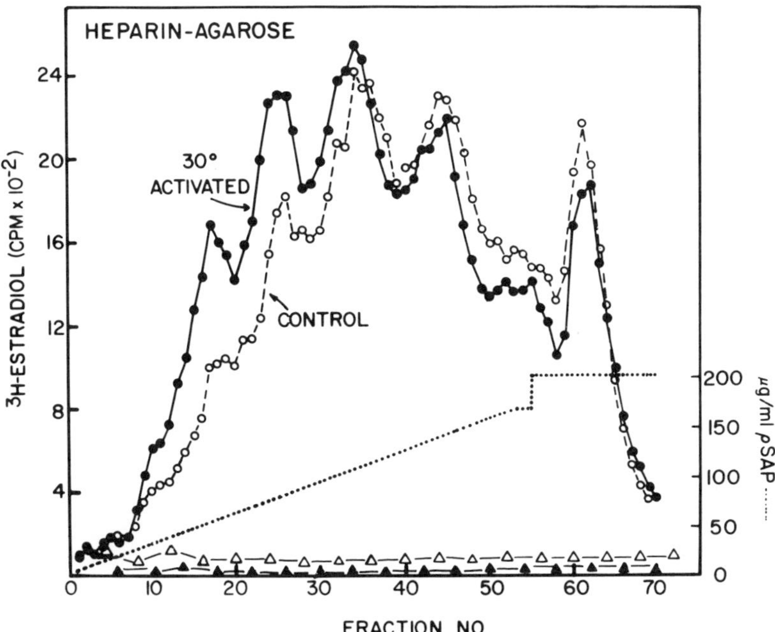

Fig. 2. *p*SAP gradient analysis of [^{3}H]estradiol-receptor complexes bound to heparin-agarose. Rabbit uterine cytosol was incubated with [^{3}H]estradiol (18.4 n*M*) for 1.5 h at 0°C. Part of the sample was incubated for an additional 30 min at 30°C while the rest of the sample was maintained at 0°C. Of the respective cytosols, 2 ml were then incubated with 0.5 ml of heparin-agarose (contained in a 0.8 × 4 cm Bio-Rad Econo column) for 1 h at 0°C with gentle mixing every 10–15 min. The nonadsorbed cytosolic components were then eluted and the columns rinsed with 25 ml TD buffer (10 m*M* Tris-HCl, 1 m*M* dithiothreitol, pH 7.4). A 40-ml linear gradient of *p*SAP (0–150 μg/ml in TD buffer) was pumped through the column at a flow rate of 0.67 ml/min. After collecting 54 fractions (0.75 ml each), a step elution of 200 μg/ml *p*SAP was used, again collecting 0.75 ml fractions. 100 μl samples from each fraction were counted directly in 5 ml of RIA-Solve II. (○) control, 0°C treatment; (●) 30°C-activated cytosol. Background, nonspecific binding of [^{3}H]estradiol was determined by the inclusion of 1.84 μ*M* (100×) unlabeled estradiol in the [^{3}H]estradiol binding step (△). Receptor-bound [^{3}H]estradiol was assayed using a hydroxyapatite procedure (▲). Results are expressed as the amount of [^{3}H]estradiol/100 μl sample. (Reprinted from Mueller et al., J. Receptor Res. 4(7), 773–785 (1984), p 776, by courtesy of Marcel Dekker, Inc.)

ically lowers the affinity of the receptors for estradiol (Kim et al., 1982). In effect, tetracaine lowers the temperature that is required to facilitate an estradiol exchange for the prebound ligand. This agent, however, clearly addresses a different region of the estrogen-binding protein since tetracaine, itself, does not directly compete with estradiol for the binding site. In contrast to *p*SAP, it requires a second estrogen molecule to displace the prebound estradiol. From studies of a series of related local anesthetics, the presence of the two positively charged nitrogens located at the opposite ends of this hydrophobic molecule are prerequisites for the observed relaxing effects on estrogen receptors.

Tetracaine, like *p*SAP, also demonstrates the presence of subsets of estrogen receptors in uterine cytosols (Van Oosbree et al., 1983–1984). As shown in Fig. 3, the washing of [^{3}H]estradiol-saturated receptors adsorbed on heparin agarose with buffers containing increasing concentra-

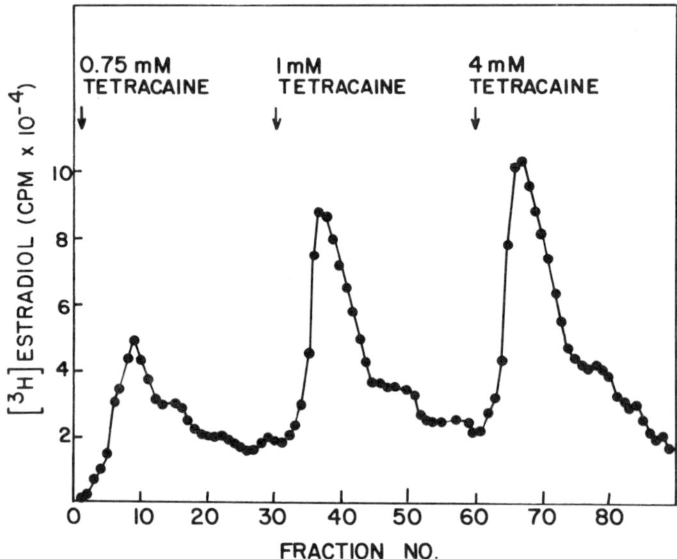

Fig. 3. Heterogeneity of estrogen receptors as revealed by tetracaine treatment. Cytosol was pretreated with 18.4 n*M* [^{3}H]estradiol for 1.5 h at 0°C. Two milliliters of the cytosol was then incubated with 1 ml of heparin-agarose for 1 h at 0°C with intermittent mixing. The slurry was poured into a Bio-Rad Econo column, and the column of heparin-agarose containing adsorbed receptors was rinsed with 25 ml of TD buffer. The column was then eluted with 22.5-ml volumes of 0.75 m*M* tetracaine (fractions 1–30), 1.0 m*M* tetracaine (fractions 31–60), and 4.0 m*M* tetracaine (fractions 61–90) by pumping the solutions through the column at a flow rate of 0.67 ml/min. Column fractions (0.75 ml) were collected and the radioactivity was determined by counting 0.5 ml from each fraction. (Reprinted from Van Oosbree et al., J. Receptor Res. 3(6), 727–743 (1983–84), p 736, by courtesy of Marcel Dekker, Inc.)

tions of tetracaine released estradiol in a stepwise, concentration-dependent manner, which is characteristic of subsets of estrogen receptors with different responses to the structure-relaxing effects of tetracaine.

A second hydrophobic molecule that influences the affinity of estrogen receptors for estradiol is arachidonic acid (Fig. 1). In this case there is a requirement for the multiple *cis* double bonds of the fatty acid. In addition, it appears necessary to have the negatively charged carboxyl group, as the substitution of a hydroxyl group or carboxyl group obviates the releasing action. Since arachidonic acid and tetracaine synergize in the releasing action, it appears reasonable that both agents may address the same hydrophobic region of the estrogen receptor, which may contain matching positive and negative charges. An interesting aspect of the arachidonic acid effect is its dependency on the presence of either KCl or NaCl. As shown in Fig. 4, the combination of 1 m*M* arachidonic acid and 0.15 *M* KCl KCl caused a dramatic reduction in the binding of [^{3}H]estradiol by estrogen receptors of uterine cytosol. Neither agent alone caused a displacement of the ligand. The dependency of the arachidonic acid-mediated response on presence of salt points to the involvement of yet a third region of the receptor molecule in the estradiol-binding process—a hydrophilic region.

Additional evidence for the participation of a hydrophilic domain in receptor function was obtained in experiments with heparin. As shown in Fig. 5, the combination of heparin and tetracaine synergize in the release of prebound [^{3}H]estradiol from receptors of uterine cytosol (Van Oosbree

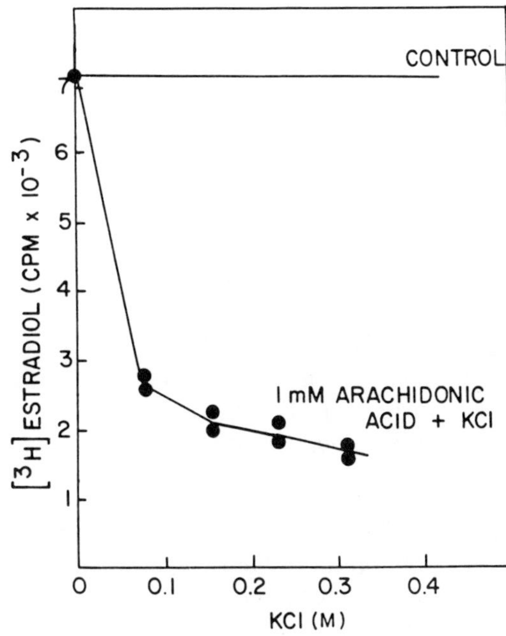

Fig. 4. Release of receptor-bound [^{3}H]estradiol by arachidonic acid and KCl. Rabbit uterine cytosol was treated with [^{3}H]estradiol (5×10^{-3} μg/ml; 18.4 n*M*) at 0°C for 1.5 h to preload the estrogen receptors. The indicated combinations of arachidonic acid and KCl were then added and the incubation continued for 3.5 h at 0°C. Hydroxyapatite assays were then performed to determine the level of receptor-bound [^{3}H]estradiol remaining.

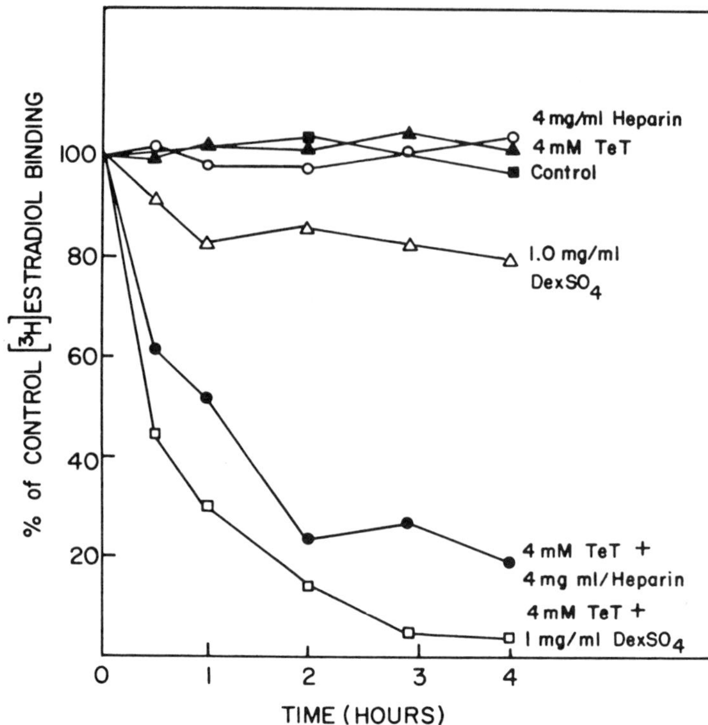

Fig. 5. Displacement of [^{3}H]estradiol from estrogen receptors by the combination of tetracaine and sulfated polysaccharides. Rabbit uterine cytosol was pretreated with [^{3}H]estradiol (18.4 n*M*) for 1.5 h at 0°C to label the estrogen receptors. At time zero, the preparation was treated with 4 m*M* tetracaine (TeT), 4 mg/ml heparin, 1 mg/ml dextran sulfate (DexSO$_4$), 4 m*M* tetracaine plus 4 mg/ml heparin, or 4 m*M* tetracaine plus 1 mg/ml dextran sulfate, and the incubation was continued at 0°C. The amount of receptor-bound [^{3}H]estradiol in each preparation was measured over the next 4 h by using the hydroxyapatite assay. Each value represents the average of duplicate assays. 100% binding = 46,900 dpm/0.1 ml cytosol. (Reprinted from Van Oosbree et al., J. Receptor Res. 3(6), 727–743 (1983–84), p 731, by courtesy of Marcel Dekker, Inc.)

et al., 1983–1984); heparin alone was ineffective. It was also observed that heparin synergized with arachidonic acid to similarly effect a release of [^{3}H]estradiol (Fig. 6).

Together, these observations are accepted as evidence that high-affinity binding of [^{3}H]estradiol by receptors involves the correct approximation of positive and negative charges in two very distinct regions of the receptor molecule. A hypothetical picture of these relationships is depicted in Fig. 7, in which the estrogen receptor is represented to contain three domains: hydrophilic, hydrophobic, and estrogen-binding. It is proposed that both the hydrophilic and the hydrophobic regions contain charged amino acid

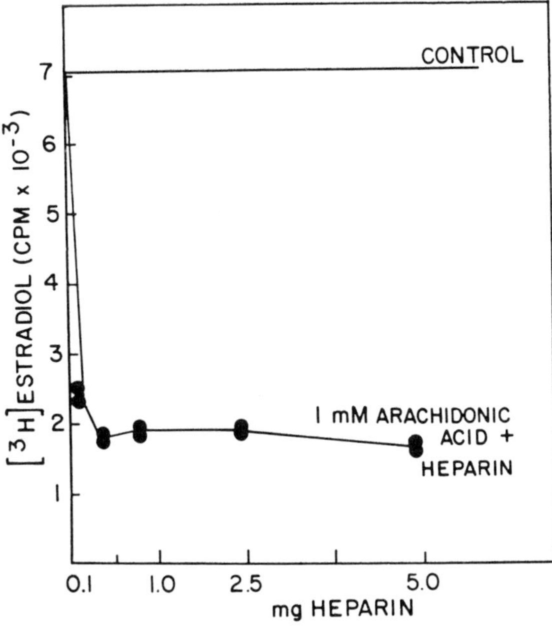

Fig. 6. Displacement of receptor-bound [^{3}H]estradiol by arachidonic acid and heparin. Receptors of rabbit uterine cytosol were preloaded with [^{3}H]estradiol for 1.5 h at 0°C. The indicated combinations of arachidonic acid and heparin were then added and the incubations continued at 0°C for an additional 3.5 h. The receptor-bound estradiol remaining was assayed by the hydroxyapatite procedure.

residues whose alignment or matching, and thus overall structure of the receptor, can be influenced by extraneous charged molecules of the appropriate polarity. It follows that the incremental interaction of such components with the receptor alters the folding of the peptide chains involved in the estrogen binding so that estrogen is either bound more tightly or released at a lower temperature. It is further suggested that an entering estrogen molecule or its analogue (i.e., an antiestrogen or simple phenol) also inflicts conformational changes on the structure of the receptor molecule so as to expose other sections of the receptor for greater or lesser degrees of interaction with chromatin or membrane components. Accordingly, the estrogen receptor is viewed as a highly dynamic molecule—acting as a transducer of conformational changes in the macromolecules involved in the assembly or disassembly of catalytically active chromatin or membranous complexes.

Preliminary evidence in support of this view of estrogen receptor action is now emerging through studies of the interaction of purified estrogen receptors with histones, cytosolic proteins, and DNA segments. For these studies it was first necessary to devise a method for the rapid isolation of functional estrogen receptors in quantities sufficient for physical in-

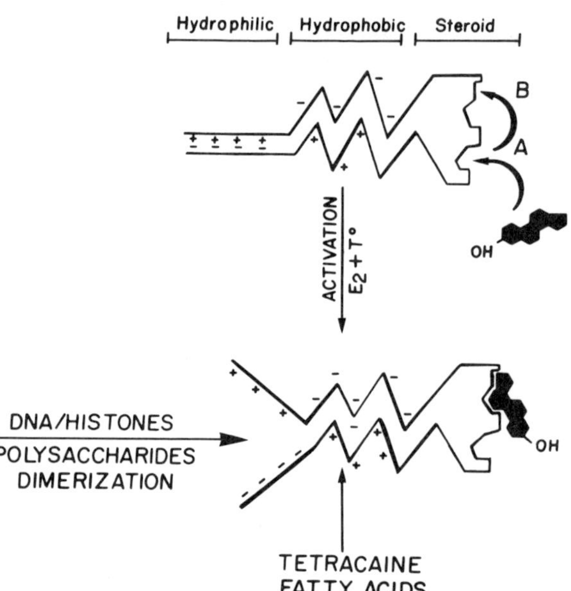

Fig. 7. Hypothetical structure of the estrogen receptor. Studies with compounds that modulate estradiol binding suggest that the receptor protein contains three separate but interactive domains. The steroid-binding domain recognizes sequentially the A-ring of the hormone (site A) with the high affinity binding occurring at site B. A second domain, noncompeting with estradiol, recognizes nonpolar, hydrophobic compounds such as tetracaine and long-chain fatty acids such as arachidonic acid. On activation with estradiol and heating at 37°, a third, hydrophilic domain becomes exposed for interaction selectively with certain DNA sequences, histones, and sulfated polysaccharides. The hydrophilic domain appears also to be involved in the dimerization of activated receptors. It is proposed that the receptor protein is highly active conformationally, responding adaptively to the simultaneous interaction of ligands at each of the three domains. In this manner, the receptor can function as a molecular transducer, inflicting conformational changes on macromolecules involved in the assembly or disassembly of functionally important chromosomal or membranous complexes. (Reprinted from Mueller, 1986)

teraction studies. To this end, a diethylstilbestrol(DES)-agarose resin was synthesized and used as an adsorbent in the purification of estrogen receptors by affinity chromatography. As demonstrated previously (Van Oosbree et al., 1984), this adsorbent is highly efficient—and with the selective elution of estrogen receptors by a solution of *p*SAP and sodium thiocyanate (NaSCN)—highly purified estrogen receptors are readily prepared (Fig. 8). The estradiol-binding activity is accounted for by two proteins that electrophorese in sodium dodecyl sulfate (SDS)-polyacrylamide gels as 65 and 50K proteins. Pretreatment of the cytosols with estradiol

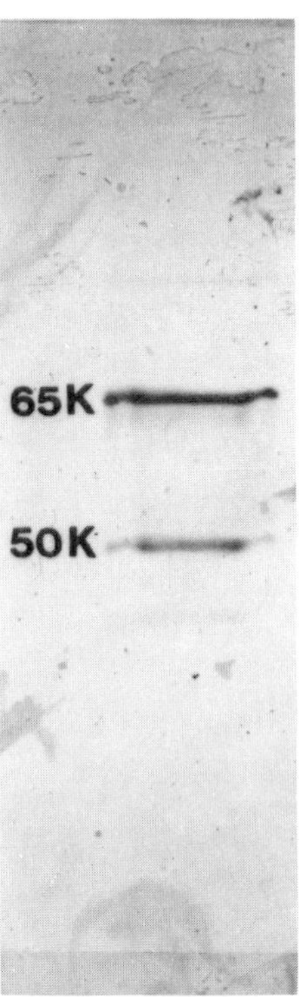

Fig. 8. Purification of M_r = 50,000 and 65,000 estrogen-binding proteins. A 0.2 ml DES-agarose affinity column was equilibrated with buffer B (10 m*M* Tris-HCl, 0.5 *M* KCl, pH 7.4) at 4°C. Ten milliliters of rabbit uterine cytosol in buffer B were applied to the column. After rinsing the column with 20 ml buffer B and 10 ml buffer A (10 m*M* Tris-HCl, pH 7.4), 1 ml of 100 μg/ml *p*SAP and 0.5 *M* NaSCN in buffer A were added. The suspension was vortexed every 5 to 15 min for 2 h at 0°C. The eluate was removed and combined with a 9 ml *p*SAP rinse of the resin. This solution was then diluted with 30 ml buffer A to yield final concentrations of *p*SAP at 25 μg/ml and NaSCN at 12.5 m*M*, and passed over a heparin-agarose column (0.2 ml) at a flow rate of 20 ml/h to adsorb the contained proteins. The heparin-agarose was then rinsed with 20 ml buffer A and the adsorbed proteins were extracted into a small volume of SDS and 2-mercaptoethanol, and resolved by SDS-gel electrophoresis. (Reprinted from Van Oosbree et al., 1984)

prevents their adsorbtion on DES-agarose, both initially and in readsorption studies. Iodination with either the lactoperoxidase or Bolton Hunter procedures, followed by digestion with V-8 protease or trypsin, yields common peptides. Accordingly, it is concluded that the proteins are structurally related. Whether they arise through posttranslational processing or through some alternative transcriptional mechanism is not known as yet. Studies in rats have revealed that the ratios of these two proteins vary in a predictable manner throughout the estrous cycle. Clearly, however, the 50K protein is not a product of proteolysis during the isolation procedure.

The finding that estrogen binding in uterine cytosols is due to two estrogen-binding proteins accounts in part for the above functional heterogeneity; however, it does not account for it all. From studies of the purified receptors, it appears, instead, that the subsets of estrogen-binding components result mostly from the estrogen receptor associations with other nonestrogen-binding components of the cytosol.

Evidence for this conclusion derives from studies of the effects of histones and cytosolic proteins on the binding of [^{3}H]estradiol by receptors purified by affinity chromatography. It was found that after the isolation process, while yielding highly purified receptor proteins, the latter had a much reduced ability to bind estrogens. When the flow-through fraction

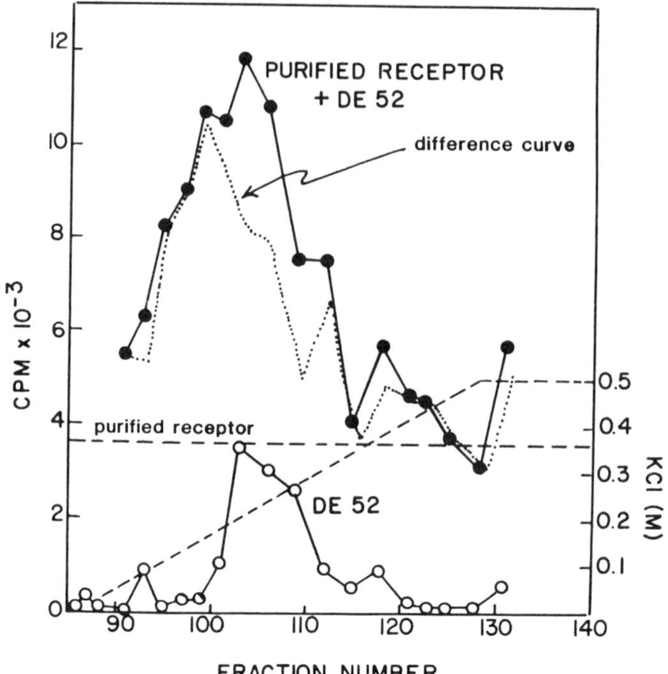

Fig. 9. Fractionation of rabbit uterine cytosol by DEAE-cellulose column chromatography. Crude cytosol (10 ml) was charged on a DE52 column (1.5 × 6.2 cm) equilibrated with 10 column volume of buffer A (10 m*M* Tris-HCl, 1.5 m*M* EDTA, 1 m*M* dithiothreitol, pH 7.4). The column was washed thoroughly and then eluted with a KCl gradient of 0–0.5 *M* in buffer A. Fractions (2 ml) were collected at a flow rate of 1 ml/4 min. Each fraction was scanned at 280 n*M*. Helper-factor activity of each fraction was assayed by adding these fractions to a mixture of purified receptors and [^{3}H]estradiol (final concentration 10 n*M*) (●). They were incubated for 16 h and an aliquot was then assayed on hydroxyapatite for [^{3}H]estradiol binding. The fractions were also assayed for receptor-binding activity (○). The difference due to the helper activity is plotted (. . . .).

of the affinity adsorption step was added back to the purified receptors, much of this binding ability was restored. This helper activity, however, was not confined to a single protein of the flow-through fraction; instead, chromatography over diethylaminoethyl (DEAE)-cellulose (Fig. 9) or gel sizing over a Toyopearl column revealed the activity to reside in multiple fractions. The isolation of the active components remains a goal of our research; however, it is clear from combinational studies that the various components influence the binding both additively and cooperatively.

Although these issues are still unresolved, it was decided to test whether any readily available purified proteins could influence the binding of

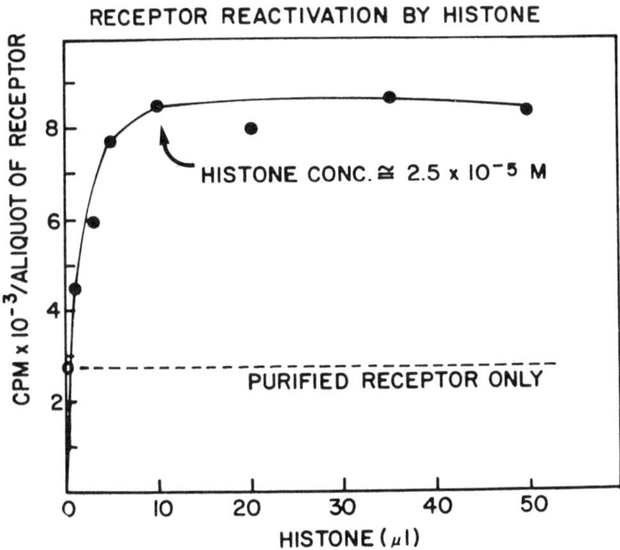

Fig. 10. Influence of mixed histones on the binding of estradiol by purified receptors. Estrogen receptors were purified by affinity chromatography over DES-agarose (Van Oosbree et al., 1984). Aliquots of the receptor solution were incubated with 10 n*M* [^{3}H]estradiol plus the indicated level of Type VII-S histones (Sigma) for 16 h at 0°C. The level of receptor-bound estradiol was assayed by the hydroxyapatite procedure. (Reprinted from Mueller, 1986)

[^{3}H]estradiol by purified receptors. Among the series of proteins tested [i.e., cytochrome C, ribonuclease, histone (type VII-S, Sigma, a mixture of histones), β-lactoglobulin B, lysozyme, ovalbumin (egg), and transferrin], only lysozyme and histone VII-S facilitated the binding. As shown in Fig. 10, this response was dose-related, with a maximal effect being reached when the ratio of receptor to histone protein was 50. Tests with the separated histones, however, revealed that the different histones influenced the binding in a distinctive manner (Fig. 11). Increasing concentrations of either histone H2A or H2B brought the binding to a plateau value that was not further influenced by the further addition of these histones. Optimal binding occurred when the molecular ratio of estrogen receptor to H2B molecules was 15. The response to histone H1, H3, and H4, however, was biphasic, with low levels facilitating the binding and higher levels counteracting this effect and even depressing the binding of [^{3}H]estradiol by the purified receptors.

These studies pointed strongly to the likely formation of histone/estrogen receptor complexes with characteristic affinities for estradiol. Preliminary evidence for this type of physical interaction was obtained through the sedimentation of histone/estrogen receptor mixtures in sucrose gradients.

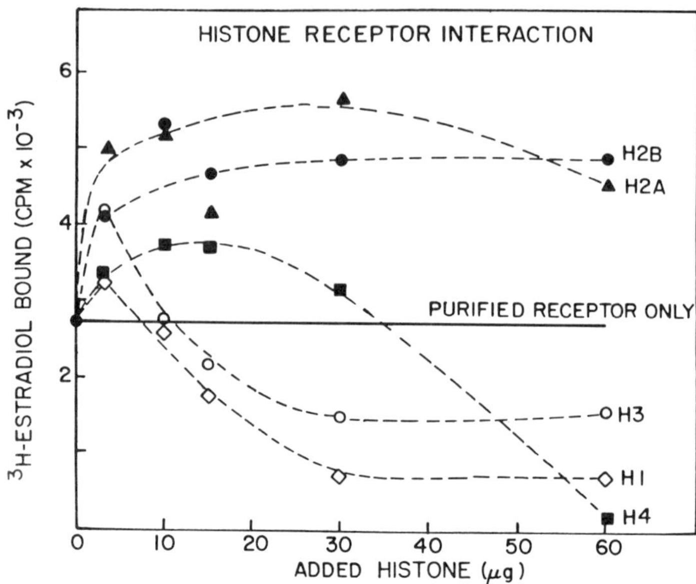

Fig. 11. Effect of individual histones on the binding of estradiol by purified receptors. Estrogen receptors were purified from rabbit uterine cytosol by affinity chromatography over DES-agarose (Van Oosbree et al., 1984). Aliquots of the receptor were mixed with the indicated levels of the chomatographically purified histones (H1, H2A, H2B, H3, and H4) and treated with 10 n*M* [^{3}H]estradiol for 16 h at 0°C. The levels of receptor-bound estradiol were then determined using the hydroxyapatite procedure.

As revealed in Fig. 12, the combination of H2A or H2B with [^{3}H]estradiol-treated receptors resulted in the sedimentation of a significant amount of receptor-bound [^{3}H]estradiol to the bottom of the gradient. Most remarkably, however, it was found that H2A and H2B synergized in this response, yielding a much greater amount of the stabilized, rapidly sedimenting receptor complexes.

Although preliminary in nature, these studies provide the first evidence that histones form specific complexes with purified estrogen receptors, which in turn influence the binding of [^{3}H]estradiol. Clearly, additional studies are needed to define the nature of the specific histone interactions and to assess the influence of nonhistone proteins on the formation and stability of such receptor-containing complexes. However, even with the studies to date, the results strongly suggest that estrogen receptors, interacting conformationally with specific histones, may guide the assembly of distinct histone receptor complexes. It follows naturally that estrogen receptors, functioning in a similar manner in a chromatin scene, and guided also by a spectrum of small molecules acting as accessory ligands, may shape the character of chromatin complexes and influence the expression of specific genes.

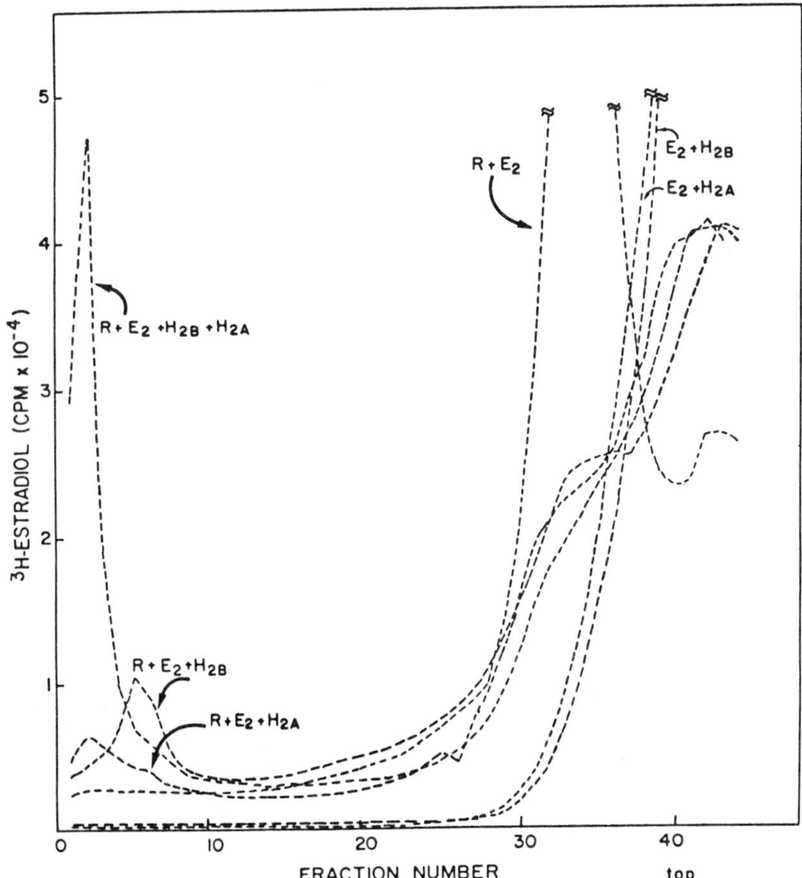

Fig. 12. Influence of histones on the sedimentation of [^{3}H]estradiol-labeled receptors. Estrogen receptors were purified from rabbit uterine cytosol by affinity chromatography over DES-agarose (Van Oosbree et al., 1984) Aliquots of the receptor were mixed with histone H2A or H2B as indicated and 20 n*M* [^{3}H]estradiol. After binding for 16 h at 0°C, the mixtures were layered 5%–20% sucrose gradients prepared in 10 m*M* Tris-HCl, 0.2 *M* KCl, 1.5m*M* EDTA (pH 7.4) and centrifuged in an SW 50.1 rotor at 38,000 × *g* for 12 h. Each gradient contained a cushion of 60% sucrose at the bottom to facilitate the recovery of sedimenting histone/receptor complexes. Fractions (5 drops) were collected from the bottom of the tubes and the amount of radioactivity in each fraction determined in a liquid scintillation spectrometer using Aquassure. Data are expressed as the counts/minute per fraction.

In summary, it is proposed that estrogen receptor molecules are conformationally highly reactive to both large and small molecules in their environment and act as molecular transducers in the guided assembly of macromolecular complexes that are functionally important in gene expression—and very likely in membrane function as well.

References

Kim UH, VanOosbree TR, Mueller GC (1982) Endocrinology 111: 260–268
Mueller GC, VanOosbree TR, Kim UH (1984) J Receptor Res 4(7): 773–785
Mueller GC (1986) In: Jordan VC (ed) 7th International Congress of Endocrinology, Satellite Symposium. University of Wisconsin Press, Madison
VanOosbree TR, Kim UH, Mueller GC (1983–1984) J Receptor Res 3(6): 727–743
VanOosbree TR, Kim UH, Mueller GC (1984) Anal Biochem 136: 321–327

Discussion of the Paper Presented by G. Mueller

MUELLER: The 65K receptor is the dominant form if you analyze uteri from ovariectomized animals. If, however, rats are followed throughout an estrous cycle, the 50K form increases dramatically toward the end of the cycle. Whether this represents a cycle-dependent processing of the receptor or an altered transcriptional event is not yet known.

CLARK: Since you just talked about the loss of activity, after the purification, and so forth, as well as elution of the column, can you not elute from the column with DES, excess DES?

MUELLER: No. Neither DES nor estradiol are able to elute the column-bound receptors very effectively; *p*SAP is by far the competing ligand of choice. The reason for this situation is not yet clear, but it very likely reflects the manner in which different estrogen or competing ligands address the receptor structure. Our view is that estradiol and other estrogens approach the receptor first with their A-ring and bind secondarily through hydrophobic interactions of the C/D-rings. In this sequence, the A-ring site on the receptor is likely to be open for approach of another phenol. The data with *p*SAP versus estradiol (or DES) suggest that the approach of another estradiol molecule to this site causes the receptor to bind the existing ligand even more tightly. In contrast, *p*SAP appears to cause the receptor to fold in a way that relaxes the interaction between the receptor and the C/D-ring of the estradiol, thus releasing this ligand at 0°C. This view of estrogen-receptor interaction is supported also by the high intolerance of substitutions, particularly polar substitutions, in the C/D-ring of the estrogen molecule.

CLARK: Well you can but it just doesn't do it as well. You kept saying that you lost activity?

MUELLER: Yes, we elute the receptor protein, but on a binding-activity test, as much as 90% of this activity has been lost. Much of the binding activity (i.e., 50%–60%), however, can be restored by the addition of the flow-through fractions of the affinity column or the purified histones.

SPELSBERG: Histones are bound to DNA as you know. They hardly come off momentarily even through DNA replication. Have you tried these types of studies in terms of activation with histones reconstituted on pieces of DNA?

MUELLER: We have not used DNA/histone combinations in our reactivation studies as yet, but this is definitely in our plans. We fully anticipate finding evidence of cooperativity in such studies. In fact, our view of estrogen action is that the receptors, acting as tranducers with bound estradiol, function in the assembly and disassembly of specific chromatin complexes or aggregations.

SCHRADER: Aside from your last comment, I think the problem in nuclei is the fact that they don't initiate RNA synthesis very well. I think the problem is structurally secondary.

MUELLER: That's the same point, Bill.

SCHRADER: When you showed the heparin columns that you eluted with *p*SAP, you did a gradient of *p*SAP, as I understood. You claimed that the radioactivity came off the receptor in individual pieces. Is that a reversible phenomenon? If you wash only part way up the column to a certain concentration with *p*SAP and then drop down to zero *p*SAP, can you relabel and show that the sites that you eluted at an early concentration of *p*SAP are specifically relabeled?

MUELLER: That's a good experiment. We have not done that, I'm sorry to say; it would be worthwhile.

SCHRADER: The second question is in the case of the heparin of the histone experiment, which you showed. You claimed that the H2A and H2B individually caused hormone binding to be improved on the receptor, but on the sucrose experiments that you showed, when you added only histone there was more free hormone on the gradient when those were present.

MUELLER: You are very observant. Yes, a significant amount of estradiol appears to have been released by the interaction of the histones with the receptors, as well as the formation of sedimentable histone/receptor/E2 complexes. We don't have an explanation for the release phenomenon except to suggest that the sedimentation experiments are of much longer duration than the hydroxyapatite binding studies. Secondly, the presence of high levels of sucrose in the sedimentation studies may also affect the receptors in some unexpected way, as it is well known that high sucrose levels do affect macromolecular structure and interaction.

SCHRADER: As I recall, interaction steroid receptors with protamine have been documented for a number of years. Why don't you think this is just a nonspecific absorption of any old basic protein to the receptor?

MUELLER: We have tested a variety of basic proteins and they don't perform in this category. For instance, we tested cytochrome C, ribonuclease, and lysozyme; these proteins are very poor performers as compared with the histones.

GREENE: I have one more question about the relationship of the 65K and the 50K proteins. I guess I'm not really quite sure where you stand on the issue from the previous discussion. From what you say, they're linked. For example, you showed the peptide mapping and really the only difference in the peptide maps is the peptide that you can add back on. Then you mentioned the McCarty work with the antibodies that they have, showing the chemical similarities. At this point one would tend to argue that one was derived from the other.

MUELLER: I think the relationship of the 65K and 50K forms of the estrogen receptor is still wide open. The different forms could arise either from an alternating transcription of the estrogen-receptor gene or by some posttranslational modification of the protein. It is even possible that some alteration in the translational process itself could explain the results. Only careful genetic experiments will define this situation. However, at this time we fell that the 50K protein does not arise from proteolysis during the isolation steps. We've tested a battery of protease inhibitors and looked deliberately for evidence of a time-dependent conversion during the isolation steps without evidence of proteolysis.

Discussants: J. CLARK, G. GREENE, G. MUELLER, W. SCHRADER, and T. SPELSBERG

Chapter 4

Type II Binding Sites: Cellular Origin and an Endogenous Ligand

B.M. MARKAVERICH AND J.H. CLARK

Introduction

Previous studies from our laboratory have described two specific nuclear estrogen-binding sites in normal (Eriksson et al., 1978; Clark et al., 1978; Markaverich and Clark, 1979; Watson and Clark, 1980) and malignant tissues. Type I sites represent the classical estrogen receptor, which has a high affinity for estradiol ($K_k = 0.1 - 1.0$ n*M*) and is present in target tissues in relatively low quantities. Nuclear type II sites appear to be a specific nuclear response to estrogenic hormones that is highly correlated with uterine hypertrophy and hyperplasia (Markaverich et al., 1981a,b,c). Likewise, nuclear type II sites are elevated in neoplastic tissues such as mouse mammary tumors (Watson et al., 1977; Watson and Clark, 1980; Watson et al., 1981) and human breast cancer (Syne et al., 1982).

At present, the precise function of nuclear type II estrogen-binding sites is unknown; however, the presence of these sites on nuclear matrix (Clark and Markaverich, 1982; Swaneck et al., 1982; Ekman et al., 1983) suggests a potential role in the regulation of DNA synthesis (Pardoll et al., 1980). This hypothesis is certainly substantiated by our studies, which show that type II sites are highly correlated with uterine hypertrophy and hyperplasia and are elevated in neoplastic tissue.

One aspect regarding the involvement of nuclear type II sites in uterotropic response that has remained an enigma is the relatively low affinity of these sites for estradiol in rat uterus (K_d 10–20 n*M*). Although the estimated K_d for type II sites in human breast cancer (~5 n*M*) is higher, this relative affinity is significantly lower than that measured for the estrogen receptor ($K_d \sim 0.1$–1 n*M*). These data would suggest type II sites are not occupied by ndogenous estradiol-17β in vivo, even though these sites show remarkable specificity for estrogenic hormones (Markaverich et al., 1980) and do not bind the classical "antiestrogens" such as nafoxidine or clomiphene (Markaverich et al., 1981c).

Recent studies from our laboratory have partially resolved this complex picture. These studies have demonstrated that nuclear type II sites are occupied in vivo by an endogenous ligand (inhibitor) that we believe is an important regulator of cell function (Markaverich et al., 1983). We have purified this material and are currently attempting to determine its precise chemical structure. This inhibitor effectively competes for ^{3}H-estradiol binding to nuclear type II sites and appears to be active in the picomolar to nanomolar range. Furthermore, preliminary studies suggest that inhibitor activity is low to nonmeasureable in rat mammary tumors, which is consistent with the high levels of nuclear type II sites observed in these neoplastic tissues. The purpose of this chapter is to review some of these studies.

Cellular Origin of Type II Sites

It has been suggested that type II sites originate in the uterus owing to eosinophil infiltration (Lyttle et al., 1984). To examine the cellular origin of type II sites, the following experiments were performed.

Estradiol Stimulation of Nuclear Type II Sites in Cultured Uterine Cells

The most direct way to determine whether estradiol stimulation of nuclear type II sites is related to eosinophil accumulation is to evaluate whether estradiol stimulation of type II sites occurs in the absence of eosinophils. For this reason we evaluated the effects of physiological levels of estradiol (10 n*M*) on the level of nuclear type II sites in cultures of isolated uterine stromal and myometrial cells (Fig. 1). These highly purified uterine cells (see methods; mini-print section) were cultured for 21 days (4 passages) and then were treated with 10 n*M* [^{3}H]-estradiol added to the tissue culture medium. Following 72 h of estrogen treatment, the cells were harvested and assayed for nuclear type II sites by [^{3}H]-estradiol exchange (see methods), and binding data were expressed as picomoles [^{3}H]-estradiol bound specifically per mg DNA. The data are representative of many experiments and clearly demonstrate that the addition of physiological quantities of estradiol directly to these cultures result in a marked stimulation of nuclear type II sites in uterine stromal (fourfold) and myometrial (80-fold) cells. In fact, the magnitude of the response of these cultured cells to estrogen and the absolute numbers of type II sites measured in these cells were not different from the stromal and myometrial cell responses obtained following in vivo estrogen stimulation (Markaverich et al., 1981a) Since these cultured stromal and myometrial cells were prepared from 21-day castrate rat uteri devoid of eosinophils (Lyttle et al., 1979),

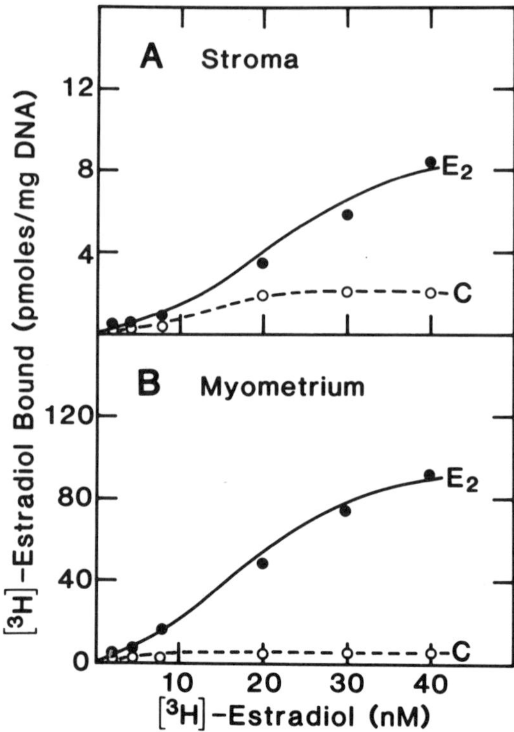

Fig. 1. Direct stimulatory effect of estradiol-17β on nuclear type II sites in rat uterine cells in culture. Highly purified rat uterine stromal (**A**) and myometrial (**B**) cells were obtained by enzymatic digestion of uteri from nonestrogenized-adult-ovariectomized rats. The isolated cell types were grown in monolayer culture for 21 days (3–4 passages). On day 22 the stromal and myometrial cell cultures were divided into groups (4 petri dishes each) and were grown an additional 72 h in the absence (control) or presence of 10 n*M* estradiol-17β E_2). Estradiol was added to the cultures daily (3 days) in 50 μl Delbecco's Modified Essential Medium (DMEM), while controls received an equivalent quantity of DMEM. Following hormone treatment the cells were harvested and assayed for nuclear estrogen binding sites by [^{3}H]estradiol exchange. Results are expressed as pmoles of [^{3}H]estradiol-bound specifically per milligrams nuclear DNA.

and since the cell isolation procedure used yields pure cultures of these respective uterine cell types, which are free of blood contaminants (McCormack and Glasser, 1980), the data demonstrate that the increase in nuclear type II sites is a direct intracellular response to estradiol. Since no eosinophils exist in these cultures or in the uterine tissue prior to cell isolation, the elevation in nuclear type II sites can not be due to eosinophil accumulation.

Distribution of Type II Sites and Peroxidase Activity in Cytosol and Nuclear Fractions of the Rat Uterus

Subsequent to the suggestion that type II sites in the rat uterus might be of eosinophillic origin, Sheehan et al. (1984) suggested that eosinophil peroxidase might be the type II estrogen-binding site. Although the direct stimulatory effect of estradiol on nuclear type II sites in cultured uterine cells (Fig. 1) suggests that this is unlikely, we decided to examine the distribution of type II sites and peroxidase activity in uterine cell fractions. Previous studies in our laboratory showed that type II sites exist in both the cytosol and nuclear fractions of the estrogen-treated rat uterus. For this reason, adult-ovariectomized rats were treated with beeswax pellets containing 20 μg estradiol-1 β, as previously described (Markaverich et al., 1981b), and cytosol and nuclear fractions were prepared. Since we had previously shown that cytosol type II sites are a soluble protein that sediments at ~4S on sucrose gradients (Clark et al., 1978) and Lyttle and DeSombre (1977) demonstrated that peroxidase activity is found in the pellet fraction following centrifugation at 39,000 *g*, we measured type II sites and peroxidase activity in the 39,000 *g* supernatant and pellet fractions obtained from low-speed (760 *g* × 20′) uterine cytosol preparations. The data show that the type II binding was measured in the 39,000 *g* supernatant fraction (Fig. 2A), and this fraction was essentially devoid of peroxidase activity (Fig. 2B), as measured by the guaiacol assay. Conversely, an aberrant [^{3}H]-estradiol binding curve was observed in the 39,000 *g* pellet fraction (Fig. 2A), which contained the peroxidase activity (Fig. 2B). [^{3}H]-Estradiol binding in this fraction may be due to peroxidase, but there is no similarity between [^{3}H]-estradiol binding in the peroxidase rich fraction. These data demonstrate type II sites and peroxidase in uterine cytosol can be separated by centrifugation and therefore they are unrelated proteins.

Similar results were also obtained with uterine nuclear fractions from these animals (Fig. 3). Saturation analysis by [^{3}H]-estradiol exchange shows that type II sites are present in the crude nuclear pellet fraction (Fig. 3A), and this fraction contains approximately 14% (0.9 units/ml) of the total uterine homogenate peroxidase activity (Fig. 3B). Furthermore, extraction of this crude nuclear pellet fraction with 0.5 *M* $CaCl_2$ had no significant effect on the level of type II sites remaining in the nuclear pellet (Fig. 3A; $CaCl_2$-resistent fraction), even though the peroxidase activity (14%) was quantitatively extracted by 0.5 *M* $CaCl_2$ (Fig. 3B). Again, as shown for uterine cytosol, the peroxidase-containing fraction (Fig. 3B; $CaCl_2$ extract) obtained by 0.5 *M* $CaCl_2$ extraction of nuclei showed an [^{3}H]-estradiol binding curve similar to that described by Lyttle et al., 1984, but lacking in similarity with the type II estrogen-binding site we routinely measure (Fig. 3A). These data demonstrate that the 0.5 *M* $CaCl_2$ extraction procedure developed by Lyttle and colleagues will quantitatively remove

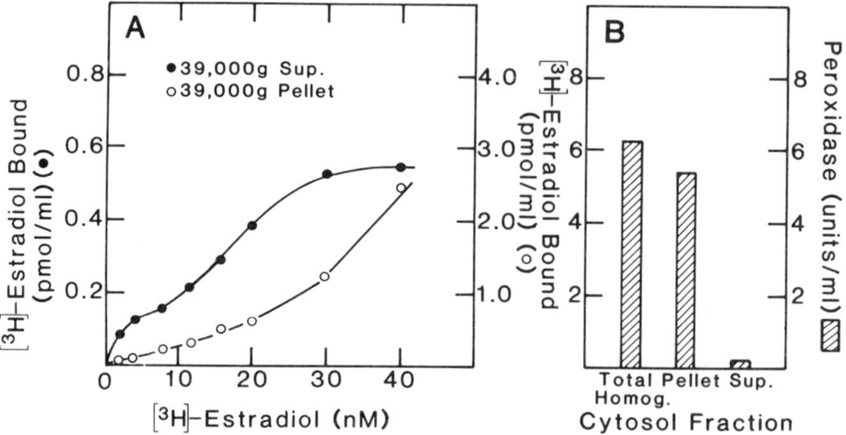

Fig. 2. Distribution of Type II estrogen binding sites (**A**) and peroxidase activity (**B**) in rat uterine cytosol fractions. Adult ovariectomized rats were implanted with 20 μg estradiol-17β or 96 h. Uteri were homogenized in TE buffer, and the homogenate was centrifuged at 760 *g* × 20′ to obtain the low-speed cytosol fraction. This cytosol was centrifuged at 39,000 *g* × 45′ to obtain the supernatant (39,000 *g* Sup.; ●) and pellet (39,000 *g* pellet; ○) fractions. The cytosol and pellet (resuspended in TE) were diluted in TE buffer in a volume equivalent to 20-mg fresh uterine wet weight/ml and were assayed for estrogen-binding sites by [^{3}H]-estradiol exchange (**A**). Aliquots of the total homogenate and the 39,000 *g* pellet and supernatant fractions were assayed for peroxidase activity (**B**) by the guaiacol assay.

the small amount of contaminating peroxidase from the low-speed nuclear pellets; however, this procedure does not solubilize nuclear type II sites. Consequently, as we demonstrated for cytosol (Fig. 2), nuclear type II sites and peroxidase are different biochemical entities.

These experiments rule out the possibility that type II sites in the uterus are of eosinophil origin or are eosinophil peroxidase. In addition, the following results and observations are presented in added proof.

1. Neonatal exposure to estradiol stimulates nuclear type II sites in the uterus (Markaverich et al., in press). Since the neonatal rat uterus does not contain significant quantities of eosinophils under the conditions (Sheehan et al., 1981), is is unlikely that type II sites are of eosinophil origin.
2. Dexamethasone treatment causes eosinophils to accumulate in the spleen; however, such treatment does not increase the number of type II sites (Markaverich et al., 1986).
3. Type II sites are present in the liver, which contains low levels of peroxidase activity (Markaverich et al., 1986).
4. Type II sites have been reported in MCF-7 human breast cancer cells in culture (Mercer et al., 1981; Oxenhandler et al., 1984) in which no eosinophils are present.

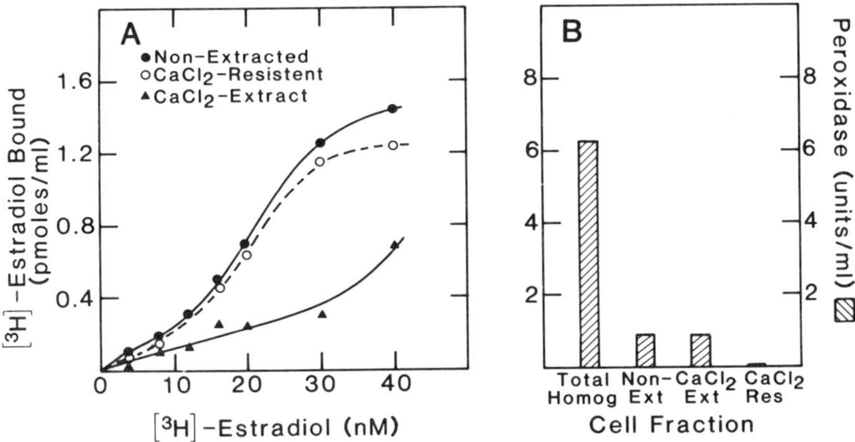

Fig. 3. Distribution of type II sites and peroxidase activity in uterine nuclear fractions. Estradiol-implanted (20 μg × 96 h) adult-ovariectomized rats were sacrificed and the uterus was homogenized in TE buffer. The whole homogenate was centrifuged at 760 *g* × 20′ to obtain the crude nuclear pellet. The pellet was resuspended (100 mg wet wt equivalent/ml) in TE buffer, split into two equal aliquots, centrifuged at 760 *g* × 7′, and the final washed pellets were resuspended in TE buffer (nonextracted; ●) or TE-0.5 *M* $CaCl_2$ ($CaCl_2$-resistant; ○) at 50 mg/ml and extracted for 45′ by resuspension every 5′ in dounce homogenizer. Following extraction, these pellets were centrifuged (760 *g* × 7′), washed 1 × in 10 ml TE buffer, resuspended (20 mg fresh wet wt/ml) in TE and assayed for estrogen-binding sites by [^{3}H]-estradiol exchange (**A**). In addition the supernantant from the 0.5 *M* $CaCl_2$ extraction ($CaCl_2$ extract; ▲) obtained following centrifugation to obtain the $CaCl_2$-resistant pellet 0) was diluted to 20 mg/ml and assayed by [^{3}H]-estradiol exchange. Aliquots of all fractions in **A**, as well as the total homogenate, were assayed for peroxidase activity (**B**) by the guaiacol assay.

5. Dilution of uterine nuclear fractions from rats 14 days after ovariectomy (Markaverich and Clark, unpublished data, 1985) shows an increase (two- to threefold) in the number of nuclear type II sites. This is consistent with the hypothesis that these sites are present in the uterus and are occupied by an endogenous ligand, which we have previously described (Markaverich et al., 1983). Since these uteri contain no eosinophils (Lyttle et al., 1979), the increase in type II sites cannot be attributed to eosinophil contamination.
6. We have shown that type II estrogen-binding sites are located on the nuclear matrix prepared from rat uterine tissue (Clark and Markaverich, 1982). Nuclear matrix preparation involves extraction of uterine nuclear fractions with 2 *M* NaCl, and type II sites are resistant to such extraction. Routine methodology for isolating peroxidase from white blood cells involves extraction with 1.5 *M* NaCl, which quantitatively solubilizes peroxidase activity (Neufeld et al., 1958). Consequently, if nu-

clear type II sites were peroxidase of eosinophilic origin, the type II site would have been solubilized from nuclear matrix preparations by 2 *M* NaCl extraction. We found this is not the case. Similarly, Simmons et al. (1984) demonstrated that type II sites are present on nuclear matrix prepared from purified chick liver nuclei.

7. No type II sites are found in uterine luminal fluid after estrogen treatment (Markaverich and Clark, unpublished observation), yet very high levels of eosinophil peroxidase accumulate in this compartment (King et al., 1981).
8. Peroxidase activity is absent in estrogen-independent mammary tumors (King et al., 1981); however, such tumors have high levels of nuclear type II sites (Watson et al., 1980).
9. When estradiol or estriol is administered by a single injection, both hormones are equally effective in causing eosinophil accumulation and other early uterotrophic responses (Tchernitchin et al., 1976; Tchernitchin and Tchernitchin, 1976). However, estriol has little effect on the elevation of nuclear type II sites and the stimulation of true uterine growth (cellular hypertrophy and hyperplasia). In contrast, estradiol stimulates both nuclear type II sites and true uterine growth (Markaverich and Clark, 1979). Therefore, elevations in type II sites and eosinophil accumulation in vivo are not correlated.

In summary, the observations and data presented in this chapter demonstrate that eosinophils are not the source of nuclear type II binding sites. The comparisons of our data with those of Lyttle et al. (1984) indicate that these investigators were not measuring nuclear type II binding sites, but, instead, other lower affinity sites that might be attributable to eosinophils and perhaps represent estradiol binding to peroxidase.

An Endogenous Ligand for Type II Sites

As explained in the introduction, type II sites appear to be occupied by an endogenous ligand. This ligand acts as an inhibitor of [^{3}H]-estradiol binding to type II sites and is an inhibitor of cell proliferation (Markaverich et al., 1983). A brief summary of our work with this inhibitor and its possible role in cell growth are presented below.

Gel Filtration Chromatography of Uterine Cytosol Inhibitor Activity

Chromatography of boiled-acid precipitated cytosol from adult ovariectomized rat uteri on Sephadex G-25 revealed two major peaks of inhibitor activity which, we designated α and β (Fig. 4). This activity was measured in the column fractions by its ability to inhibit the binding [^{3}H]-estradiol to nuclear type II. A minor peak of activity was also seen in the void

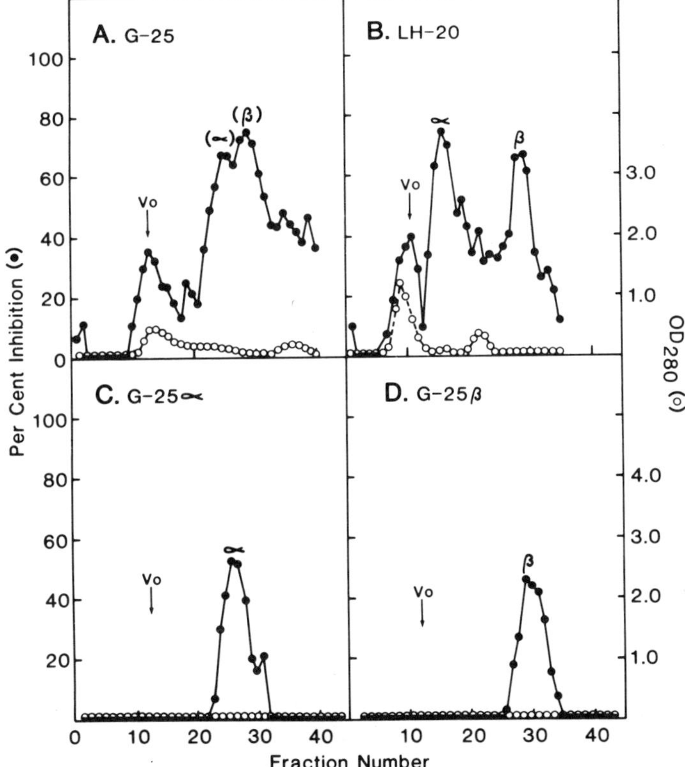

Fig. 4. Chromatography of rat uterine cytosol inhibitor preparations on Sephadex G-25 (**A**, **C**, **D**) or LH-20 columns (**B**).

volume of the column, which is apparently associated with protein, as there was a significant OD_{280} reading in these fractions. In addition, incubation of the cytosol inhibitor preparation with 0.4 *M* KCl for 60 min at 4°C prior to chromatography on Sephadex G-25 (TE buffer containing 0.4 *M* KCl) dissociated the inhibitor activity from the void volume fractions (data not shown).

On the basis of sizing experiments using tryptophan (MW 204.2) and adenosine triphosphate (ATP) (MW 507.2) as markers, we estimated that the molecular weight of the α and β inhibitor components is in the range of 300–400. Surprisingly, chromatography study of an aliquot of this same cytosol preparation on LH-20 (Fig. 4B) shows that the two components are more clearly resolved. To determine if the order of elution of the α- and β-peaks on LH-20 chromatography was analagous to their behavior on Sephadex G-25, we collected the individual α- and β-peak fractions from LH-20 and repeated the chromatography study on these fractions on the Sephadex G-25 column (Fig. 4C,D). The results demonstrate that the β-peaks from LH-20 columns correspond to the elution of this material on Sephadex G-25. Chromatography of cytosol preparations on larger preparative LH-20 columns facilitates complete separation of the α- and β-peak fractions owing to the selective retention of the β-material and has greatly facilitated purification of this inhibitor activity (see below).

Comparison of LH-20 Elution Profiles of Cytosol Obtained From Rat Uterus, Normal Lactating Rat Mammary Gland, and Estrogen-Induced Rat Mammary Tumors

Since nuclear type II sites may be involved in the mechanisms by which estrogens cause cell growth, we reasoned that perhaps this inhibitor may play a role in modulating these cellular responses. If this were the case, then rapidly proliferating neoplastic tissues that respond to estrogen might show deficiencies in inhibitor activity. To examine this possibility we prepared boiled-acid precipitated cytosol from rat uterus, normal lactating rat mammary gland, and estrogen-induced rat mammary tumors and conducted LH-20 chromatography studies on aliquots of these cytosols. Individual fractions were assayed for inhibitor activity. These data show that the larger LH-20 column (1.0 × 50 cm: 75 ml bed volume) used for these experiments clearly separated the α- and β-inhibitor peaks (Fig. 5) in rat uterine cytosol. Furthermore, although rat mammary tumor cytosol contained approximately equivalent quantities of the α-peak material, the tumors contained little of the β-component, and in some cases the β-inhibitor component was nonmeasurable.

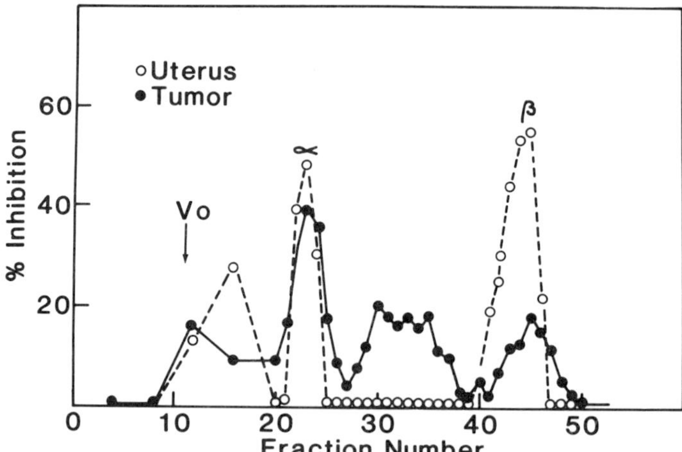

Fig. 5. Comparison of LH-20 elution profiles from acid-precipitated-boiled cytosol preparations (100 mg fresh tissue equivalents/ml) from rat uterus (○) or an estrogen-induced rat mammary tumor (●).

This observation has also been extended to mouse mammary tumors and human breast cancer (Markaverich et al., 1984). Detailed studies by dilution analysis and LH-20 chromatography showed that mouse mammary gland contained 20-fold more inhibitor activity than mammary tumors in the same strain of animals. This difference in activity appears to result from a primary deficiency in the β-inhibitor peak material (Fig. 6). In addition, we have completed a series of mixing experiments where cytosol inhibitor preparations from uterus and tumor were mixed prior to LH-20

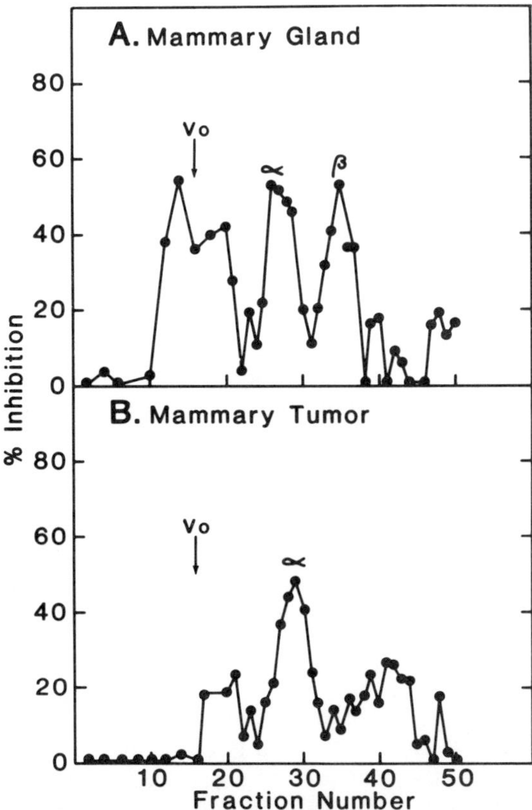

Fig. 6. LH-20 chromatography of mouse mammary gland (**A**) and mouse mammary tumor (**B**) cytosol. Aliquots (1 ml) of the cytosol preparations (80 mg fresh tissue equivalents/ml) were loaded on an LH-20 column and the columns eluted with TE (10 m*M* Tris; 1.5 m*M* EDTA) buffer. Fractions (0.5 ml) were collected and assayed for inhibitor activity as described in methods. Results are plotted as [^{3}H]-estradiol-binding to nuclear type II sites (% inhibition) where buffer controls (0% inhibition) contained approximately 45,000 cpm bound. The results were obtained from a single experiment. However, we have analyzed 15–20 separate tumor cytosol preparations and very similar results were obtained.

chromatography. The results of these experiments demonstrated that the β-inhibitor component in uterine cytosol was quantitatively recovered following chromatography. These results suggest that the tumor cytosol is indeed deficient in the β-inhibitor component and that the low levels of this molecule in tumor cytosol (as compared with uterus) is not due to some intrinsic degradation during the preparation. At present we do not know whether there is a precursor-product relationship between α- and β-inhibitor peaks (since they are of very similar molecular weight) and the tumor cannot readily form the β-material, or if perhaps the tumors metabolize this β-inhibitor component. Certainly the inhibitor activity in fractions 28–38 (Fig. 5) suggests that there is an altered form of inhibitor in tumors which is not observed in the uterus. Whether this activity (fractions 28–38) represents an altered or metabolized form of the β-peak material in tumors (Fig. 5) remains to be resolved. However, preliminary experiments indicate that the presence of the β-peak material in crude inhibitor preparations is associated with biological activity. Acid-precipitated-boiled cytosol inhibitor preparations from rat uterus of liver (containing α- and β-inhibitor components) inhibit the growth of rat mammary tumor cells in culture by approximately 80%–90% in 4–7 days. Conversely, preparations from rat mammary tumors (containing α, but lacking the β-inhibitor component) had no significant effect on cell growth in identical experiments continued for 3–4 weeks. These results suggest that the absence of the β-inhibitor peak material in tumors is correlated with rapid cell proliferation in these populations.

Purification and Partial Characterization of Inhibitor Activity

As stated earlier, positive structural identification of the inhibitor activity (LH-20 β-peak component) remains to be established. However, high-pressure liquid chromatography (HPLC) analysis showed that a major peak of inhibitor activity can be eluted from the silica column. This experiment has been repeated a number of times on three to four separate liver preparations, and the activity consistently elutes 5–6 min following injection. The samples at this point are very clean since we observe a single peak of UV absorbance at 254 n*M*, which is coincident with the inhibitor activity. The peak is somewhat broad and appears to have shoulders, suggesting multiple inhibitor components that are not separated. To date we have tried a number of HPLC procedures to separate these components (various elution conditions, reinjection of peak fractions), and we cannot further separate these components on a straight phase column.

This heterogeneity of putative inhibitor molecules was supported by gas chromatography–mass spectrometric (GC-MS) analysis. The GC-MS analysis of the pooled HPLC fractions (5.4–5.8 min) positively identified a number of fatty acids as major components in the sample inhibitor preparations. However, these are unlikely candidates for the inhibitor since the authentic compounds did not inhibit [^{3}H]-estradiol binding to type II sites over a wide range of concentrations (0.001 nM–100 μM). The putative inhibitor activity appears to be associated with two remaining components in the sample which have a molecular weight of 302 and 304. Comparison of sample spectra with those of known compounds in the National Institutes of Health (NIH) Bureau of Standards library suggests the inhibitor is very similar to phenanthrene-like molecules.

Discussion and Conclusions

These experiments demonstrate that the adult ovariectomized rat uterus and a variety of rat tissues contain an inhibitor that interferes with [^{3}H]-estradiol binding to type II sites in uterine nuclei. This inhibitor is specific for nuclear type II sites and does not interfere with estrogen binding to cytoplasmic or nuclear estrogen receptor. Consequently, if this inhibitor is involved in the modulation of estrogenic response in target tissues, its effects are expressed through an interaction with nuclear type II sites. We currently feel that this molecule represents an endogenous ligand for type II sites (Markaverich et al., 1983).

Preliminary characterization of this inhibitor in rat uterine cytosol demonstrates this molecule(s) is stable to heat (100° × 60′), and 0.1 N HCl, and therefore it is unlikely to be protein in nature. In addition, trypsin and proteinase K do not destroy its activity (data not shown), and the inhibitor activity chromatographs on Sephadex G-25 or LH-20 (Fig. 2) as two major peaks with an estimated molecular weight of 350. We have purified the β-peak material from rat liver (which appears identical to that seen in the uterus) by thin layer chromatography and HPLC, and it appears to consist of two nearly identical phenanthrene-like compounds with molecular weights on the basis of mass spectrometry of 302 and 304. Proof that these are in fact the inhibitor molecules awaits purification to homogeneity, structural identification and demonstration that the "identified" material has equivalent biological activity. At the present time we feel that these phenanthrene derivatives are good candidates for the inhibitor activity since the only other measureable compounds in the sample preparation (free fatty acids) did not inhibit [^{3}H]-estradiol binding to nuclear type II sites. Although one could argue that if the putative inhibitor competes for [^{3}H]-estradiol binding to nuclear type II sites it should also compete for [^{3}H]-estradiol binding to the estrogen receptor, this is not necessarily the case. Nuclear type II do not bind triphenylethylene derivatives (antiestrogens) even though these compounds bind to the estrogen receptor (Markaverich et al., 1981b). Likewise, nuclear type II sites also appear to bind this inhibitor with amazing specifity, whereas we have been unable to show this inhibitor interacts with the estrogen receptor (Markaverich et al., 1983). Certainly, if the inhibitor were associated with the estrogen receptor in vivo we would have observed a dilution effect on binding (Fig. 1) or direct inhibition of [^{3}H]-estradiol binding to type I sites in the direct competition experiments. It is also possible that if we were able to obtain milligram amounts of the inhibitor, competition for [^{3}H]-estradiol binding to type I sites would be observed with pharmacological concentrations (m*M*). Our data demonstrate that at physiological concentrations this interaction is unlikely.

Since we have not been able to directly assess the effects of this inhibitor activity in vivo, it is very difficult at this early time to describe any direct role for this compound in estrogen action. These experiments await chemical identification of the inhibitor. Once a positive identification has been made, determination of its biological significance in vivo should be straightforward. Preliminary in vitro experiments, however, are very promising. It appears that there is a deficiency in the β-inhibitor component in rat mammary tumor cytosol as compared with normal uterus (Fig. 5), and lactating mammary gland (Fig. 6). We have observed that inhibitor preparations from rat mammary tumors that were deficient in the β-peak material did not inhibit growth of uterine stromal and myometrial cells, or rat mammary tumor cells in culture. In contrast, inhibitor preparations from uterus or liver that contain the β-aterial reduced cell numbers by approximately 80% following 4–7 days of treatment. Whether or not this inhibition of cell growth in culture results from an acceleration of cell death or an inhibition of cell division or both remains to be resolved.

Although the physiological significance of this inhibitor remains to be resolved, we speculate at this time that the inhibitor may act to modify or regulate uterotropic responses to estrogen or perhaps act in a "protective" capacity in cases of hyperestrogenization. Such hypotheses are consistent with our current knowledge concerning a possible role for nuclear type II sites in estrogen action. We have shown that these secondary nuclear estrogen binding sites are only activated or stimulated in the nucleus under conditions that cause uterine hypertrophy, hyperplasia, and DNA synthesis (Markaverich et al., 1981b,c). Furthermore, dexamethasone and progesterone antagonism of uterine growth in the rat is associated with an inhibition of estrogen stimulation of nuclear type II sites, and these antagonists do not affect the normal functions of the estrogen receptor. On the basis of these experiments we have suggested that nuclear type II sites may be involved in estrogen action. Since nuclear type II estrogen binding sites appear to be localized on the nuclear matrix (Clark and Markaverich, 1982), which has been implicatedin DNA replication (Pardoll et al., 1981), we feel that this inhibitor activity may modulate or block estrogen-induced DNA synthesis by inhibiting estrogen stimulation of these secondary nuclear estrogen binding sites.

Perhaps the failure of estrogen to stimulate cell growth (hyperplasia; DNA synthesis) in estrogen target tissues such as the pituitary and hypothalmus (Kelner and Peck, 1981) is related to the inability of estrogen to modulate the activity of this inhibitor in these tissues. Certainly, the failure of estrogen to stimulate nuclear type II sites in these estrogen target organs makes this a tenable hypothesis. Our findings that rat and mouse mammary tumors and human breast cancer (Markaverich et al., 1984) contain significantly lower levels (~ 15–20-fold) of this inhibitor activity which is correlated with a deficiency in the β-peak component is consistent with this hypothesis. Likewise, nuclear type II sites appear to be permanently activated in ovarian-dependent (Watson and Clark, 1980) or independent (Watson et al., 1980) mouse mammary tumors and human breast cancer (Syne et al., 1982), regardless of the endocrine status. Therefore these higher levels of nuclear type II sites in malignant tissues are correlated with this inhibitor deficiency. Likewise, we have measured basal levels of nuclear type II sites in a variety of tissues that do not normally respond to estrogen via hypertrophy and hyperplasia (diaphragm, spleen, liver) and these tissues do contain significant quantities of inhibitor activity (Markaverich et al., 1983). Therefore, our hypothesis is that this inhibitor may be a component of all tissues, as are nuclear type II sites. In tissues that do not normally respond to estrogen in a proliferative manner, type II sites are complexed with this inhibitor and consequently the functions of these sites are not expressed. Conversely, in tissues that do respond to estrogens, the association of the receptor-estrogen complex with target cell nuclei may result in a dissociation of the inhibitor from nuclear type II sites. Under these conditions cellular hypertrophy and hyperplasia are observed. Consistent with this hypothesis is our observation that in estrogen-treated nuclei additional nuclear type II sites are observed following dilution. Since this effect is not observed in uterine nuclei from ovariectomized animals, we feel that this dissociation of the inhibitor from nuclear type II sites is estrogen dependent. Obviously, the lower levels of inhibitor activity in neoplastic tissues is consistent with the elevated levels of type II sites measured in tumors and the rapid proliferation rate in these cell populations. Although only tentative at this point in time, we feel that this is a reasonable model for potential regulation of cell proliferation by type II binding inhibitor.

Acknowledgments. This work was supported by NIH grants HD-08436, CA-26112 and CA-35480.

References

Clark JH, Hardin JW, Upchurch S, Eriksson H (1978) J Biol Chem 253:2630–2634
Clark JH, Markaverich BM (1982) In: Maul G (ed) The nuclear envelope and the nuclear matrix. Alan R Liss Inc, New York, pp 259–264
Ekman P, Barrack ER, Greene GL, Jensen EV, Walsh PC (1983) J Clin Endocrinol Metab 57:166–176
Eriksson H, Upchurch S, Hardin JW, Peck EJ, Clark JH (1978) Biochem Biophys Res Commun 81:1–7
Kelner KL, Peck EJ Jr (1981) J Receptor Res 2:47–62
King WJ, Allen TC, DeSombre ER (1981) Biol Repro 25:859–870

Lyttle CR, DeSombre ER (1977) Proc Natl Acad Sci USA 74:3162–3166
Lyttle CR, Garay RV, DeSombre ER (1979) J Steroid Biochem 10:359–363
Lyttle CR, Medlock KL, Sheehan DM (1984) J Biol Chem 259:2697–2700
Markaverich BM, Clark JH (1979) Endocrinology 105:1458–1462
Markaverich BM, Roberts RR, Finney RW, Clark JH (1983) J Biol Chem 44:1575–1579
Markaverich BM, Roberts RR, Finney RW, Clark JH (1983) J Biol Chem 258:11663–11671
Markaverich BM, Upchurch S, McCormack SA, Glasser SR, Clark JH (1981a) Biol Repro 24:171–181
Markaverich BM, Upchurch S, Clark JH (1981b) J Steroid Biochem 14:124–132
Markaverich BM, Williams L, Upchurch S, Clark JH (1981c) Endocrinology 109:62–69
Markaverich BM, Roberts RR, Alejandro MA, Clark JH (1986) J Biol Chem 261:142–146
McCormack SA, Glasser SR (1980) Endocrinology 106:1634–1649
Mercer WD, Edwards DP, Chammnes GC, McGuire WL (1981) Cancer Res 41:4644–4652
Neufeld HA, Levay AN, Lucas V, Martin AP, Stotz E (1958) Proc Natl Acad Sci USA 233:209–211
Oxenhandler RW, McCune R, Subtelney A, Truelove C, Tyrer HW (1984) Cancer Res 44:2516–2523
Sheehan DM, Branham OO, Medlock KL, Olson ME, Zehr DR (1981) Endocrinology 109:76–82
Pardoll DM, Vogelstein B, Coffey DS (1980) Cell 19:527–536
Sheehan D, Medlock KL, Lyttle CR (1984) 7th Inter Congr Endocrinol Abst No 2418, p 1469
Simmons RCM, Means AR, Clark JH (1984) Endocrinology 115:1197–1202
Swaneck GE, Alvarez JM, Sufrin G (1982) Biochem Biophys Res Commun 106:1441–1445
Syne JS, Markaverich BM, Clark JH, Panko WB (1982) Cancer Res 42:4443–4448
Tchernitichin A, Tchernitichin X (1976) Experientia 32:1249–1250
Tchernitchin A, Tchernitchin X, Galand P (1976) Differentiation 5:145–150
Watson CS, Clark JH (1980) J Receptor Res 1:91–111
Watson CS, Medina D, Clark JH (1977) Cancer Res 37:3344–3348
Watson CS, Medina D, Clark JH (1981) Endocrinology 107:1432–1437

Discussion of the Paper Presented by J. Clark

CLARK: It's a small molecule with a molecular weight between 300 and 400 daltons.
O'MALLEY: If this model is correct, it would seem reasonable that other steroid hormones may have these types of antagonistic affects too. I remember when I first went to NIH years ago, Roy Hurst was just retiring there and he used to talk about hormones and an evolving class of antihormones that was going to come, and they never came. But it may be that there are many types of things that have been missed. What do you see as a potential of this to apply to one other steroid hormone or other proliferative responses?
CLARK: At first we thought each steroid hormone might have an endogenous an-

tagonistic ligand. We have performed preliminary experiments with the progesterone system but have not found a similar progestin antagonist.

O'MALLEY: Did you look at a tissue in which androgens cause proliferation?

CLARK: Yes. The prostate does contain a binding site for estradiol which can be stripped by charcoal exposure to reveal type II-like sites. We presume these may be occupied by a ligand.

LIAO: We showed a similar type of binding that is going up and we later isolated our protein. We didn't know what it was until one day we extracted other proteins with aceonte. We also did charcoal, which will remove the binder. Our protein binder was steroid. But you have to remove the endogenous binder before you can show the binding of the steroid to our protein. Two years ago we published that the one we removed was cholesterol. So my question is, is it cholesterol or not?

CLARK: No, it's not cholesterol. However, we had a similar experience just recently with the antiestrogen binding site in which we thought there was an endogenous antiestrogen for that site and set about to isolate, purify, and identify it, and, in fact, it came out to be cholesterol.

LIAO: By the way, in the same extract, we isolated phospholipid and the ratio 1:1. I'm not going to ask you if it was phospholipid.

CLARK: No, it is not a phospholipid.

BAXTER: Jim [Clark], I take it that at this stage you don't have any evidence that would exclude the possibility that this thing might be having its growth-inhibiting properties through some type of other interactions, some interaction other than through the estrogen receptor. I mean through the type II.

CLARK: Do you mean do we have any other evidence to exclude it? No.

BAXTER: Have you looked to see whether this substance, along those lines, has any affect on the growth of tumors that may contain no estrogen receptors?

CLARK: Yes, they will in fact inhibit any cell type that we've tested, whether they contain estrogen receptor or not.

BAXTER: So that would raise the possibility that they could be acting primarily through some locus other than type II?

CLARK: No, it does not because the type II sites are present in all cells that we have examined. They're elevated and inducible in estrogen target tissues.

BAXTER: I see, so you mean the type II sites are ubiquitous and are present in other cell types?

CLARK: Yes, as far as we know.

BAXTER: There is another point that confused me somewhat and that was the interrelationship between the estrogen receptor, which I guess is the type I and these sites. Was that based on some direct experimental evidence that I missed or is that just hypothesis?

CLARK: That was based on fantasy.

HAUSSLER: Jim, do you think that this class of compounds does more than inhibit proliferation? That is, do they cause differentiation of some cells for instance?

CLARK: I can't think of a paper on differentiation.

BARDIN: What were the void volumes of α and intermediate peaks?

CLARK: We don't know that yet.

Discussants: W. BARDIN, J. BAXTER, J. CLARK, M. HAUSSLER, S. LIAO, and B. O'MALLEY

Chapter 5

Human Progesterone Receptors Have Two Intracellular Hormone Binding Proteins That Are Covalently Modified in Nuclei

K.B. HORWITZ, M.D. FRANCIS, AND L.L. WEI

Introduction

Progesterone receptors (PR) are believed to be gene-regulatory proteins by virtue of their ability to bind chromatin and DNA. Neither the structural organization nor the nuclear sites of action of these proteins is resolved. The questions that persist include the size and number of hormone-binding subunits of the holoreceptors and the means by which receptors are transformed* to tight chromatin-binding proteins in response to hormone treatment. With regard to the number of hormone-binding forms, two proteins of different molecular weight have been described for chick oviduct and human breast cancer PR, and two molecular weight forms are occasionally seen for estrogen, glucocorticoid, and Vitamin D_3 receptors as well (reviewed in Horwitz et al., 1985a). However, it is generally argued that for all these receptors, smaller molecular weight proteins are proteolytic artifacts formed during in vitro incubations (Loosfelt et al., 1984). Questions also persist about the mechanisms by which progesterone converts the receptors from loose to tight chromatin-binding proteins, and virtually nothing is known about the means by which the actions of the nuclear hormone-receptor complexes are terminated.

To study the structure of PR in intact cells and to avoid the artifacts that may arise during in vitro hormone-receptor binding reactions, we developed an in situ photoaffinity labeling method for PR that permits study of the proteins with minimal incubations (Horwitz and Alexander, 1983). The strategy is to use [^{3}H]R5020, a synthetic photoactive progestin, and suitable incubation temperatures, to move receptors to various intracellular sites. The cells, still intact, are then irradiated with UV rays

*Defined according to Sherman and Stevens (1984): "Activated" receptors are potentially able to bind hormone; "untransformed" hormone-receptor complexes are formed by hormone treatment in the cold and are characterized by low affinity for nuclei, chromatin, or DNA; "transformed" hormone-receptor complexes have high affinity for nuclei in intact cells, as a result of hormone treatment plus warming.

to yield covalently linked hormone-receptor complexes at any intracellular location. This method has been used to show, first, that PRs consist of two equimolar hormone-binding proteins that share considerable homology; second, that they are dissimilarly covalently modified in their untransformed state; third, that after progestin binding and transformation—acquisition of tight nuclear-binding capacity—the two PR proteins undergo further covalent modifications consistent with phosphorylation; and, fourth, that nuclear receptor action may be terminated by a processing reaction that leads to the simultaneous decrease of both proteins from the

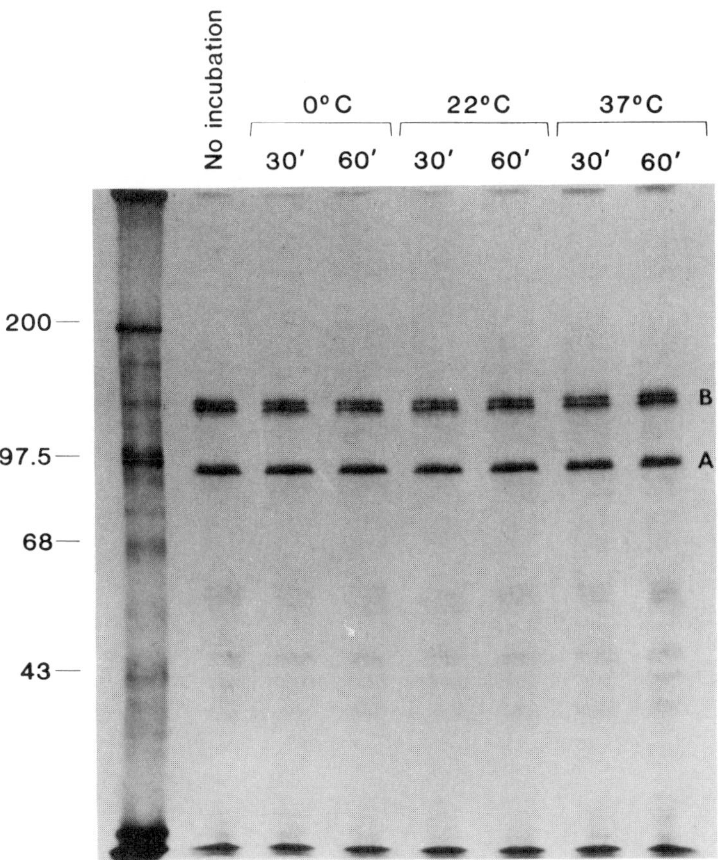

Fig. 1. Structural stability of in situ photolabeled untransformed PR incubated in vitro. T47D cells were incubated with [^{3}H]R5020 for 3 h at 0°C, irradiated 2 min with 300 nm UV, then homogenized and a cytosol was prepared. Aliquots of the photolabeled cytosol were denatured immediately (no incubation), or incubated for 30 or 60 min at 0, 22, or 37°C. These cytosols were then denatured and, together with [^{14}C] labeled molecular weight standards, subjected to SDS-PAGE on a 7.5%–19% acrylamide gradient gel. A fluorogram of the gel is shown.

cell (Horwitz et al, 1985b). A model is presented to describe the role of phosphorylation in receptor action.

The "Doublet B-Singlet A"

Figure 1 illustrates the nature of the photolabeled untransformed (cytoplasmic) proteins seen after T47D cells have been incubated with [^{3}H]R5020 at 0°C for 2h, irradiated with UV rays for 2 min, and then homogenized. If the cytosol prepared from the postnuclear supernatant is subjected to electrophoresis on gradient gels under denaturing conditions, two bands are seen. The heavier B protein is a doublet of M_r 117–120,000 daltons. The smaller A protein is a singlet of M_r 94,000 daltons. Figure 1 also shows the remarkable stability of these two proteins during further in vitro incubations that would be expected to promote proteolytic degradation. Even incubation for 1 h at 37°C fails to degrade the B protein either to A or to any other fragments. The T47D cells have unusually low endogenous protease activity, as shown also by the stability of the protein products of transfected genes (B. O'Malley, personal communication, 1985). We have previously shown that proteins A and B bind only progestins specifically (Horwitz et al., 1985a) and that they are present in equimolar amounts (Horwitz and Alexander, 1983). Figure 2 shows that they can be demonstrated even if homogenization and centrifugation are eliminated, that is, by in situ photoaffinity labeling of intact cells, solubilization in buffer containing detergent and protease inhibitors, denaturation, and immediate electrophoresis. Inspection of Fig. 2 shows that under these conditions both A and B are present (H) and that they bind R5020 specifically (H+C).

The "doublet B-singlet A" is characteristic of activated but untransformed receptors, having low affinity for nuclei. This structure is seen not only in 0°C in situ photolabeled receptors but also in receptors from cytosols incubated with [^{3}H]R5020 at 0°C in vitro (Fig. 3). Both proteins form part of the structural organization of the native hormone-free aporeceptors, but are not bound together as subunits (unpublished material, 1986).

Covalent Modifications in the Nucleus

When the cells are treated with hormone and also warmed, 95% of receptors become transformed and acquire tight nuclear-binding capacity within 5 min. In this state receptors can no longer be recovered in cytosols, but instead must be extracted from nuclei with buffers of high ionic strength. The transformation reaction does not alter the receptors. Acutely transformed nuclear receptors (1–5 min after hormone addition) struc-

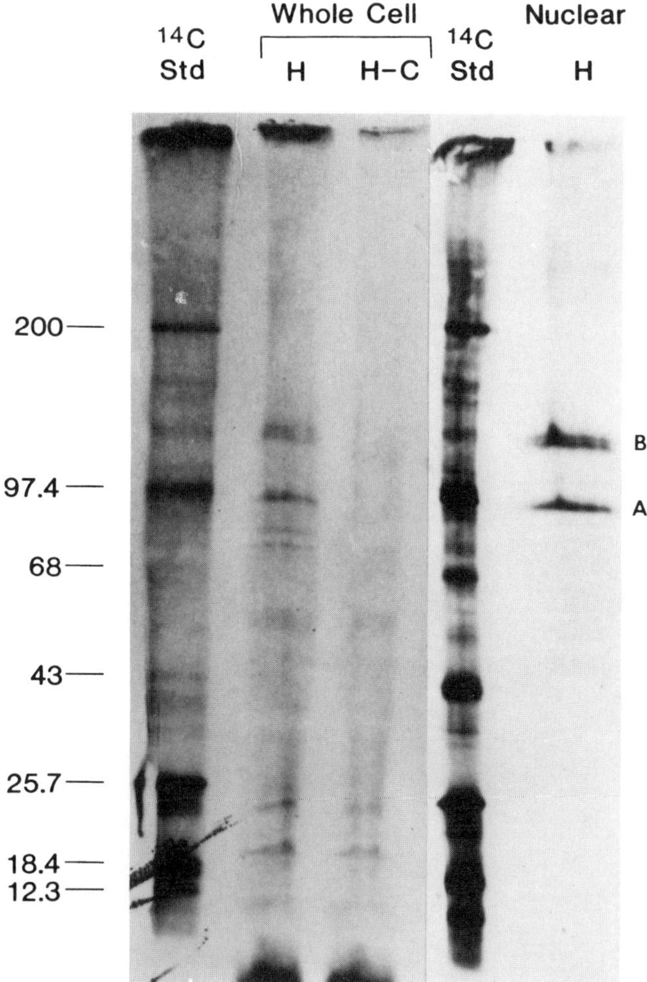

Fig. 2. The two PR subunits solubilized from intact cells after in situ labeling with [^{3}H]R5020. T47D_{co} cells growing in petri dishes were treated 5 min at 37°C with 80 n*M* [^{3}H]R5020 only (H) or with labeled R5020 plus a 100-fold excess of unlabeled R5020 (H + C). The cells were then cooled and irradiated 2 min at 0°C with 300 nm UV. The whole cells were then lysed from the dishes with a solubilization buffer containing 1% SDS, 0.4 *M* KCl, and protease inhibitors. The lysate was heated at 100°C for 2 min and resolved on 7.5%–19% gradient gels. The standard salt extracted nuclear receptors are shown on the right for comparison.

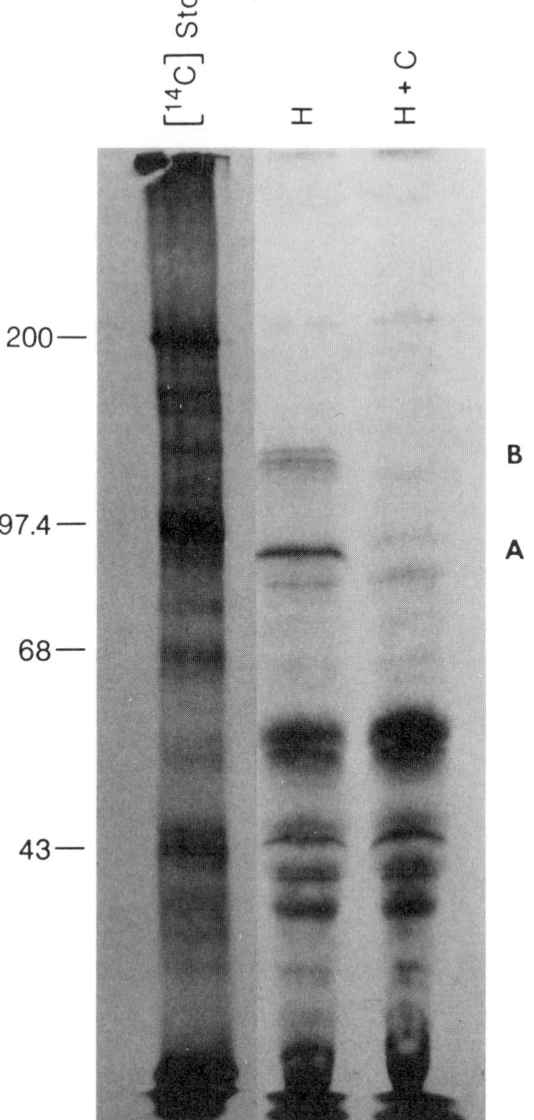

Fig. 3. Structure of in vitro photoaffinity labeled untransformed PR. Untreated T47D cells were homogenized and a cytosol was prepared. Aliquots were incubated 2 h at 0°C with 20 n*M* [^{3}H]R5020 alone (H) or together with a 100-fold excess of unlabeled R5020 (H + C). The labeled cytosols were then irradiated 2 min with 300 nm UV. After addition of SDS, dithiothreitol and bromphenol blue, samples were heated 2 min at 100°C, then subjected to SDS-PAGE on 11.5% acrylamide gels. [^{14}C] Molecular weight standards were run in a parallel lane. A fluorogram of the gel is shown.

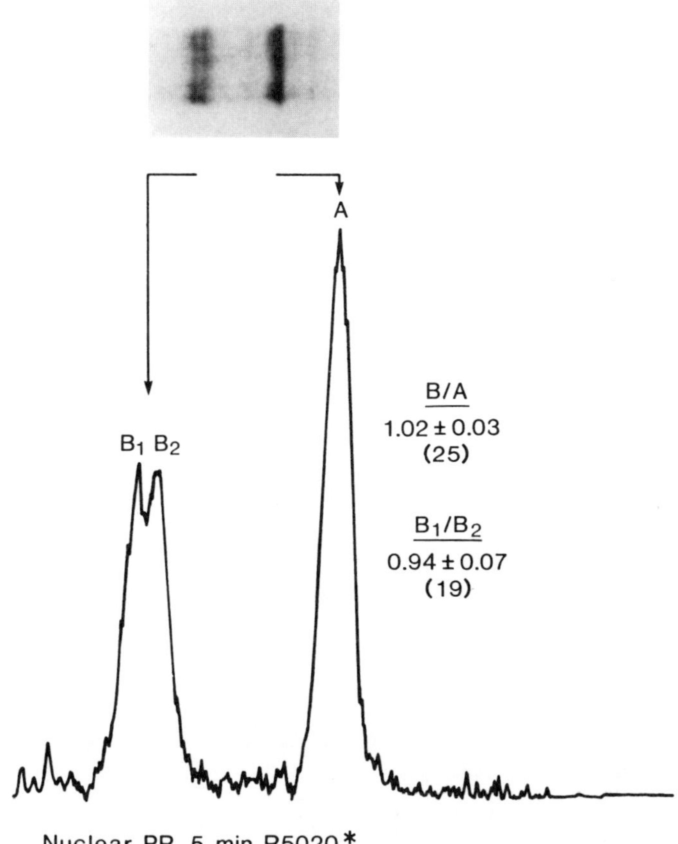

Fig. 4. Structure of transformed nuclear receptors photoaffinity labeled in situ. T47D cells growing in petri dishes were incubated 5 min at 37°C with 80 m*M* [^{3}H]R5020 to transform receptors. The dishes with the cells intact were then irradiated 2 min with 300 nm UV. Cells were harvested, homogenized, nuclei were pelleted, and incubated 1 h at 0°C with buffer containing 0.4 *M* KCl. A nuclear extract was prepared by ultracentrifugation. Proteins in the extract were precipitated with ammonium sulfate, and the redissolved pellets were denatured by boiling in SDS and subjected to SDS-PAGE on 7.5%–19% gradient gels. Gels were fixed, treated with fluor, dried, and used to expose x-ray film. The film was scanned with a densitometer and analyzed by an integrator program. The ratio of peaks B to A for 25 lanes and also the ratio of B_1 to B_2 in the "B doublet" of 19 scans are shown on the right. Numbers are averages ± SEM.

turally resemble the untransformed cytoplasmic receptors; the "B doublet-A singlet" structure prevails (Fig. 4). This can also be concluded from Fig. 1, where transformation was accomplished in vitro by the 60-min, 37°C incubation. There also the transformed receptors resemble the untransformed ones (0°C treated), and we would conclude that transformation itself leaves the primary structure of each of the receptor proteins unchanged, as far as can be determined by electrophoresis.

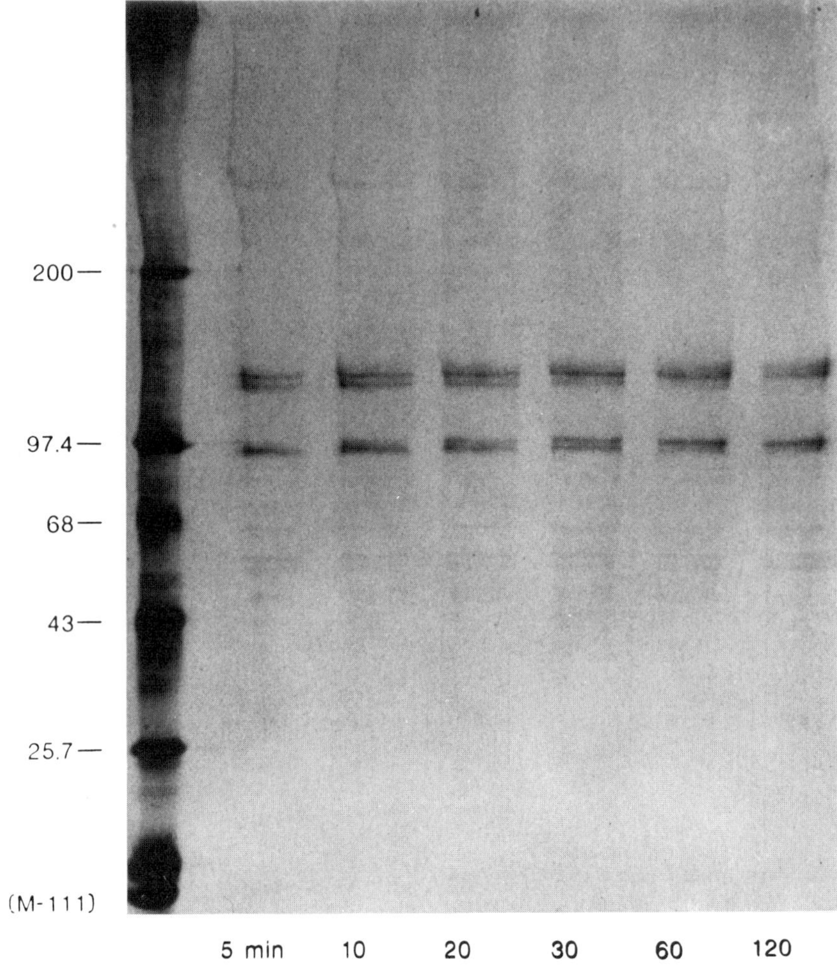

Fig. 5. Hormone-dependent covalent modification of nuclear PR photolabeled in situ. T47D cells grown in petri dishes were treated with [^{3}H]R5020 for 5 min to 2 h at 37°C. The dishes were then cooled and irradiated 2 min with UV. The cells were harvested, homogenized, and a nuclear extract was prepared, concentrated by ammonium sulfate precipitation and subjected to SDS-PAGE followed by fluorography exactly as described for Fig. 4.

The holoreceptors, as well as the acutely transformed nuclear receptor complexes, have equal amounts of A and B, as well as equimolar levels of the B_1 and B_2 isoforms (Fig. 4). However, as nuclear residence time increases, the equimolar relationship between the two B isoforms begins to shift. Figure 5 shows that the lower molecular weight band of the protein B doublet disappears with time and is entirely converted to the heavier form of the doublet approximately 30 min after the start of hormone treatment. Meanwhile protein A, which is a singlet at 5 min, splits to become a doublet by 30 min, and then the lower molecular weight form of the doublet shifts entirely to the heavier form by 60 min. Thus, both B and A display reproducible 2,000–3,000 dalton decreases in electrophoretic mobility 30–60 min after acquisition of tight nuclear-binding capacity. Such shifts occur when proteins become covalently modified, usually as a result of phosphorylation (Wegener and Jones, 1984; Pike and Sleator, 1985).

Figure 6 shows the changes in nuclear receptors during 24 h of R5020 treatment. The "doublet B-singlet A" condition at 5 min is replaced at 4 h by the heavier, covalently modified form of both proteins. Prolonging hormone treatment even further leads to the simultaneous down-regulation or "processing" of both proteins. It is clear that A and B are simultaneously lost by a mechanism that generates no hormone-binding fragments of molecular weight greater than 10,000 daltons. The nature of this reaction is unknown, although it occurs both for estrogen receptors (Horwitz and McGuire, 1980), and PRs (Mockus and Horwitz, 1983) and probably serves to terminate their action. Following R5020 treatment, the receptors lost from nuclear compartments do not reappear in cytoplasmic compartments (not shown) so that processing represents an absolute decrease of total cellular receptors as measured by hormone binding. The receptor loss is not due to inactivation of [^{3}H]R5020 because the high levels of a nuclear matrix progestin-binding protein measured to control for this possibility remain unchanged during the entire 24-h period. In contrast to R5020, after progesterone treatment, nuclear receptor loss is followed by cytoplasmic replenishment.

The Size of the Human Receptors

The preceding studies show that the molecular weights we have assigned to hormone-binding proteins A and B of human cells are considerably higher than have previously been reported for chick oviduct and rabbit uterine PRs. Protein A migrates just ahead of the 97.4 M_r phosphorylase B marker (Fig. 1, e.g.), and, by immunoblot analysis (not shown), just behind the 90-kdalton nonhormone-binding protein described by Toft and collaborators (Riehl et al., 1985). Thus, we have assigned protein A a molecular weight of 94,000 daltons. The molecular weight of protein B was more difficult to assess since no [^{14}C] labeled standard was used in

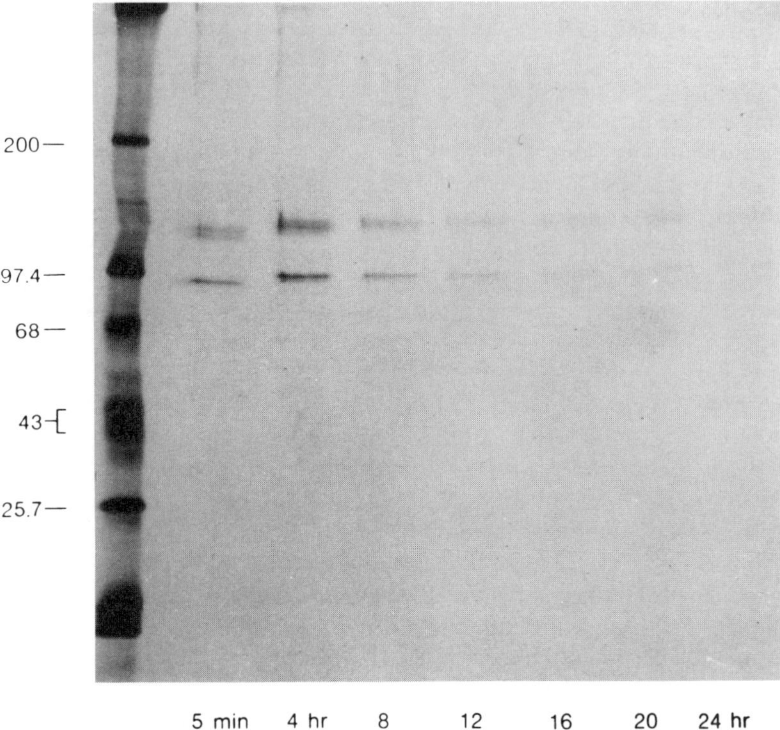

Fig. 6. Hormone-dependent covalent modification and processing of nuclear PR photolabeled in situ. T47D cells growing in petri dishes were treated with [^{3}H]R5020 for 5 min to 24 h. They were then UV irradiated before being homogenized. Nuclei were pelleted at low speed, extracted with 0.4 *M* KCl-containing buffer, and a soluble extract and salt-resistant pellet were obtained. The postnuclear supernatant from the low speed was saved. Salt-resistant nuclear pellets and cytoplasm were solubilized by 30 min incubation at room temperature in SDS and protease inhibitors, then they and nuclear extracts were denatured by boiling in SDS, subjected to SDS-PAGE, followed by fluorography. Results shown are the PR in the nuclear extracts. No receptors were seen in the other cell compartments.

the 110–120-kdalton range. Instead, we compared photoaffinity labeled human and chick receptors on the same electrophoretic gel (Fig. 7) and found that the human B protein is indeed 10–15 kdalton heavier than its chick counterpart. This was confirmed by immunoblot analysis using the 9G10 antibody that binds to a nonhormone-binding heat shock protein of M_r 106–108 kdalton homologous to chick protein B (Peleg et al., 1985). As Fig. 8 shows, the heat shock protein is present in T47D cytosols; however, its molecular weight is identical to that of chick protein B (Fig. 7) but at least 10 kdaltons less than that of human protein B. Based on these

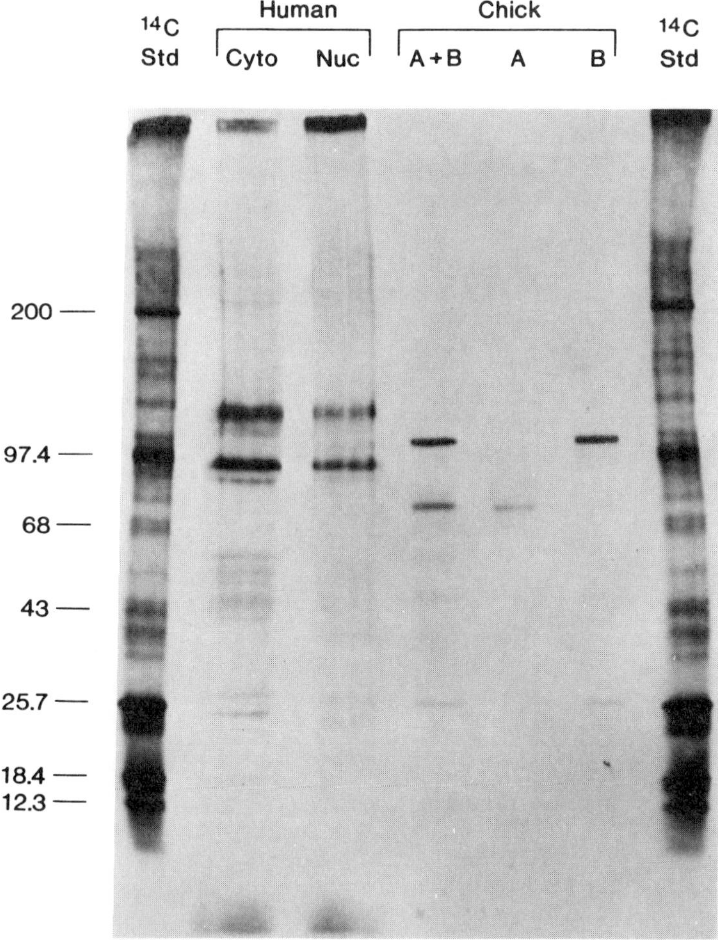

Fig. 7. An electrophoretic comparison of cytoplasmic and nuclear human PR subunits and of purified chick oviduct A and B proteins. *Human PR:* T47D cells were grown to confluence in petri dishes, then treated for 5 min at 37°C with 80 n*M* [^{3}H]R5020 to label receptors in nuclei (nuc), or treated for 2 h at 0°C with 80 n*M* [^{3}H]R5020 to label untransformed cytoplasmic receptors (cyto). Cells were then washed with ice-cold buffer, the buffer was removed, and the uncovered petri dishes were inverted for 2 min at 0°C on the surface of a 300-nm UV transilluminator. Cells were dislodged and homogenized; nuclei were sedimented and extracted for 1 h at 0°C with 0.4 *M* KCl-containing buffer. The nuclear extract was concentrated with ammonium sulfate. Cytosol was prepared from the postnuclear supernatant. This and the redissolved nuclear extracts were heated in SDS at 100°C, subjected to SDS-PAGE, and visualized by fluorography. *Chick PR:* Purified and photolabeled chick oviduct receptor proteins A and B were gift of M. Birnbaumer and W. Schrader. They were heated in SDS, and then pooled (A + B) and/or separately subjected to chromatography.

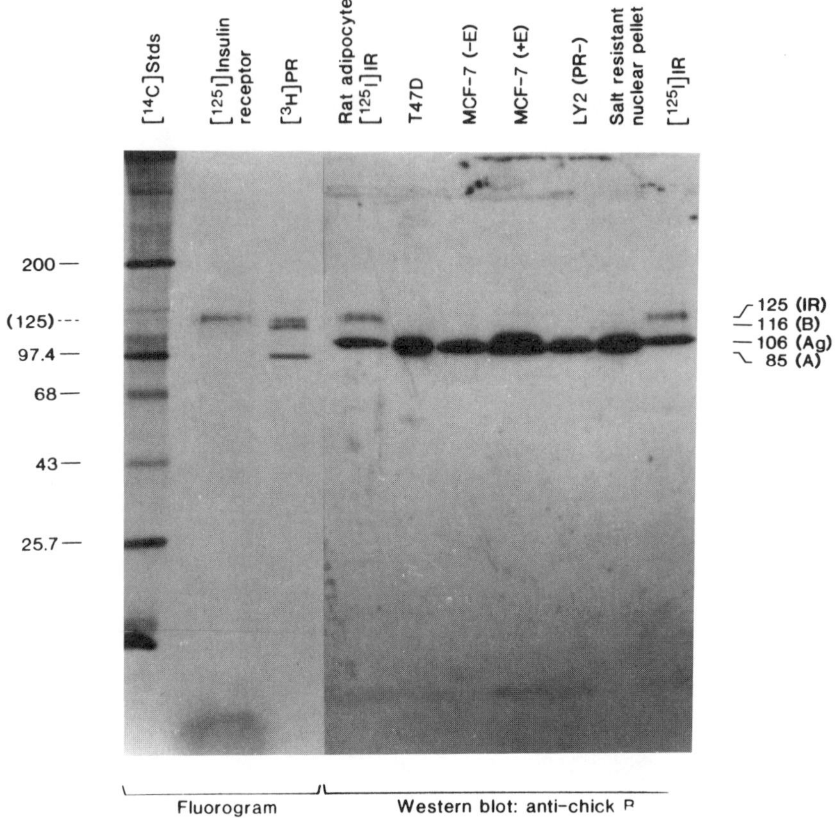

Fig. 8. Size comparison of human proteins A and B measured by fluorography with the heat shock protein homologue of chick protein B, measured by immunoblotting. *Immunoblot:* Untreated T47D and LY-2 cells were harvested from growth medium; MCF-7 cells were treated 5 days ± 10 n*M* 17β-estradiol in medium containing charcoal-stripped serum. Cytosol was prepared from all cells. The salt-resistant nuclear pellet was obtained from T47D cells treated 5 min with [^{3}H]R5020 and UV irradiated. Rat adipocyte membranes covalently labeled at the α-subunit of insulin receptors with a photosensitive [^{125}I] insulin derivative were a gift of P. Berhanu (Denver, Colo). Cytosols, membranes, and nuclear pellets were solubilized in detergent, subjected to SDS-PAGE, and the separated proteins were electrophoretically transferred to nitrocellulose paper. The paper was blocked, incubated 18 h at 4°C with the 9G10 antibody prepared against the 106–108 kDa heat shock protein homologue of chick protein B, then incubated with anti-rat IgG, followed by [^{125}I] protein A. The nitrocellulose was air dried and used to expose x-ray film for 3 days at −70°C. *Fluorogram:* A parallel gel with [^{14}C] labeled molecular weight markers, [^{125}I] insulin covalently labeled to insulin receptor (IR) from rat adipocytes, and in situ photoaffinity labeled progesterone receptors (PR) from T47D cells, was impregnated with En^{3}hance, dried, and placed on X-Omat XAR-5 film at −70°C for 20 days. The insulin receptors covalently labeled with [^{125}I] insulin were used as a marker to line up the fluorogram and western blot. Molecular weights for the IR, protein B of human PR, the antigen recognized by 9G10, and the A protein of human PR are shown on the right side of the figure.

studies we have assigned to human protein B an $M_r \sim$ 117–120,000 daltons to encompass the molecular weight of the doublet. Parenthetically, Fig. 8 also shows that the 108-kdalton heat shock protein is ubiquitous—it is present also in MCF-7 cells where it is estrogen inducible; it is present in PR-negative, antiestrogen-resistant LY-2 cells developed by M. Lippman; it is enriched in nuclei and may be on the nuclear matrix; and it is even found in rat adipocyte membranes containing [^{125}I] labeled insulin receptor α-subunit (M_r 125,000 daltons), which we included as a molecular weight marker, and to align the fluorogram and blot.

Human PRs Contain Two Hormone-Binding Proteins

The argument that A is formed in vitro as an artifactual proteolytic fragment of B is buoyed by several studies. First, despite repeated demonstration of two progestin-binding proteins in chick oviduct and human breast cancer cells, two hormone-binding proteins are usually not demonstrated for other steroid receptors (Schrader et al., 1981; Horwitz et al., 1985a). On occasions when two proteins are seen, the proteolysis explanation is invoked (Loosfelt et al., 1984). Second, there are studies demonstrating structural similarity between chick oviduct and human proteins A and B by peptide mapping of photolabeled receptors (Gronemeyer et al., 1983; Birnbaumer et al., 1983a; Horwitz et al., 1985b) and by immunoreactivity (Gronemeyer et al., 1985). These are compelling arguments but inconclusive, since they also fail to reveal the circa 25-kdalton fragment that must be generated from B but not A if the proteolysis argument is true. Nor do the studies prove a proteolysis mechanism, since other explanations for the biosynthesis of partially homologous proteins are possible (see below). [Although it has been reported that iodinated tryptic peptides from proteins A and B are different, these results have not yet been confirmed using the hormone-binding forms of A and B (Birnbaumer et al., 1983b)]. Third are the immunoblot studies of rabbit uterine PR. These studies show that if homogenization is done quickly, little or no protein A is generated. Unfortunately, the antibody used for these experiments has not been fully characterized—receptor purification data are unpublished to our knowledge (Logeat et al., 1983; Logeat et al., 1985), and the antibody used for blotting studies failed to bind to 50% of native PR in salt-containing sucrose gradients (Logeat et al., 1983), suggesting that it may have very different affinities for proteins A and B. Thus, the protease hypotheses has not, in our opinion, been conclusively established.

Our studies of human breast cancer PR suggest that protein B ($M_r \sim$ 120,000) and protein A ($M_r \sim$ 94,000) are integral intracellular proteins: (1) Regardless of the site from which they are extracted (cytoplasmic or nuclear), proteins A and B are found in equimolar amounts (Horwitz and Alexander, 1983 and Fig. 4). If a protease is degrading protein B, it must

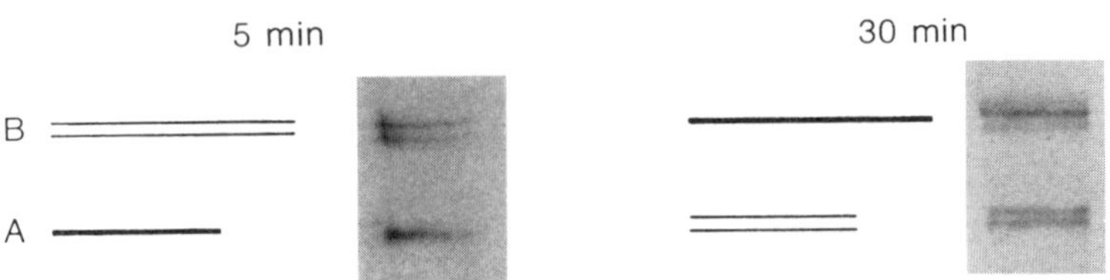

Fig. 9. Comparison of human PR proteins A and B extracted from nuclei 5 or 30 min after hormone treatment at 37°C. Experimental details are exactly as described in Fig. 5, and the figure is discussed in the text.

be present in cytosols as well as in nuclear extracts, be equally active in buffers of different ionic strengths and pH, and its activity must stop after degrading half of the B molecules. (2) Degradation of B to A can not be demonstrated during in vitro incubations (Fig. 1), forcing the conclusion that the putative enzyme acts instantaneously on half of the B molecules when the cells are first broken. (3) Both receptor proteins are seen in cells photolabeled in situ, then lysed directly in buffer-containing detergent and protease inhibitors, and immediately subjected to electrophoresis (Fig. 3). This suggests that the putative protease acts intracellularly. (4) The two receptor proteins are dissimilarly modified in their untransformed state. Since B is a doublet while A is a singlet in holoreceptors, one would have to postulate that if A is a proteolytic artifact, then the offending protease clips out a domain of B containing the phosphorylation site(s). However, such a protease can not be invoked for the 30-min resident nuclear receptors, where protein B is a singlet and protein A a doublet (Figs. 5 and 9). Here different proteases would have to be involved, that either clip the phosphorylated B at two sites 3,000 daltons apart to generate the A doublet, or clip B at a single site then dephosphorylates half the newly generated molecules. Thus we are left to conclude either that B is subject to degradation by a series of unusual intracellular proteases, or that A is a protein closely related to B, which is formed intracellularly by other mechanisms. Possible mechanisms include gene duplication, alternative transcription from a single gene, or alternative processing of a single precursor RNA. Examples of all these have been described (Nabeshima et al., 1984; Amara et al., 1984).

The Role of Phosphorylation in PR Function

Both the doublet structure of apoprotein B and the mobility shifts of proteins B and A seen after hormone treatment are most likely due to phosphorylation for several reasons: Such shifts are characteristically seen for phosphoproteins on sodium dodecyl sulfate–polyacrylamide gel electrophoresis (SDS-PAGE) (Wegener and Jones, 1984); there is increasing evidence that steroid receptors are phosphoproteins (Pike and Sleator, 1985; Migliaccio and Rotondi, 1984; Kurl and Jacob, 1984; Dougherty et al., 1982); a similar shift seen after hormone treatment of vitamin D_3 receptors is due to phosphorylation (Pike and Sleator, 1985). With this in mind, our data conform to the model shown in Fig. 10. Newly synthesized receptors are phosphorylated in a hormone independent step. This reaction is catalyzed by Kinase I. Does the initial phosphorylation confer hormone-binding activity to previously inactive receptors? Does it promote binding of receptors to nonhormone-binding proteins to generate the 8S hetero-oligomers? (Our unpublished data suggest that A and B-receptors form independent 8S oligomers.) Both phosphorylated protein molecules of the "doublet B-singlet A" structure are able to bind hormone and, after warming, can be transformed to the tight chromatin-binding state. This reaction is rapid (occurs in less than 5 min), requires no phosphorylation, and is unaccompanied by major modifications of the primary receptor structure. We postulate that it is only after acquisition of tight nuclear-binding capacity that an intranuclear phosphoprotein kinase(s) (Kinase II) modifies both proteins further. This second phosphorylation is not related to transformation since it occurs after that step. In fact, since it

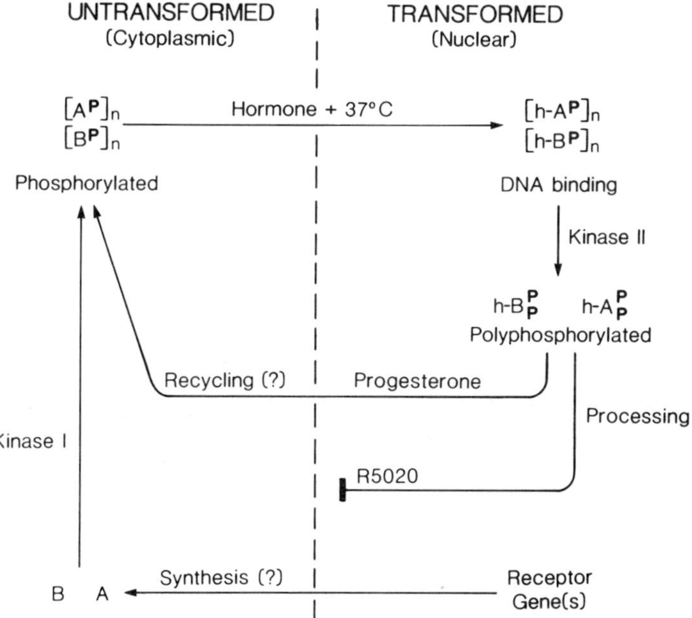

Fig. 10. A model describing the role of phosphorylation in progesterone receptor action. Details are described in the text.

occurs 30-60 min after acquisition of tight chromatin binding capacity, the actions of Kinase II may serve to inactivate the nuclear receptors leading to their subsequent "processing". Receptor levels are chronically down-regulated after R5020 treatment, but replenish quickly after progesterone treatment. Whether this is due to recycling or new synthesis is unknown.

That both A and B proteins may be important for the physiological actions of progesterone is suggested by the studies of Spelsberg et al. (Boyd and Spelsberg, 1979; Boyd-Leinen et al., 1984) who show in chick oviducts that seasonal and developmental failures to respond to progesterone are correlated with the absence of one or the other of the two molecular species of PR.

The model leads to several predictions: The first is that at least two kinases act on PR; one that phosphorylates receptors in their untransformed state, and another that phosphorylates A and B on chromatin after transformation. Whereas the second kinase is undoubtedly intranuclear, the location of the first depends on the intracellular site of the untransformed receptors—a site that remains unclear. The second prediction is that receptors that have been in the nucleus long enough to have been polyphosphorylated by Kinase II will have lower affinity for specific DNA sequences than do freshly translocated receptors, or be unable to regulate transcription. The third deals with receptor processing (Fig. 6). The nuclear receptors down-regulate without prior down-shift in molecular weight. Therefore, we would postulate either that dephosphorylation is not required or that dephosphorylation is accompanied by instantaneous loss of hormone-binding capacity; a reversal of the Kinase I mediated activation step. Clearly, some mechanism must exist to terminate the action of PR on DNA.

Acknowledgments. This work was funded by a National Institutes of Health grant (CA26869), a National Science Foundation grant (PCM8318063), and by the National Foundation for Cancer Research. Dr. Horwitz is supported by a Research Career Development Award from the National Cancer Institute (CA06694). We are grateful to our generous colleagues for some of the reagents used in these studies: Marc Lippman, the LY-2 cells; David Toft, the AC88 anti-*Achlya* antibody; Dean Edwards and Bill McGuire, the 108 kDa heat-shock protein antibody (9G10); Bill Schrader and Mariel Birnbaumer, the photolabeled chick receptors; and Paulos Berhanu, the photolabeled rat adipocyte insulin receptors.

Our chapter was originally presented at the 1984 Laurentian Hormone Conference, and in abstract form at the June 1985 National Meeting of the Endocrine Society.

References

Amara SG, Evans RM, Rosenfeld MG (1984) Mol Cell Biol 4: 2151–2160
Birnbaumer M, Schrader WT, O'Malley BW (1983a) J Biol Chem 255: 1637–1644

Birnbaumer M, Schrader WT, O'Malley BW (1983b) Biol Chem 258: 7331–7337
Boyd PA, Spelsberg TC (1979) Biochemistry 18: 3685–3690
Boyd-Leinen P, Gosse B, Rasmussen K, Martin-Dani G, Spelsberg TC (1984) J Biol Chem 259: 2411–2421
Dougherty JJ, Puri RK, Toft DO (1982) J Biol Chem 257: 14226–14230
Gronemeyer H, Harry P, Chambon P (1983) FEBS Lett 156: 287–292
Gronemeyer H, Govindan MV, Chambon P (1985) J Biol Chem 260: 6916–6925
Horwitz KB, Alexander PS (1983) Endocrinology 113: 2195–2201
Horwitz KB, McGuire WL (1980) J Biol Chem 255: 4699–9705
Horwitz KB, Mockus MB, Lessey BA (1982) Cell 28: 633–642
Horwitz KB, Wei LL, Sedlacek SM, d'Arville CN (1985a) Recent Prog Horm Res 41: 249–316
Horwitz KB, Francis MD, Wei LL (1985b) DNA 4: 451–460
Kurl RN, Jacob ST (1984) Biochem Biophys Res Commun 119: 700–705
Logeat F, Hai MTV, Fournier A, Legrain P, Buttin G, Milgrom E (1983) Proc Natl Acad Sci USA 80: 6456–6459
Logeat F, Pamphile R, Loosfelt H, Jolivet A, Fournier A, Milgrom E (1985) Biochemistry 24: 1030–1035
Loosfelt H, Logeat F, Hai MTV, Milgrom E (1984) J Biol Chem 259: 14196–14202
Migliaccio A, Rotondi A, Auricchio F (1984) Proc Natl Acad Sci USA 81: 5921–5925
Mockus MB, Horwitz KB (1983) J Biol Chem 258: 4778–4783
Nabeshima Y, Fujii-Kuriyama Y, Muramatsu M, Ogata K (1984) Nature 308: 333–338
Peleg S, Schrader WT, Edwards DP, McGuire WL, O'Malley BW (1985) J Biol Chem 260: 8492–8501
Pike JW, Sleator NM (1985) Biochem Biophys Res Comm 131: 378–385
Riehl RM, Sullivan WP, Vroman BT, Bauer VJ, Pearson GR, Toft DO (1985) Biochemistry 24: 6586–6591
Schrader WT, Birnbaumer ME, Hughes MR, Weigel NL, Grody WW, O'Malley BW (1981) Recent Prog Horm Res 37: 583–632
Sherman MR, Stevens J (1984) Ann Rev Physiol 46: 83–105
Wegener AD, Jones LR (1984) J Biol Chem 259: 1834–1841

Discussion of the Paper Presented by K. Horwitz

MUELLER: Has anybody looked for homogenous staining sections in the karyotype?

HORWITZ: Not to my knowledge.

MUELLER: Because again it could be a reiterated gene. If you had a reiterated gene, then for the production of two proteins, you would have to consider some kind of alternative transcriptional process to be a very likely event in getting two kinds of gene products.

HORWITZ: There are several ways to obtain two similar proteins other than by proteolytic degradation. Reiterated genes are one way. Alternative transcription or translation sites with a single gene are other ways. As I discussed in my presentation, proteolysis seems to be, in my opinion, the least likely mechanism.

CLARK: In the experiments when the receptor disappeared, those were all done with a photoaffinity light bulb?

HORWITZ: Right. Those were all photolabeled receptors.
CLARK: If you try to do those experiments with a nonphotoaffinity light, that's not what you see right?
HORWITZ: We see the same kind of disappearance if the receptors are measured by conventional ligand-binding assays except that the rate appears to be faster because we do those experiments with progesterone-saturated receptors instead of with R5020-saturated receptors.
CLARK: There is no return?
HORWITZ: There is no return of cytoplasmic receptors only if the ligand is R5020. If the ligand is progesterone, and we remove the hormone and allow the cells to recover, then we do see restoration of the cytoplasmic receptors. Replenishment takes 24–48 h, and appears to be protein-synthesis dependent.
CLARK: That's similar for both the photoaffinity label as well as the nonphotoaffinity?
HORWITZ: Progesterone cannot be photolinked to receptors. For R5020 there is no replenishment with or without photolinking. I hope it's clear that in our studies, all the UV treatment is done at the end of the hormone pulse; once we irradiate the cells, they're damaged. Therefore, we cannot and do not, photolink first and then do the experiment, if that is what you're getting at. The UV pulse is the last treatment done just before the cells are harvested, and therefore it has no effect on receptor kinetics or cell viability.
SPELSBERG: Kate [Horwitz], I can't recall from your one slide with the antichick oviduct B antibody, did that react with both species or just one of your species?
HORWITZ: It reacted with the "B antigen." This is a protein present in the cells that's neither A nor B receptor, but is a 108K heat shock protein.
SPELSBERG: One last question. Antiprogestins, what do they do to your cells? Enhance growth or what?
HORWITZ: RU38 486 acts like a progestin in that it inhibits cell growth. However, another biological response that I did not show you here is that progestins increase the number of insulin receptors on the cells; the antiprogestin blocks that increase. So, RU38 486 has differential effects. In the case of specific protein synthesis, it acts like an antagonist, whereas in the case of growth, it acts like an agonist. I think its actions are very complex and it will take a lot more experiments to sort out its mechanisms.
SCHRADER: I have two questions. If you grow cells chronically in progesterone and then add R5020, what is the steady-state receptor apparent molecular weight? What do you see? You said that if you pulse with R5020, molecular weight goes up and then eventually those bands just disappear. What would happen in a cell that instead of being grown in either zero-progesterone or very high progesterone, is pulsed with progesterone?
HORWITZ: The first experiment, a short-term time course, showed the upward shift in molecular weight of A and B that occurs 30–60 min after the start of continuous R5020 treatment. The second study, a long-term time course showed the processing of receptors at 8–12 h after the start of continuous R5020 treatment. A pulse of R5020 gives essentially the same results as continuous R5020, because R5020 is not washed out of cells and is not metabolized. We can not do the experiment with pulsed progesterone because it can't be photolinked.
SCHRADER: So there is no replenishment of progesterone receptors?
HORWITZ: After R5020 there is no replenishment as long as there is hormone

around. R5020 is not metabolized and cannot be removed. In contrast, progesterone can be removed, so its withdrawal, whether by physical removal or by metabolism, does lead to replenishment. The two hormones have different effects on replenishment of progesterone receptors.

SCHRADER: My second question is with respect to the heterogeneity of the lipid droplet induction in the T47Dco cells. As I understand it, this is not a clonal line.

HORWITZ: That's right, it's not cloned.

SCHRADER: Do you know whether or not all the progesterone receptors might be present in only those cells that make lipid?

HORWITZ: We don't know whether all or only some of the cells have progesterone receptors and have not yet tried to clone the T47D cells ourselves. I'm not sure how easy they are to clone. Chuck McGrath tried to clone MCF-7 cells, which consist of two different cell populations, a larger and a smaller cell type. He was able to get a pure cell population of the large-cell type, but eventually after a few passages, some of the large cells reverted back to small-cell type, so it was not a permanent selection. There is a subclone of T47D called "clone 11" that Henri Rochefort has used, but I don't know how these were derived.

SCHRADER: But these will just float?

HORWITZ: I suppose the lipid-containing cells would float, and as you imply, this property could be used to select for receptor-positive cells. Immunohistochemistry should give us more information about the receptor distribution.

GREENE: I have a couple of comments. First of all in regard to what Jim [Clark] was asking. For estrogen receptor we, as well as Benita Katzenellenbogen, have looked at various times to processing a receptor with tamoxifen receptor as well as immunochemically. We see the same thing as you describe. You just see a disappearance of receptor eventually over a period of time with no appearance anywhere else. You don't see degraded products or anything else, it's just gone. I can't clarify any issue here, but I can say that in terms of relationship with DNA, we now have eight monoclonal antibodies against the human progesterone receptor and at least seven of these are unique, and in every case the epitope is represented both in B and A. Whatever these two components, they are certainly closely related to each other.

MILGROM: Did you look on the DNA-binding properties of both forms?

HORWITZ: No, we have not. We have only done Cleveland mapping and find that A and B are very closely related.

MILGROM: Did you look on other cell types? Have you done any similar experiments?

HORWITZ: We are planning to measure PR by in situ photolabeling in primary cultures of rabbit uterine cells. We see both A and B proteins in MCF-7 cells.

———: You cited that you had no apparent estrogen receptors here. With such a dominance of progesterone receptors, maybe you got that process all suppressed. I wondered if you were to take and treat with progesterone for a period of time, and then took away, whether or not you might possibly begin to show an estrogen role?

HORWITZ: That's an interesting experiment. One thing that I did not show you here but is published is that we can demonstrate a small number of estrogen receptor sites that are of the tight chromatin-binding form. They have to be extracted from the nuclei with salt and then measured by an exchange type of assay. So, they resemble estrogen-occupied nuclear receptors. We considered that T47D cells synthesize estrogen endogenously and measured the biosynthesis of estradiol in these cells; we find that they do not make their own estrogen. So, it is possible that what we are dealing with, is an abnormal estrogen receptor that is permanently occupying its nuclear binding site in the absence of hormone and constitutively turning-on the transcription of the progesterone receptor gene.

Discussants: J. Clark, G. Greene, K. Horwitz, E. Milgrom, G. Mueller, W. Schrader, and T. Spelsberg

Chapter 6

The Two Phosphorylation Reactions of the Progesterone Receptor

F. LOGEAT, M. LE CUNFF, R. PAMPHILE, AND E. Milgrom

It is generally admitted that the early action of steroid hormones consists of two steps (Jensen et al., 1968). After entering the target cell the hormone interacts with a specific receptor. The second step is the tight binding of the steroid-receptor complex to chromatin, where it modulates the transcription of specific genes. We have recently observed, in the case of the progesterone receptor, the existence of a third step involving a hormone-dependent phosphorylation of the receptor (Logeat et al., 1985b).

Experimental Evidence of Receptor Phosphorylation

Labeling of the Receptor

Rabbit uterine slices were incubated with ^{32}P. One batch of the uterine slices was also incubated in the presence of the progestin R5020. At the end of the incubation the tissue was homogenized. Cytosolic and nuclear fractions were prepared. Nuclei were extracted by high salt. The receptor was purified by immunoaffinity chromatography (Logeat et al., 1985a) from both the cytosol and the nuclear extract. Purified receptor was then submitted to polyacrylamide gel electrophoresis in the presence of sodium dodecyl sulfate (SDS) and the gels were autoradiographed (Fig. 1).

In the uterine slices, incubated in the absence of hormone, the receptor was located in the cytosol and, as is usually the case after purification, the 110,000-dalton native receptor had been partially proteolyzed into the 79,000-dalton species (Logeat et al., 1985a; Loosfelt et al., 1984). Both forms of receptor had been labeled by ^{32}P, showing that the receptor is a phosphoprotein.

In the uterine slices incubated in the medium containing hormone, the receptor was found in both the cytosol and the nuclear extract. It was fourfold more phosphorylated than in the absence of hormone. Moreover the electrophoretic mobility of the receptor that had been exposed to hor-

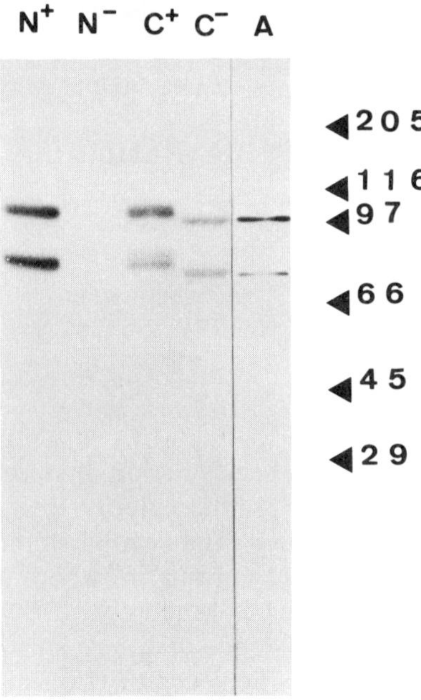

Fig. 1. Effect of the hormone on the phosphorylation of the progesterone receptor. Two batches of uterine slices were incubated with ^{32}P. In one batch the progestin R5020 was added to a final concentration of 0.1 μM after 20 min of incubation. After 40 min of incubation the tissues were homogenized, and cytosol and nuclear extracts were prepared from which receptors were purified by immunoaffinity chromatography (Logeat et al., 1985a). Aliquots were electrophoresced in 9% polyacrylamide gels (Logeat et al., 1985; Loosfelt et al., 1984) from which autoradiographs were made. N+ = nuclear receptor, incubation in the presence of hormone; N− = nuclear receptor, incubation in the absence of hormone; C+ = cytosolic receptor, incubation in the presence of hormone; C− = cytosolic receptor, incubation in the absence of hormone; Lane A = Western blot of an immunopurified cytosolic receptor.

mone had decreased in both nuclear extract and cytosol. Such differences of electrophoretic mobility, on polyacrylamide-SDS gels, between the phosphorylated and the nonphosphorylated form of a protein have been described previously in several instances (Wegener and Jones, 1984). They were ascribed to variations in the binding of SDS to each form.

The progesterone receptor is a phosphoprotein. In the presence of the hormone it becomes further phosphorylated at some supplementary site(s) as shown by the change of its electrophoretic mobility.

The effect of the hormone on receptor phosphorylation was also confirmed in vivo: rabbits received intraperitoneal injections of ^{32}P and the hormone was administered subcutaneously 30 min before they were killed.

Again, increased phosphorylation was observed under the effect of the progestin (data not shown).

Phosphorylation Studied Through Changes of Electrophoretic Mobility of Receptor

The hormone-dependent phosphorylation of the receptor provokes a change in its electrophoretic mobility. It thus appears possible to study receptor phosphorylation by directly observing this change in mobility.

The Western blot method employing monoclonal antibodies raised against the receptor (Logeat et al., 1985a; Loosfelt et al., 1984) was used to compare receptor obtained from rabbits that had or had not received hormone. In the latter case the receptor was present in the cytosol and clearly migrated ahead of the receptor present in the nuclear extract of rabbits that had received progestin (Fig. 2). Moreover, the small proportion of receptor recovered in the cytosol after progestin administration also had a decreased electrophoretic mobility (data not shown).

Immunoaffinity chromatography allowed to purify progesterone receptor from nuclear extracts and cytosols of rabbits that had or had not received the hormone. With purified receptors the same difference in electrophoretic mobility was seen (Logeat et al., 1985b).

When purified nuclear receptor was incubated with alkaline or acidic phosphatase, its electrophoretic mobility was reversed to that of the hormone-free cytosolic receptor (F. Logeat, unpublished experiments, 1985).

These experiments show that nearly all the receptor molecules had a

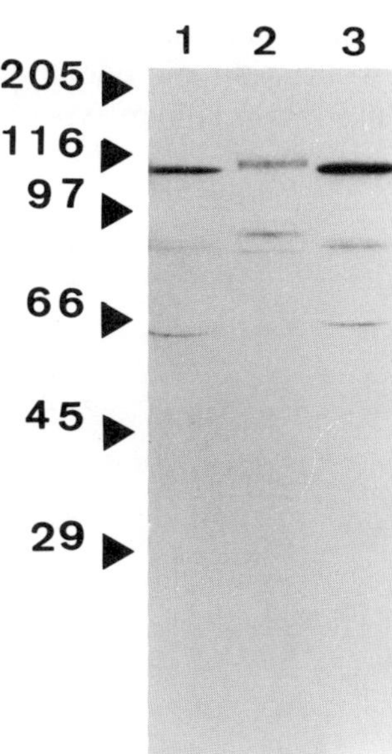

Fig. 2. Differences in the electrophoretic mobility of the progesterone receptor provoked by hormone-dependent phosphorylation. The Western blot method (Logeat et al., 1985a; Loosfelt et al., 1984) was used to study the uterine progesterone receptor present in crude nuclear extract (lane 2) prepared from hormone-treated rabbits and in crude cytosol prepared from control rabbits (lanes 1 and 3). The group of hormone-treated rabbits received an injection of 10 mg of the progestin R5020 30 min before sacrifice.

different electrophoretic mobility under the conditions used, in which the receptor was saturated with hormone, as compared with the uncomplexed receptor. Thus, this hormone-dependent phosphorylation reaction is not a partial reaction involving only a fraction of the receptor. Moreover, the hormone acts very rapidly, since these observations were made 30 min after administration of progestin to the rabbits.

The Two Phosphorylation Reactions of the Progesterone Receptor

The experiments described above show that the progesterone receptor undergoes two successive phosphorylation reactions (Fig. 3). The first one is independent of hormone administration and thus is probably a step in the biosynthesis of the receptor. Only the second one is hormone dependent in vivo, and it is this one that modifies the electrophoretic mobility of the receptor.

The polyphosphorylated form is the species of receptor that is tightly attached to chromatin and that probably modulates gene transcription.

Discussion: Possible Significance of the Hormone-Dependent Phosphorylation of Receptor

The biological significance of the hormone-dependent receptor phosphorylation is not clearly understood at the present time. It may be a necessary step in hormone action: the polyphosphorylated receptor being the active species in the modulation of gene expression. If this is the case, then all in vitro acellular experiments involving receptor interactions with regu-

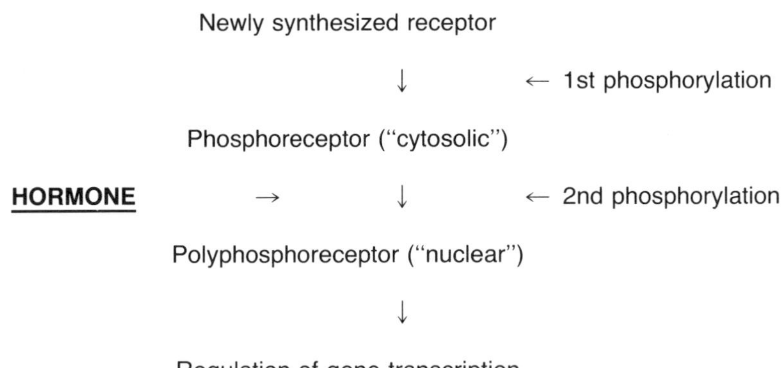

Fig. 3. The progesterone receptor undergoes two phosphorylation reactions; only the second one is hormone dependent.

latory regions of genes (Von der Ahe et al., 1985; Bailly et al., 1983) or changes in gene transcription (Jost et al., 1985) should be reassessed using the nuclear form of receptor.

Alternatively, it is possible that hormone-dependent phosphorylation does not change the gene-regulatory properties of the receptor but is a mechanism involved in target-cell desensitization. It has been shown in several cases that phosphorylation of proteins decreases their half-life (Engstrom et al., 1981). This may explain the decrease in concentration (down-regulation or processing) of progesterone receptor when the hormone is administered (Milgrom et al., 1973).

It is generally admitted that steroid hormones that act in the nucleus have completely different mechanisms of action from those of hormones or growth factors acting on the cell membrane. However, several of the latter (insulin, EGF, PDGF, etc.) have receptors that are phosphorylated upon binding of their ligand (Kasuga et al., 1983; Cohen et al., 1980). In this respect the progesterone receptor resembles these receptors for polypeptidic hormones and growth factors. Moreover, the latter receptors are hormone-regulated autokinins. We are presently examining the possibility that this might be the case for the progesterone receptor. A protein kinase activity copurifies with the receptor, but it is not clear at this time if it is the same kinase that, in vivo, phosphorylates the receptor under the effect of the hormone.

Another question that remains unanswered is how universal is the mechanism described here? Will it apply to the other steroid receptors and perhaps to other receptors of intranuclear acting hormones (thyroid hormones, metabolites of vitamin D_3). If changes of electrophoretic mobility of receptors are also observed in these cases, the question could be solved by using affinity labeling methods even in cases where lack of adequate antibodies impedes the direct study of receptor phosphorylation.

Finally, the mechanism of action of several classes of steroid antagonists (for instance the antiestrogen tamoxifen, the antiprogestin RU486) is presently unknown (Rauch et al., 1985). These compounds bind to receptors, activate them, and provoke their binding to DNA and chromatin. The reason why they are antagonists and not agonists has not been established to date. It would be interesting to analyze possible differences in receptor phosphorylation under the influence of hormones and antihormones.

Acknowledgments. The manuscript was typed by N. Malpoint. This work was supported by the Institut National de la Santé et de la Recherche Médicale (INSERM), the Fondation pour la Recherche Médicale, the UER Kremlin-Bicêtre, the Association pour la Recherche sur le Cancer (ARC).

References

Bailly A, Atger M, Atger P, Cerbon MA, Alizon M, Vu Hai MT, Logeat F, Milgrom E (1983) J Biol Chem 258: 10384–10389

Cohen S, Carpenter G, King L (1980) J Biol Chem 255: 4834–4842
Engstrom L, Ragnarsson U, Zetterqvist O (1981) In: Rosen OM, Krebs EG (eds) Protein phosphorylation. Cold Spring Harbor Conferences on Cell Proliferation, Vol 8. Cold Spring Harbor Laboratory, New York, pp 561–574
Jensen EV, Suzuki T, Kawashima T, Stumpf WE, Jungblut P, DeSombre ER (1968) Proc Natl Acad Sci USA 59: 632–638
Jost JP, Greiser M, Seldran M (1985) Proc Natl Acad Sci USA 82: 988–991
Kasuga M, Fujita-Yamaguchi Y, Blithe DL, Kahn CR (1983) Proc Natl Acad Sci USA 80: 2137–2141
Logeat F, Pamphile R, Loosfelt H, Jolivet A, Fournier A, Milgrom E (1985a) Biochemistry 24: 1029–1035
Logeat F, Le Cunff M, Pamphile R, Milgrom E (1985b) Biochem Biophys Res Commun 131: 421–427
Loosfelt H, Logeat F, Vu Hai MT, Milgrom E (1984) J Biol Chem 259: 14196–14202
Milgrom E, Luu Thi MT, Atger M, Baulieu E (1973) J Biol Chem 248: 6366–6374
Perrot-Applanat M, Logeat F, Groyer-Picard MT, Milgrom E (1985) Endocrinology 116: 1473–1484
Perrot-Applanat M, Groyer-Picard MT, Logeat F, Milgrom E (1986) of Cell Biol 102: 1191–1199
Rauch M, Loosfelt H, Philibert D, Milgrom E (1985) Eur J Biochem 148: 213–218
Von der Ahe O, Janich S, Scheidereit C, Renkawitz R, Schutz G, Beato M (1985) Nature 313: 706–709
Wegener AD and Jones LR (1984) J Biol Chem 259: 1834–1841

Discussion of the Paper Presented by E. Milgrom*

GREENE: We have done a lot of work on T 47 D human breast cancer cells, and Kate Horowitz has done some work as well, looking at receptor both labeled with R5020 covalently and in our case, with antibodies. What we find are two components. We have not been able to find any condition under which we do not observe these two components. That does not mean that A is not derived from B. I am not questioning that, but what I would like to ask is: Have you looked at human tissue with your antibodies? Do you see anything different from what you would see in the rabbit? Is there anything peculiar about the rabbit?

MILGROM: We have looked in other species and we usually see, as you do, several bands, the main ones being 110,000 and 79,000 daltons. However, the proportion varies; in some human breast cancer tumors we find mainly the 110,000-dalton receptor. We have also observed in some tumors kept in liquid nitrogen for a long time, partially thawed several times, the progressive disappearance of the 110,000-dalton band and the appearance of smaller forms. We thus think that there is nothing peculiar about the rabbit, except that having a long experience of this

---

*In the oral presentation by E. Milgrom, data were also discussed on the number of receptor subunits (see Logeat et al., 1985a or Loosfelt et al., 1984), on imunocytochemical studies at the light microscope and ultrastructural level (see Perrot-Applanat et al., 1985, 1986).

species we have defined the optimal conditions to decrease proteolysis, which has not been done in other species. Our opinion is that there is a unique steroid-binding receptor of 110,000 molecular weight in both rabbit and human, the smaller forms being due to proteolysis.

HORWITZ: I really regret that my talk is not tonight instead of tomorrow, because I am going to address all of these questions tomorrow night, so maybe we can wait until then and go over it all at that time.

HAUSSLER: That was a beautiful talk. I think you asked whether the hormone-dependent phosphorylation is general to other steroid-hormone receptors. I'd like to add one more to the list and that is the vitamin D receptor, because, as we have observed in our laboratory, it is phosphorylated. My talk is tomorrow night and I do not want to give it now. Basically we see exactly what you see by the Western blot. The upshift, as you pointed out, is due to phosphorylation. We have seen it using ^{35}S methionine internal-labeled receptor and through direct phosphorylation. Our receptor is much smaller, it is about 55,000 daltons, so it is half the size of your progesterone receptor. But it seems to be doing all the same things. The question is, where is the kinase? You seem to imply that it could be a nuclear chromatin-associated protein kinase, or is it the receptor itself?

MILGROM: May I ask one question and then I shall answer. Do you see also the two types of phosphorylation (one "basal," nonhormone-dependent and the hormone-dependent one)?

HAUSSLER: No. In fact, that is different. Without hormone treatment, we see no phosphorylation.

MILGROM: We are now working on the problem of the kinase. There is a kinase activity that copurifies with the receptor, but I would not swear now that it is the receptor itself.

SPELLBERG: My talk is tomorrow night too. I had two quick questions. One is: you know the condensed chromatin, the heterochromatin, is at least tenfold more concentrated than euchromatin. Are you considering that maybe you are not really seeing any kind of localization other than just a per unit mass DNA phenomenon?

MILGROM: This could perhaps explain the situation in the absence of hormone. However, the DNA in condensed chromatin is in major part not accessible, and there is no evidence of receptor binding to DNA in absence of hormone. In any case, this explanation does not hold for the observations made after hormone administration.

SPELSBERG: I was intrigued by the fact that some cells do not contain receptor. Have you looked at different physiological states to see if this changes?

MILGROM: No. We have not.

O'MALLEY: I am talking tonight but I want to save my questions for tomorrow night if you don't mind. There might be another way of looking at the EM nuclear localization. If, in the absence of hormones, receptor did not have a preference for specific sites, then it might fall on the bulk of the chromatin, which is the condensed chromatin. When you give hormone, it now tends to move to specific sites. Then you might have localization on the specific sites, and there are likely to be euchromatin type sites. That could be a mechanism for the shift.

MILGROM: This may be one of the possible mechanisms.

O'MALLEY: I have three questions: One, when you get the receptor off the immunoaffinity column, what percentage will bind DNA? Two, can you demonstrate with the receptor preparations specific interactions in genes? And, three, with step 1 or 2 phosphorylation is the relative affinity for DNA changed?

MILGROM: First, about 60%–90% (depending of the batch) of immunoaffinity-purified receptor will bind to DNA. Second, this purified receptor shows sequence-specific interactions with regions of the uteroglobin gene. And, third, we are presently studying the effect of phosphorylation on receptor binding to genes. I have no answer to this question at the present time.
SCHRADER: I want to ask you a question about the phosphorylation. To label the receptor by incubation with ^{32}P you must have had to turn over the phosphate on receptor protein. What fraction of the receptor in your extracts do you feel contains ^{32}P and what sort of conditions did you use to label?
MILGROM: Most of the results that I showed were obtained through incubation of uterine slices with ^{32}P. In these conditions it is very difficult to measure the molarity of incorporated ^{32}P since the specific activity of intracellular adenosine triphosphate (ATP) is not known. We have also studied kinase activity in purified receptor preparations but, as I have said before, I am not sure that we are observing in this situation the same hormone-dependent kinase that works in vivo.
SCHRADER: Okay, one last question. If you run a 2 D gel on the N^+ or the C^+ lanes, how many phosphorylated spots do you see?
MILGROM: We have experienced difficulties in running 2 D gels with purified receptors.
LIAO: Just a short question on this ribosome localization of the receptor seen by ultrastructural immunocytochemistry. Have you compared it before and after hormone injection?
MILGROM: Yes. It seems to be more clear-cut after hormone injection, but this is really difficult to assess owing to the small amount of receptor.
BRANER: In your light microscope immunocytochemical studies you pointed out cellular heterogeneity and you showed as an example myometrial tissue. I have used the same antibodies as Geff Greene to study monkey uterus. With estrogen receptor all the smooth-muscle cells are positive. The negative cells are fibroblasts, other connective tissue cells, and cells of the lining of capillary vessels.
MILGROM: You are right. Many (but not all) of the nonmuscular cells are negative. However, in the case of the progesterone receptor, you also have heterogeneity among muscular cells.
GREENE: When we look at reproductive tissues, in general, the uterus, we find fairly homogenous staining for estrogen receptor, not that there isn't any variation in intensity, but we don't usually see extremes. When we get into cultured cells like MCF-7, or if we look at breast cancers, then we can see tremendous heterogeneity, which may be related to various factors. In terms of precise localization of receptor, our experience has been a little different from yours. At the EM level we have seen receptor associated exclusively with dispersed chromatin. I'm not sure if we can guarantee that there is no hormone around, since we have studied postmenopausal uterine cells. I'll show a couple of slides tomorrow night, but that has been our experience so far.

Discussants: R. BRANER, G. GREENE, M. HAUSSLER, K. HORWITZ, S. LIAO, E. MILGROM, B. O'MALLEY, W. SCHRADER, and T. SPELSBERG

Chapter 7

Receptor-Mediated Action of the Vitamin D Hormone

M.R. HAUSSLER, D.J. MANGELSDORF, C.A. DONALDSON, S.L. MARION, N.M. SLEATOR, AND J.W. PIKE

1,25-Dihydroxyvitamin D_3 (1,25$(OH)_2D_3$) is now considered to be the active hormonal sterol derived from the sunlight vitamin, vitamin D_3. As depicted in Fig. 1, 1,25$(OH)_2D_3$ is formed in the kidney according to the calcium and phosphorus needs of the organism (Haussler and McCain, 1977). Its main functions are the stimulation of intestinal calcium and phosphate absorption as well as bone remodeling. In addition to its mineral conservation effects in the kidney, 1,25$(OH)_2D_3$ induces a 24-OHase enzyme that appears to initiate a catabolic cascade for side chain cleavage and metabolic elimination of both the hormone (Fig. 1) and its 25$(OH)D_3$ precursor (Chandler et al., 1984). Thus the 1,25$(OH)_2D_3$ hormone is dynamic in the sense that its production is controlled by calcium/parathyroid hormone (PTH) and phosphate status, and its biodegradation is self-initiated.

The actions of 1,25$(OH)_2D_3$ are apparently mediated by a classical steroid hormone receptor system that localizes the hormone in the target-cell nucleus where the sterol-receptor complex controls DNA transcription (Haussler and McCain, 1977). Based on the presence of 1,25$(OH)_2D_3$ receptors, it is now clear that this hormone functions in a variety of targets beyond those involved in mineral and bone homeostasis, including those of the endocrine system, fibroblasts, and hematopoietic cells (Fig. 1). It is probable that 1,25$(OH)_2D_3$, like retinoic acid and several steroids, plays a basic role in cell differentiation (Mangelsdorf et al., 1984). The present challenge is to characterize the molecular mechanism of action of 1,25$(OH)_2D_3$ in this spectrum of target cells and to discover and integrate individual biological functions into a scenario of vitamin D-mediated development and physiologic adaptation. The molecular key to approaching this problem is the 1,25$(OH)_2D_3$ receptor protein that modulates the expression of vitamin D-regulated genes in various target cells.

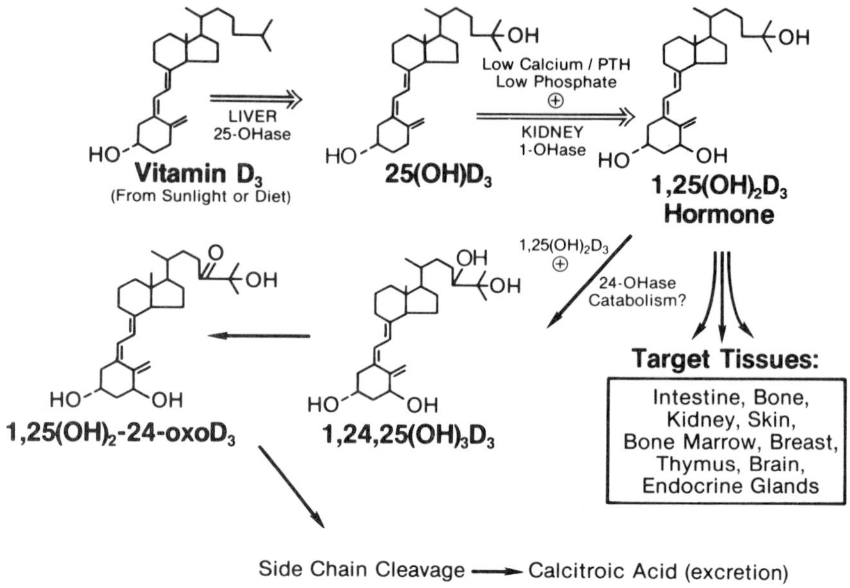

Fig. 1. Vitamin D endocrine system. The 24-OHase initiated series of reactions that culminates in side chain cleavage also utilizes 25(OH)D_3 as a starting substrate (not shown). This series of catabolic reactions is induced by 1,25$(OH)_2D_3$ and occurs primarily in kidney but also probably to some degree in all vitamin D target cells.

Immunochemical Identification of the 1,25$(OH)_2D_3$ Receptor

The 1,25$(OH)_2D_3$ receptor protein has been extensively studied since its original discovery, in vivo, as a chromosomal protein in chick intestinal nuclei (Haussler and Norman, 1969). It was biochemically and pharmacologically characterized as a 3.3S macromolecule that binds 1,25$(OH)_2D_3$ with high affinity and specificity by Brumbaugh and Haussler (1973, 1974) and was later observed to have the important property of DNA binding (Pike and Haussler, 1979). However, further characterization of the receptor molecule was not possible until monoclonal antibodies to the protein were developed (Pike et al., 1983). These monoclonal antibodies against chick intestinal receptor displaced the native hormone-receptor complex in a sucrose gradient and, most importantly, were found to interact with both avian and mammalian 1,25$(OH)_2D_3$ receptors with very high affinity ($K_d = 10^{-11} M$). The latter property enabled us to develop immunoblot methodology capable of identifying the molecular mass of receptor proteins. Figure 2 illustrates immunoblots from chick intestine and liver high-salt cytosols. This procedure results in the detection of a major immu-

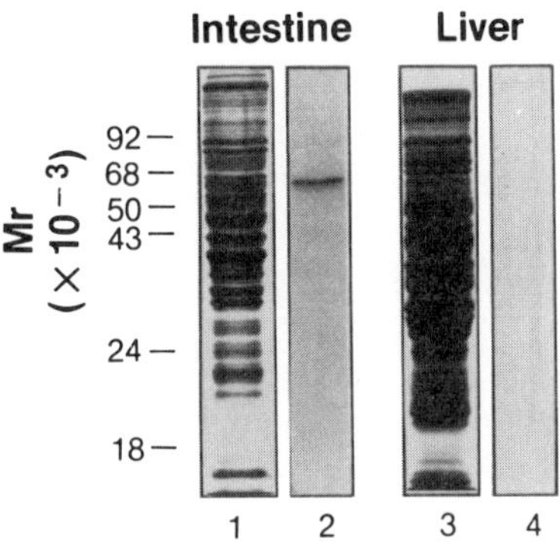

Fig. 2. Western blot immunodetection of the chick receptor for $1{,}25(OH)_2D_3$. Tissue extracts were prepared by homogenization in 0.3 *M*KCl, 1 m*M* EDTA, 0.01 *M* Tris-HCl, pH 7.4, 5 m*M* dithiothreitol (0.3-KETD) followed by ultracentrifugation (high-salt cytosol). The supernatants were eletrophoresed in sodium dodecyl sulfate (SDS) -11% polyacrylamide gels (Pike et al., 1983) and either stained with Coomassie blue (lanes 1 and 3) or transferred to nitrocellulose (lanes 2 and 4) and immunoblotted. Electrophoretic transfer to nitrocellulose membranes was accomplished in 3 h at 50V or overnight at 20V, with a modified elution buffer containing 25 m*M* Tris, 192 m*M* glycine, and 0.015% SDS. Membranes were then incubated with 0.02 *M* Na_2HPO_4, pH 7.2, 0.15 *M* KCl-phosphate buffered saline (PBS) /3% bovine serum albumin (BSA) for 2–4 h at 22°C, and then transferred to PBS/1% BSA containing 4 μg/ml pure 9A7 monoclonal antibody (Pike et al., 1983) and shaken overnight at 4°C. Washing was achieved by gently shaking the membranes for 90 min at 22°C in 50 ml 0.05 *M* Tris-HCl, pH 7.5, 0.2 *M* NaCl-Tris buffered saline (TBS)/0.05% Tween 20 with four changes. Following a short preincubation in PBS/3% BSA (10 min), the sheet was developed for 2 h in 25 ml PBS/1% BSA containing 1.5×10^6 cpm of purified rabbit anti-9A7 previously iodinated using chloramine T. After additional washing exactly as above, the membranes were dried and then autoradiographed at −70°C for various lengths of time (24–72 h), using Kodak X-Omat RP or AR film and a Dupont Cronex Hi-plus intensifying screen. The sensitivity of this Western blot procedure is approximately 0.5 ng of avian $1{,}25(OH)_2D_3$ receptor.

noreactive protein in intestine at 60,000 daltons and a minor band at 58,000 daltons. The liver, which is not considered to be a vitamin D target in the adult chicken, possesses no immunoreactive proteins (Fig. 2). We therefore conclude that the avian $1{,}25(OH)_2D_3$ receptor consists of two monomeric forms, a major species of 60,000 daltons and a minor form of 58,000 daltons. This conclusion is consistent with the original observation that several proteins between 50,000 and 65,000 daltons are evident following denaturing gel electrophoresis of purified chick receptor (Pike and Haussler, 1979) and, with gel filtration estimates of the molecular weight for the native monomer of approximately 60,000 daltons (Pike et al. 1983). It is not known if both immunoreactive species in the avian intestine bind the $1,25(OH)_2D_3$ hormone and it is conceivable that the minor form of 58,000 daltons arises through proteolytic degradation of the 60,000 dalton receptor. Recent experiments involving the in vitro translation of chick intestinal mRNA similarly reveal a 60,000/58,000-dalton doublet of immunoprecipitable receptor (Mangelsdorf DJ, Pike JW, Haussler MR, unpublished data). Thus, these two forms of the receptor could result from alternative mRNA splicing or differing transcription start/termination sites, although proteolysis cannot be ruled out during the in vitro translation or subsequent immunoprecipitation procedures.

We have also identified mammalian $1,25(OH)_2D_3$ receptors by immunoblot analysis. In these experiments, cultured target cells are labeled in suspension with $1,25(OH)_2[^3H]D_3$ and the nuclei then isolated and extracted with 0.3 *M* KCl. The nuclear extract is then chromatographed on DNA-cellulose and eluted with a gradient of KCl. Figure 3 illustrates such an experiment when human acute promyelocytic leukemia (HL-60) cells are incubated with $1,25(OH)_2[^3H]D_3$. The labeled hormone-receptor complex elutes from the DNA-cellulose column at approximately 0.23M KCl and, when individual fractions are immunoblotted (see inset to Fig. 3), a 52,000 dalton immunoreactive protein is evident which elutes with the receptor. We conclude that the human $1,25(OH)_2D_3$ receptor is a single protein of 52,000 daltons. This protein displays the three key properties of the $1,25(OH)_2D_3$ receptor: hormone binding, DNA binding, and interaction with monoclonal antibody. Similar analysis of other mammalian cell lines (Pike, 1985), as well as in vitro translation of their respective mRNAs (Mangelsdorf DJ, Pike JW, Haussler MR, unpublished data), indicate the following molecular weights for mammalian $1,25(OH)_2D_3$ receptors: mouse fibroblasts (3T6) = 55,000; rat osteosarcoma (ROS 17/2.8) = 54,000; porcine kidney (LLC-PK_1) = 54,000. Thus, mammalian $1,25(OH)_2D_3$ receptor monomers appear as single proteins ranging in molecular weight from 52,000 to 55,000 daltons. The $1,25(OH)_2D_3$ receptor monomer is similar in size to the thyroid (Casanova et al., 1984) and estrogen (Walter et al., 1985) receptors, which have molecular weights of 57,000 and 65,000 daltons, respectively. Like all thyroid and steroid hormone-binding proteins, the $1,25(OH)_2D_3$ receptor is a DNA-binding protein that now appears to be predominantly associated with the nucleus, even in the unoccupied state (King and Greene, 1984). Walters et al. (1980) provided the first detailed biochemical evidence that the unoccupied $1,25(OH)_2D_3$ receptor was localized in the nucleus, and we recently confirmed this employing immunocytochemistry. We observed nuclear localization in human breast cancer tissue, mouse osteoblasts, mouse kidney, and rat hippocampus. Although these preliminary findings require in-depth confirmation, the concept is emerging that, like the thyroid receptor, the $1,25(OH)_2D_3$ receptor is a loosely associated chromosomal protein. It is hypothesized that the receptor's affinity for DNA increases when occupied with the $1,25(OH)_2D_3$ hormone (Pike and Haussler, 1983), and the receptor subsequently binds to upstream regulatory regions of certain genes.

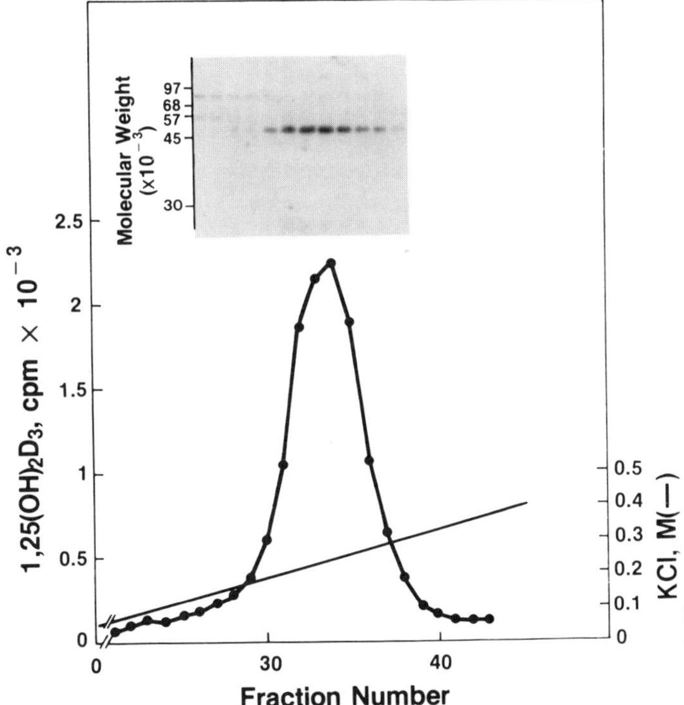

Fig. 3. Analysis of HL-60 cell nuclear 1,25(OH)$_2$D$_3$ receptor by DNA cellulose chromatography and immunoblotting. HL-60 cells (8×10^8) were incubated in suspension with 2 n*M* 1,25(OH)$_2$[^{3}H]D$_3$ (79 Ci/mmol) for 90 min at 37°C in RPMI with 1% fetal bovine serum (FBS). Nuclei were isolated, extracted with 0.3 *M* KCl, and the extract subjected to DNA cellulose chromatography (Mangelsdorf et al., 1984) The receptor was eluted in 3.0 ml fractions with a linear KCl gradient, with 0.25 ml aliquots counted for tritium (33% efficiency), and the balance precipitated with 6.5% trichloroacetic acid (TCA) for immunoblot analysis *(inset)*. Immunoblotting was carried out on a portion of the precipitated fractions solubilized in sample buffer and processed as described in the legend to Fig. 2. Only fractions within the salt gradient are illustrated, and the lanes of the immunoblot insert correspond to the DNA cellulose column fractions on the abscissa. The peak fraction contained 0.129 pmoles of receptor by hormone-binding quantitation.

1,25(OH)$_2$D$_3$ Action in the Hematopoietic System

As introduced in Fig. 1, the nonclassical sites for the 1,25(OH)$_2$D$_3$ receptor include cells derived from bone marrow and thymus. These data suggest that like the glucocorticoids and other steroids, 1,25(OH)$_2$D$_3$ may be an immunomodulator. The most intensively investigated model system within the cell-mediated immune system is the differentiation of HL-60 leukemia cells. This leukemic line differentiates into macrophage-like cells when treated with $1,25(OH)_2D_3$ in culture (Mangelsdorf et al., 1984; Bar-Shavit et al., 1983). The HL-60 cells also contain significant quantities of the $1,25(OH)_2D_3$ receptor (2,000–4,000 copies/cell) as demonstrated previously (Mangelsdorf et al., 1984) and confirmed via immunoblot analysis in Fig. 3. We have also shown that saturation of the HL-60 nuclear receptor by $1,25(OH)_2[^3H]D_3$ in complete culture medium correlates with the kinetics of $1,25(OH)_2D_3$-induced FMLP (chemotaxin) receptors (Mangelsdorf et al., 1984), indicating that this expression of the differentiated phenotype can be mediated by the $1,25(OH)_2D_3$ receptor protein. However, suppression of proliferation and differentiation of HL-60 cells into macrophages require at least 72 h, even though the saturation of nuclear receptors with $1,25(OH)_2D_3$ can be achieved as early as 4 h in suspended cells. One rapid action of $1,25(OH)_2D_3$ in HL-60 cells that has been observed by Reitsma et al. (1983) is the attenuation of c-*myc* oncogene mRNA levels. We also have studied c-*myc* mRNA levels in HL-60 cells treated for short periods (12 h) with $1,25(OH)_2D_3$. As illustrated in Fig. 4, dot blot hybridization analysis of c-*myc* mRNA utilizing a labeled cDNA probe reveals a marked depression in the apparent expression of this mRNA in cells exposed to $1,25(OH)_2D_3$. Although it is tempting to speculate that the $1,25(OH)_2D_3$ receptor complex regulates c-*myc* transcription via direct binding in the c-*myc* gene regulatory region, it is plausible that this regulation is secondary and that the hormone-receptor complex controls the transcription of a key gene(s) whose product(s) in turn controls c-*myc* mRNA levels. Since $1,25(OH)_2D_3$ regulates calcium-binding protein (CaBP) in traditional target cells, it is equally possible that intracellular calcium fluctuations

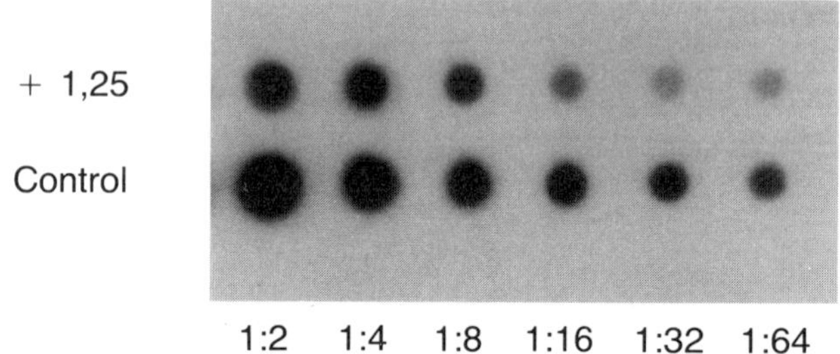

Fig. 4. Suppression of c-*myc* oncogene expression by $1,25(OH)_2D_3$ in HL-60 cells as assayed by dot blot hybridization. Total cellular RNA was extracted with guanidinium isothiocyanate and precipitated through CsCl from HL-60 cells grown in the absence (control) or presence (+1,25) of $5 \times 10^{-8}M$ $1,25(OH)_2D_3$ for 12 h. For each experiment 400 μg of RNA was serially diluted, then spotted and baked on to nitrocellulose paper (Thomas, 1983). The bound RNA was then probed with 1.7×10^7 dpm of $[^{32}P]$-labeled *myc* DNA (Oncor) and washed extensively as described (Thomas, 1983). After autoradiography of the hybridized filter, each spot was cut out and counted to quantitate radioactivity. The presence of $1,25(OH)_2D_3$ provokes a 40% reduction in the levels of c-*myc* RNA.

modulated by CaBP represent a second messenger in $1,25(OH)_2D_3$-induced differentiation of HL-60 cells. It should also be noted that a number of other compounds that differentiate HL-60 cells, such as retinoic acid and a phorbol ester tumor promotor (TPA, or 12-O-tetradecanoylphorbol-13-acetate), elicit suppressions in c-*myc* mRNA levels similar to that produced by $1,25(OH)_2D_3$ (Grosso and Pitot, 1984). Thus, this effect on c-*myc* expression may reflect either a declining proliferation rate or possibly the general triggering of differentiation.

We next studied the influence of HL-60 cell differentiation on the concentration of $1,25(OH)_2D_3$ receptors. Receptors are assessed via the dual properties of $1,25(OH)_2D_3$ and DNA binding as depicted in Fig. 5. Cells are differentiated to macrophages with the potent inducer, TPA, instead of $1,25(OH)_2D_3$ so that subsequent $1,25(OH)_2[^3H]D_3$ binding assays can

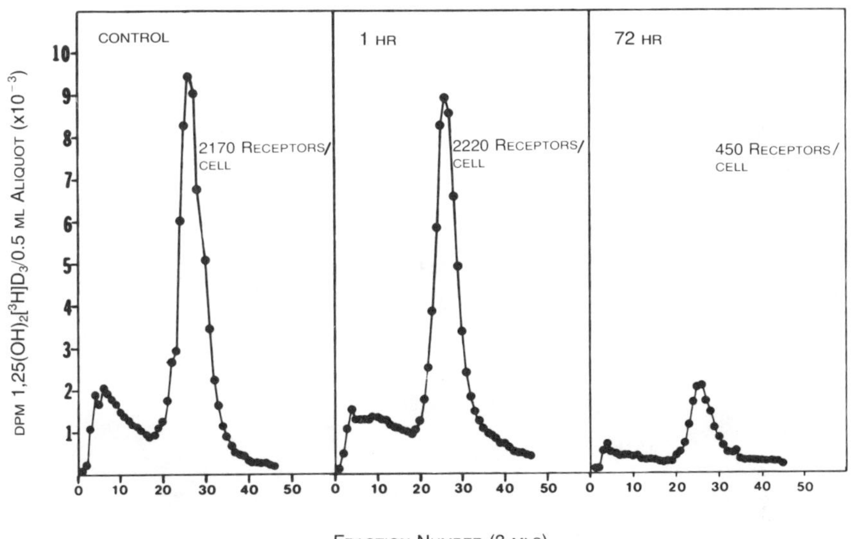

Fig. 5. Tumor promoter (TPA) regulation of the $1,25(OH)_2D_3$ receptor in HL-60 cells as assessed by DNA-cellulose chromatography of $1,25(OH)_2[^3H]D_3$ binding in nuclear extracts. HL-60 cells were incubated in the absence (control) or presence of TPA for 1 h ($5 \times 10^{-8} M$ TPA) and 72 h ($10^{-9} M$ TPA for 48 h, then $4 \times 10^{-9} M$ for 12 h). The lower doses of TPA for the 72 h experiment were necessary to keep the cells viable. For each experiment, 2.75×10^8 cells were harvested after TPA treatment and incubated with 2 n*M* $1,25(OH)_2[^3H]D_3$ for 90 min at 37°C in 10 ml RPMI with 1% FBS. Preparations of nuclear extracts and DNA-cellulose chromatography were performed as previously described (Mangelsdorf et al., 1984). After samples were applied to columns and a consistent baseline established, DNA-binding components were eluted with linear salt gradients and counted for radioactivity. In each case the peak of receptor-bound radioactivity eluted at 0.23 *M* KCl.

be performed. Short-term treatment with TPA (1 h) does not influence the level of $1,25(OH)_2D_3$ receptor, which remains similar to that of control HL-60 cells (~2200 copies/cell). This indicates that short-term activation of protein kinase C, which occurs upon TPA binding, probably does not affect either the hormone or DNA-binding functions of the $1,25(OH)_2D_3$ receptor. However, after 72 h the TPA-treated cells become adherent to the culture flask and macrophage-like and, as shown in Fig. 5, there is an 80% reduction in the concentration of $1,25(OH)_2D_3$ receptors. This precipitous fall in receptor number accompanies the process of macrophage differentiation and may relate to the proposed role of monocytes and macrophages as osteoclast precursors (Ko and Bernard, 1981). $1,25(OH)_2D_3$ increases osteoclast number, in vivo (Holtrop et al., 1981), and is proposed to mediate bone resorption, in part, by increasing the differentiation of osteoclast progenitors as illustrated in Fig. 6. Osteoclasts do not possess $1,25(OH)_2D_3$ receptors (Merke et al., 1985) and therefore it is their number but not their activity that is directly affected by $1,25(OH)_2D_3$. Our observation that $1,25(OH)_2D_3$ receptor concentrations are dramatically reduced upon differentiation of HL-60 cells to macrophages could reflect the first stage in the attenuation of receptor expression that is presumably complete upon fusion of macrophages to multinucleate osteoclasts. As

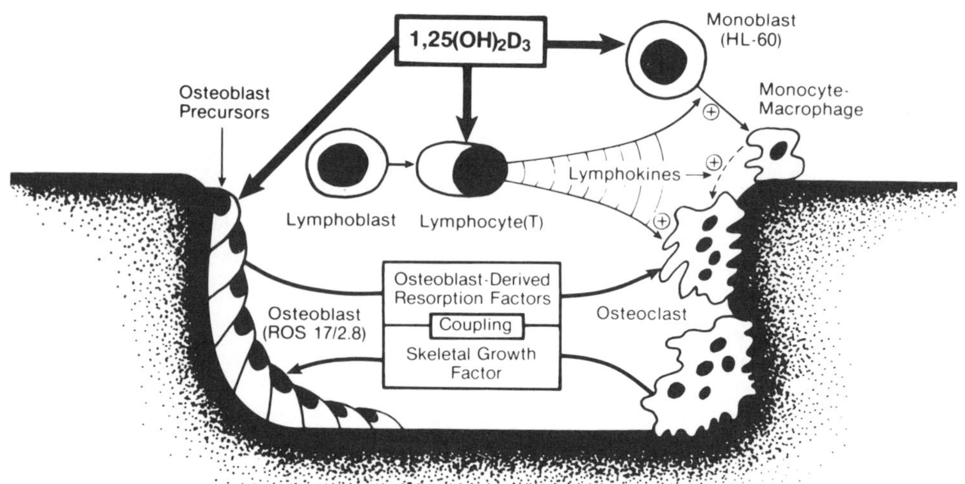

Fig. 6. Proposed function of $1,25(OH)_2D_3$ and its receptor in bone remodeling and immunomodulation. The diagram is intended to represent the normal bone remodeling process, but certain transformed cell lines are listed with their normal counterparts. Thus, HL-60 serves as a good model system of a monocyte/macrophage precursor and the ROS 17/2.8 osteogenic sarcoma cell serves as an osteoblast-like line.

illustrated in Fig. 6, it is important to note that $1,25(OH)_2D_3$ indirectly augments the final stages of osteoclast differentiation through a newly recognized operation on T-lymphocytes. T-cells contain $1,25(OH)_2D_3$ receptors (Provvedini et al., 1983), and both $1,25(OH)_2D_3$ receptor number and lymphokine production are modulated by $1,25(OH)_2D_3$ (Tsoukas et al. 1984); at least one of the regulated lymphokines has been shown to elicit macrophage fusion to giant multinucleate osteoclast-like cells (Abe et al., 1983). Finally, $1,25(OH)_2D_3$ could cause bone resorption by binding to its well-known receptors in osteoblasts (Dokoh et al., 1984) and bringing about the release of osteoblast-derived resorption factors as depicted in Fig. 6. $1,25(OH)_2D_3$-stimulated osteoblasts also remineralize bone during the remodeling cycle. Therefore, it is clear that along with its mediating receptor, $1,25(OH)_2D_3$ accomplishes a complex yet elegant regulation of the bone remodeling cells. $1,25(OH)_2D_3$ responsive cells of the immune system appear to play a central role in the function of $1,25(OH)_2D_3$ on bone resorption, but this may be only one facet of a more general action of the hormone as a novel immunomodulator.

Regulation and Covalent Modification of the $1,25(OH)_2D_3$ Receptor

As detailed above, the concentration, and presumably the expression, of the $1,25(OH)_2D_3$ receptor is modulated by the state of differentiation of HL-60 cells (Fig. 5). It is also known that the receptor level increases dramatically in normal human T-lymphocytes when they are activated (Provvedini et al., 1983). In our hands, the receptor copy number per T-cell rises from 200 to 2,000 when the cells are activated for 72 h with concanavalin A (Haussler et al., 1986). Thus, it is evident that the concentration of the $1,25(OH)_2D_3$ receptor is critically dependent on the state of differentiation and/or activation of certain vitamin D target cells—probably to restrict or amplify the action of the hormone at various stages in the physiology of the cell.

The $1,25(OH)_2D_3$ hormone also appears to alter the level of its receptor in certain cultured cell lines. Sher et al. (1985) reported that in T47D breast cancer cells, $1,25(OH)_2D_3$ induces processing of the receptor; however, only the labeled hormone was followed as a measure of the receptor in these experiments. Conversely, Costa et al. (1985) observed that analogues of $1,25(OH)_2D_3$, like $24,25(OH)_2D_3$, cause a significant up-regulation of $1,25(OH)_2D_3$ receptor-binding activity in cultured kidney cells as well as in other cell types. In our own studies involving immunoblot detection of the receptor in $1,25(OH)_2D_3$-treated 3T6 mouse fibroblasts (Pike JW, Sleator NM, Haussler MR, unpublished data), we have found an enhancement in receptor levels as early as 4 h and as high as tenfold within 48–72 h. Our data in 3T6 cells and those of Costa et al. (1985) may be

explained by invoking the principle that 1,25$(OH)_2D_3$ influences receptor levels via an alteration of the proliferation and differentiation state of the cultured cells. However, in view of the rapidity of this event, it is more likely that the hormone is capable of altering the turnover rate of the receptor, either by enhancing its biosynthesis or by decreasing its degradation, or both. In addition, 1,25$(OH)_2D_3$ could conceivably facilitate efficient translation of the receptor mRNA as well as stabilize the newly synthesized protein. The mechanism whereby 1,25$(OH)_2D_3$ apparently upregulates the level of its receptor in mouse 3T6 and other cells is currently under study in our laboratories. We are also investigating the possible relevance of this finding to the biology of the receptor in mammalian systems, in vivo.

Recently we discovered in the receptor a striking qualitative modification that occurs when 3T6 cells are exposed to the 1,25$(OH)_2D_3$ hormone (Pike and Sleator, 1985). As depicted in Fig. 7, when a high-salt cytosol preparation from untreated 3T6 cells is immunoblotted, the receptor appears as a band at 54,500 daltons. The higher molecular weight bands are proteins that are crossreactive with the second antibody and thus are independent of the antireceptor monoclonal antibody. If this 3T6 cytosol preparation is incubated with 1,25$(OH)_2D_3$, in vitro, prior to immunoblotting, no difference is seen in the migration of the receptor (Fig. 7, lane 2). In contrast, after treatment of intact 3T6 cells with 1,25$(OH)_2D_3$ in culture for 90 min, the receptor extracted from the nucleus exhibits a small but significant reduction in electrophoretic mobility (Fig. 7, lane 3). The upshifted form of the 3T6 cell 1,25$(OH)_2D_3$ receptor displays an approximate molecular weight of 55,000. These data indicate that the occupied receptor isolated from the nuclear fraction has undergone covalent modification that causes it to migrate more slowly upon denaturing gel electrophoresis. Although this modification occurs only in intact cells treated with 1,25$(OH)_2D_3$ hormone, current efforts are underway to duplicate this event, in vitro, and to characterize the enzyme responsible.

To study further this hormone-dependent alteration in the receptor, we labeled the protein internally with [^{35}S]methionine in 3T6 cells. Initial experiments (Pike and Sleator, 1985) demonstrated the feasibility of metabolically labeling the receptor in 3T6 cells and suggested that it was maximally labeled 2 h after introduction of [^{35}S]methionine. Metabolic labeling, immunoprecipitation, and denaturing electrophoresis confirmed the hormone-dependent upshift in electrophoretic mobility (Pike and Sleator, 1985) that was seen previously in immunoblots (Fig. 7). Figure 8 illustrates ^{35}S-labeled receptor from 3T6 cells analyzed 4 h after exposure to both [^{35}S]methionine and 1,25$(OH)_2D_3$. An intense doublet appears at 55,000 daltons, representing the residual unmodified receptor and its upshifted derivative in the 3T6 total cell lysate (Fig. 8). When the cells are instead incubated with [^{32}P]orthophosphate in the presence or absence of 1,25$(OH)_2D_3$, a striking pattern of phosphorylation is revealed for the im-

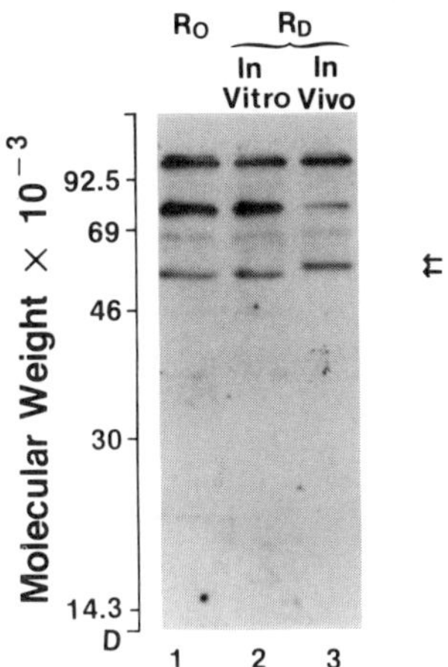

Fig. 7. Apparent modification of the $1,25(OH)_2D_3$ receptor in mouse 3T6 fibroblasts treated with $1,25(OH)_2D_3$. Lane 1: Untreated 3T6 cells (2.5×10^7/ml) were sonicated in (0.3-KETD) and the supernatant (cytosol) equivalent to 8×10^5 cells submitted to immunoblotting by a procedure similar to that detailed in the legend to Fig. 2. Conditions of immunoblot analysis were modified such that incubations were carried out with primary antibody (4 μg/ml), secondary antirat IgG antibody (1:10,000), and finally, iodinated protein A (100,000 cpm/ml; ICN). Lane 2: Cytosol equivalent to lane 1 was incubated for 2 h at 4°C with 2 n*M* $1,25(OH)_2D_3$ to form in vitro occupied hormone-receptor complexes prior to immunoblottiing. Lane 3: Intact 3T6 cells in culture were pretreated (in vivo) with 2 n*M* $1,25(OH)_2D_3$ for 90 min at 37°C, nuclei isolated as described elsewhere (Pike and Haussler, 1983), and an 0.3 *M* KCl extract of the nuclei immunoblotted. R_o represents unoccupied receptor, and R_D is used to designate $1,25(OH)_2D_3$-occupied receptor. Arrows at the right mark the unmodified receptor and the slower migrating derivative formed in the nucleus after treatment of cells with $1,25(OH)_2D_3$.

munoprecipitated receptor (Fig. 8). No ^{32}P is detected in the receptor without addition of $1,25(OH)_2D_3$ to the cells, but significant and selective phosphorylation of the upshifted band occurs when 3T6 cells are treated for 4 h with $1,25(OH)_2D_3$. Thus, the anomalous migration of the $1,25(OH)_2D_3$ receptor is due to phosphorylation. Preliminary data indicate that phosphorylation in the presence of $1,25(OH)_2D_3$ under these conditions (i.e., no phosphatase inhibitors) occurs on a serine residue(s) (Pike JW, Sleator NM, Haussler MR, Weigel N, unpublished data). Results of in vitro translation of 3T6 cell mRNA followed by immunoprecipitation of biosynthesized receptors (Mangelsdorf DJ, Pike JW, Haussler MR, unpublished data) are consistent with the upper receptor band being a posttranslationally modified form; only the lower nonphosphorylated receptor band at 54,500 daltons appears during in vitro translation. A similar phosphorylation of the nuclear progesterone receptor in uterine slices has re-

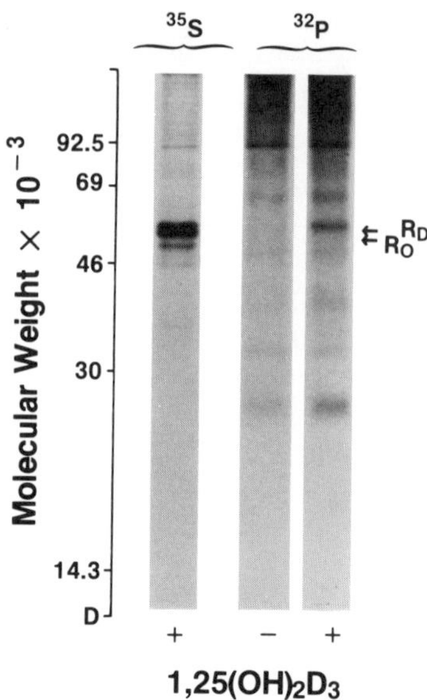

Fig. 8. Metabolic labeling of 3T6 fibroblasts with [^{35}S]methionine or [^{32}P]-orthophosphate followed by immunoprecipitation of 1,25(OH)$_2$D$_3$ receptor. Confluent 3T6 cells in monolayer culture were incubated with [^{35}S]methionine (10^6 cells) in the presence of 5 n*M* 1,25(OH)$_2$D$_3$ (left lane) for 4 h at 37°C or with [^{32}P] orthophosphate (10^7 cells) in the absence (center lane) or presence (right lane) of 5 n*M* 1,25(OH)$_2$D$_3$ for 4 h at 37°C. To accomplish this, confluent cultures of 3T6 cells (10^6–10^7 cells) were first washed twice under sterile conditions with either methionine-free or phosphate-free DMEM, respectively. Cells were then incubated for 4 h in methionine-free medium supplemented with 2% fetal bovine serum (FBS) and [^{35}S]methionine (250 μCi/ml; 1,100 Ci/mmol, Amersham International) or for 4 h with phosphate-free DMEM supplemented with 2% FBS and [^{32}P]orthophosphate (0.5 mCi/ml, carrier-free, Amersham International). Cells were then lysed in 0.3-KETD plus 0.5% Triton X-100 and immunoprecipitated as follows. Lysates of metabolically labeled 3T6 cells were subjected to immunoprecipitation by the addition of antireceptor monoclonal antibody coupled directly to Sepharose-4B via a CNBr-activated intermediate. The derivatized beads (~ 10 μl) were incubated with the lysate (0.1–0.2 ml) overnight at 4°C with gentle shaking. Following incubation, the derivatized beads were washed by repeated resuspension in 0.5-KETD (0.5 *M* KCl) containing 0.5% Tween 20, 0.1% sodium dodecyl sulfate (SDS), 0.5% NP-40, and 1% sodium deoxycholate. After sedimenting the beads through 0.5 ml of the above buffer containing 1*M* sucrose and following a final wash in KETD (without KCl), the beads were boiled for 3 min in SDS-denaturing buffer and electrophoresed as described in the legend to Fig. 2. The polyacrylamide gels were fixed in 40% methanol-10% acetic acid, treated with Enhance (New England Nuclear), and fluorographed overnight at −70°C on Kodak Omat XAR film. R_o = unoccupied receptor; R_D = receptor prepared from cells treated with 1,25(OH)$_2$D$_3$.

cently been reported by Logeat et al. (1985). Therefore, the $1,25(OH)_2D_3$ receptor and probably other steroid receptors become phosphorylated upon hormone binding in culture and probably in vivo. The interaction of the $1,25(OH)_2D_3$ hormone with the receptor could alter the conformation of the protein, rendering it a good substrate for a nuclear protein kinase. Because phosphorylation is an early, hormone-dependent event coinciding with nuclear localization, it is likely of functional importance. The phosphorylated receptor (occupied or perhaps with hormone dissociated) could be the form of the receptor that binds most avidly to upstream regulatory regions of vitamin D-controlled genes, or perhaps that alters transcription in another fashion. In addition to functional activation, phosphorylation may act as a signal for receptor processing to inactive forms. Regardless of the role of $1,25(OH)_2D_3$ receptor phosphorylation, this qualitative modification of the receptor is probably a key biochemical event in the mechanism of action of $1,25(OH)_2D_3$.

A Proposed Model for the Molecular Action and Biological Significance of the Vitamin D Receptor

Our current working hypothesis regarding the biochemical mechanism of $1,25(OH)_2D_3$ action is shown in Fig. 9. In the integrated model of $1,25(OH)_2D_3$ receptor biochemistry and biology, we show a fraction of the unoccupied receptor in the cytoplasm, but the equilibrium probably favors the nuclear compartment where the receptor exists as a loosely associated chromosomal protein. When complexed with the $1,25(OH)_2D_3$ hormone, the receptor is phosphorylated—possibly at multiple sites. The phosphorylated receptor is proposed to be the form that locates upstream regulatory regions of vitamin D-modulated genes in order to control transcription. These upstream activating sequences are analogous to those found for other steroid hormone receptors (Renkawitz et al., 1984; Dean et al., 1984; Jost et al., 1984). Altered DNA transcription then results in enhanced or repressed levels of various mRNAs (Fig. 9). The altered mRNA concentrations depicted in Fig. 9 are hypothetical (i.e., extrapolated from $1,25(OH)_2D_3$ elicited protein changes) except for CaBP, which is increased (Perret et al., 1985), and c-*myc* and collagen, which are decreased (Fig. 4; Rowe and Kream, 1982). The concept that the $1,25(OH)_2D_3$-receptor complex binds to regulatory/promotor sequences for each of the putative-induced mRNAs is only one possible mechanism. Regardless of the exact sequence of events, there is little doubt that control of mRNA levels is the key manner in which $1,25(OH)_2D_3$ ultimately orchestrates the concentration of its bioactive proteins. As summarized in Fig. 9, there are a number of $1,25(OH)_2D_3$-induced proteins that occur in various target cells. Individual cell types are presumably preprogrammed to modulate certain of the induced proteins, which then carry out the

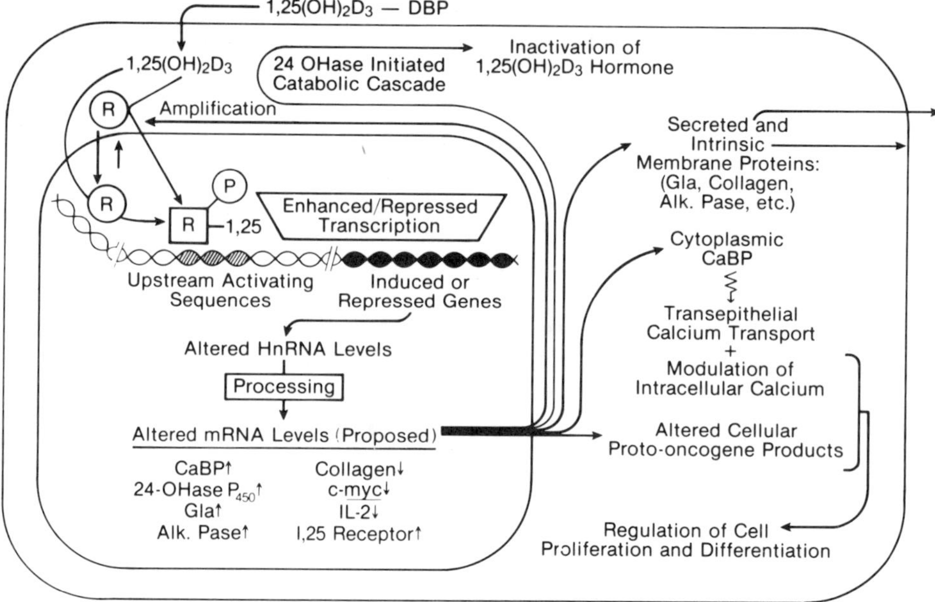

Fig. 9. Working hypothesis for $1,25(OH)_2D_3$ mechanism of action in target cells: 1986. DBP = serum vitamin D-binding protein; R = receptor (R circled represents unoccupied receptor while R in a square depicts the hormone occupied or "activated" receptor); P = phosphorylation of the receptor.

appropriate altered cell function. 3T6 fibroblasts would be an example of a cell type where receptor up-regulation or amplification occurs. Other target cells like those of the kidney may primarily inactivate $1,25(OH)_2D_3$ via the induced 24-OHase-initiated cascade and still carry out transepithelial transport of minerals. Bone cells apparently respond to $1,25(OH)_2D_3$ by inducing or repressing an array of proteins (e.g., bone Gla containing protein, collagen, alkaline phosphatase, etc.) required for the complex process of bone mineralization and remodeling. CaBP undoubtedly functions as a part of intestinal calcium transport, but could also modulate intracellular calcium in other targets. Along with control of oncogene products, intracellular calcium may act to regulate cell proliferation and differentiation. This is particularly true in certain hematopoietic and transformed cells that appear to respond to $1,25(OH)_2D_3$ by differentiation and suppression of the embryonic or malignant phenotype (Haussler et al., 1986).

In conclusion, what was traditionally a calcium absorption/antirachitic vitamin, has now been unveiled as a steroid hormone biological modifier capable of inducing cellular alterations ranging from specific protein induction to differentiation. The $1,25(OH)_2D_3$ hormone and its receptor protein are individually and elegantly regulated as they cooperate in controlling gene expression. While there is little doubt that the receptor protein, perhaps in its phosphorylated form, mediates the functions of $1,25(OH)_2D_3$ at the level of DNA, the goal for the future is to elucidate the molecular details of this protein-DNA interaction.

Acknowledgments. This work was supported by National Institutes of Health grants AM-15871 and AM-33351 to M.R. Haussler and AM-32313 and AM-34750 to J.W. Pike. We thank Elizabeth Allegretto and Michael Kelly for insightful discussions and Paula Leece for preparation of the manuscript.

References

Abe E, Miyaura C, Tanaka H, Shiina Y, Kuribayashi T, Suda S, Nishii Y, DeLuca HF, Suda T (1983) Proc Natl Acad Sci USA 80:5583–5587

Bar-Shavit Z, Teitelbaum SL, Reitsma P, Halal A, Pegg LE, Trial JA, Kahn AJ (1983) Proc Natl Acad Sci USA 80:5907–5911

Brumbaugh PF, Haussler MR (1973) Life Sci 13:1737–1746

Brumbaugh PF, Haussler MR (1974) J Biol Chem 249:1258–1262

Casanova J, Horowitz ZD, Copp RP, McIntyre WR, Pascual A, Samuels HH (1984) J Biol Chem 259:12084–12091

Chandler JS, Chandler SK, Pike JW, Haussler MR (1984) J Biol Chem 259:2214–2222

Costa EM, Hirst MA, Feldman D (1985) Endocrinology 117:2203–2210

Dean DC, Gope R, Knoll BJ, Riser ME, O'Malley BW (1984) J Biol Chem 259:9967–9970

Dokoh S, Donaldson CA, Haussler MR (1984) Cancer Res 44:2103–2109

Grosso LE, Pitot HC (1984) Biochem Biophys Res Commun 119:473–480

Haussler MR, McCain TA (1977) N Engl J Med 297:974–983, 1041–1050

Haussler MR, Norman AW (1969) Proc Natl Acad Sci USA 62:155–162

Haussler MR, Donaldson CA, Marion SL, Allegretto EA, Kelly MA, Mangelsdorf DJ, Pike JW (1986) In: Gotto AM, O'Malley BW (eds) The role of receptors in biology and medicine. Raven Press, New York, pp 91–104

Holtrop ME, Karen AC, Clark MB, Holick M, Anast CS (1981) Endocrinology 6:2293–2301

Jost JP, Seldran M, Geiser M (1984) Proc Natl Acad Sci USA 81:429–433

King WJ, Green GL (1984) Nature 307:745–747

Ko JS, Bernard GW (1981) Am J Anat 161:415–425

Logeat F, LeCunff M, Pamphile R, Milgrom E (1985) Biochem Biophys Res Commun 131:421–427

Mangelsdorf DJ, Koeffler HP, Donaldson CA, Pike JW, Haussler MR (1984) J Cell Biol 98:391–398

Merke J, Klaus G, Waldherr R, Ritz E (1985) Proceedings, American Society for Bone and Mineral Research. Abstract 362

Perret C, Desplan C, Dupre JM, Thomasset M (1985) In: Norman AW, Schaefer K, Grigoleit H-G, Herrath Dv (eds) Vitamin D. A chemical, biochemical and clinical update. Walter de Gruyter & Co, Berlin, New York, pp 365–366

Pike, JW (1985) In: Norman AW, Schaefer K, Grigoleit H-G, Herrath Dv (eds) Vitamin D. A chemical, biochemical and clinical update. Walter de Gruyter & Co, Berlin, New York, pp 97–105

Pike JW, Haussler MR (1979) Proc Natl Acad Sci USA 76:5485–5489
Pike JW, Haussler MR (1983) J Biol Chem 258:8554–8560
Pike JW, Sleator NM (1985) Biochem Biophys Res Commun 131:378–385
Pike JW, Marion SL, Donaldson CA, Haussler MR (1983) J Biol Chem 258:1289–1296
Provvedini DM, Tsoukas CD, Deftos LJ, Manolagas SC (1983) Science 221:1181–1183
Reitsma PH, Rothberg PG, Astrin SM, Trial J, Bar-Shavit Z, Hall A, Teitelbaum SL, Kahn AJ (1983) Nature 306:492–494
Renkawitz R, Schutz G, von der Ahe D, Beato M (1984) Cell 37:503–510
Rowe DW, Kream BE (1982) J Biol Chem 257:8009–8015
Sher E, Frampton RJ, Eisman JA (1985) Endocrinology 116:971–977
Thomas PS (1983) Methods Enzymol 100:255–266
Tsoukas CD, Provvedini DM, Manolagas SC (1984) Science 224:1438–1440
Walter P, Green S, Green G, Krust A, Bornert J-M, Heeltsch J-M, Staub A, Jensen E, Scrace G (1985) Proc Natl Acad Sci USA 82:7889–7893
Walters MR, Hunziker W, Norman AW (1980) J Biol Chem 255:6799–6805

Discussion of the Paper Presented by M. Haussler

MILGROM: I have two questions. First, do you have any kinase activity with your purified receptor? Did you try this?
HAUSSLER: Yes we have tried it. Based on the in vitro experiments, we believe the kinase activity is not present intrinsically in the receptor.
MILGROM: The second question. You showed you could increase your smaller form of receptor in the chick system by using DNA chromatography and you said that there were two possibilities: either it was appearing during the chromatography, or it was enriched. But to distinguish between both possibilities, you simply have to look at what comes through the column without binding. Did you look at that? If one form is enriched, it should be depleted in the fraction passing through the column.
HAUSSLER: In other words, we should be losing A form at the expense of gaining B form? It's possible that A form was in fact also proteolyzed to a fragment that we couldn't see any more with our antibodies. I don't know the answer to that. I realize that the second explanation may not be totally rigorous, but when you look carefully across a DNA cellulose profile, it seems that the A and B do not exactly comigrate. So there is some suggestion that they bind a little differently to DNA and we may be able to somehow enrich the forms. The compelling thing to us is the in vitro translation, which shows the A and B forms rather reproducibly. We have only been able to show phosphorylation in cultured mammalian cells. It's still possible that the avian A or B forms are somehow modified or proteolyzed, but this would have to occur in the in vitro translation system.
CLARK: That was really a nice talk. The diagram of the osteoblast (Fig. 6) was beautiful. I'd like a copy of that and also of your last slide (Fig. 9), but I'm interested in whether you did the work on the osteoblast measurements of vitamin D receptors?
HAUSSLER: Yes.
CLARK: How did you do that? How do you handle osteoblasts and are there cultures of osteoblasts?

HAUSSLER: There are two ways, neither of which is totally satisfactory. At least osteoblasts are easier to work with than osteoclasts, so if you want to do research on osteoblasts with estrogen or whatever, the two best systems are either fetal calvaria or rat osteosarcoma cells. The problem is there could be other kinds of cells in the fetal calvaria system. What we've used is the osteosarcoma line, which is a pure osteoblast-like population, to study the vitamin D receptor and response. We have observed modulation of alkaline phosphatase, growth, and changes in the shape of cells. These cells appear not to have estrogen or testosterone receptors, although we thought about the possibility that osteoblasts could be regulated in that fashion. I suppose I'm probing into your question, but if you want to know where the sex steroids might fit into the model (Fig. 6), then we have some ideas. There are reports of estrogen receptors in T-cells, I believe in the $OKT8^+$ lymphocytes, so it's possible that sex hormones enter the bone remodeling scenario at the lymphocyte. It is also possible that estrogen controls differentiation of the osteoblast from its precursor. It is an open question as to where the sex hormones might fit in our model for bone turnover.

BARDIN: That was such a nice talk I hate to ask a question like this, but I will do it anyway. In the first part of your talk you used in vitro translation to show that you had multiple translation products. You said that there wasn't proteolysis. I know that in a lot of translations there are proteases and a lot of proteins do turn over and you can show degradation of protein. In the second part of your talk you used the same system to show that you obtained a single band. How do you resolve those two different translation results?

HAUSSLER: I suppose the worst interpretation is that all of our smaller mammalian receptors are proteolyzed versions of larger receptors similar to the avian A form. Hopefully that is not true. We have never seen any reproducible evidence for a larger form of the mammalian receptor so we tend to believe the single band of molecular weight 54,000. It seems that multiple avian receptor forms are back in vogue. But if you really look at the data, my interpretation would be in agreement with yours, namely, that the avian B form results from proteolysis of the A form, even in the in vitro translation and subsequent immunoprecipitation.

LIAO: According to your scheme, the phosphorylated steroid hormone receptor does not bind hormone again unless phosphate is removed, is that right?

HAUSSLER: That is true according to the scheme, but we don't feel that phosphorylation/dephosphorylation regulates hormone binding, although we don't know for sure. Probably more arrows are needed in the scheme, perhaps one showing that the phosphoreceptor can rebind hormone.

LIAO: Do you care to comment about Horwitz's report on estrogen receptor phosphorylation, Also I think Dr. Pratt's group has demonstrated glucocorticoid receptor phosphorylation. Are they different?

HAUSSLER: Many of the phosphorylations shown have been in vitro by exogenous protein kinases, which we think may be supplemental phosphorylations for the one that is intrinsically important. Undoubtedly one can phosphorylate receptors with exogenous kinases. Since our work is done in cell culture, this may, in fact, be normal phosphorylation going on in the nucleus. I don't know whether it is the same that is seen exogenously. There is a nice paper by Ron Kahn in the *Journal of Biological Chemistry* in which he showed the insulin receptor kinase yielded different patterns of phosphorylation, depending on whether it was done in vitro or in vivo. Purified insulin receptor (which will self-phosphorylate) produces a totally different pattern of phosphorylation than when cells are incubated with insulin. Some of Pratt's work has been done with whole animals and tissues and is probably similar to ours, but most of the other work has not been in vivo phosphorylation. I suspect that everyone is going to start visualizing receptor doublets like those originally seen by Horwitz in cultured cells, and pointed out by Geoff Greene at this meeting, when they obtain steroid receptors from animals that have seen hormone.

SARKAR: Can you tell us how much is really known about the vitamin D receptor and the decreased expression of protein gene products?

HAUSSLER: Of oncogene products?

SARKAR: Right.

HAUSSLER: We've looked at several other oncogenes (*fos* and *sis*) that we thought might be likely candidates and they appear not to be regulated by vitamin D. Our positive data are limited to a suppression of *myc*, which is usually associated with decreased cell proliferation. At the moment we are not certain that 1,25$(OH)_2D_3$ is an important regulator of oncogene expression. We have independent studies using the soft agar technique showing that 1,25$(OH)_2D_3$ inhibits colony formation in a number of tumor cell lines. Whether this is through oncogene regulation or other kinds of differentiation, we really don't know.

GREENE: I may have missed it, but on your mapping you assigned the DNA-binding domain toward the carboxyl end. What was your evidence?

HAUSSLER: No firm evidence at all, other than some preliminary data that come from an unsubstantiated clone of the receptor gene. We are only certain that the monoclonal site is near the DNA-binding domain, and we have arbitrarily oriented it near the C-terminus.

MUELLER: With respect to your phosphorylation, do you have any ideas as to the amino acid target for that phosphorylation?

HAUSSLER: We have only preliminary data that it's serine, but I must add that there were no phosphatase inhibitors in the system.

MUELLER: Did you test vanadate yet?

HAUSSLER: We haven't tested it yet. Without vanadate, we can't exclude the dephosphorylation of a tyrosine residue(s).

MUELLER: In the second situation in your cultures, to induce terminal differentiation, what is the relationship of pulses of vitamin D? For instance, take it away, and then establish the calcium-binding protein induction or oncogene depression phenomena?

HAUSSLER: That's a good question and a good experiment. We have added the 1,25$(OH)_2D_3$ and left it in for the duration of the experiment. We have not done the withdrawal or wash-out experiments.

MUELLER: In many cases I think when we do terminal differentiation experiments we seem to lose track of the fact that there is a very tight window in starting cells to grow. For instance, in the erythroid culture differentiation the cell commits itself to doing certain things terminally, which I classify as last gasp efforts. In other words, when they get down to the point that their nucleus has fallen apart, then the cells spew out some information for making glucagon, for example, and that's the end of the cell. In terms of the calcium-binding protein, it would be interesting to stage that experiment so that you could see the window and determine if there is any carry over. It might not be directly related to your receptor mechanism.

Discussants: W. BARDIN, J. CLARK, G. GREENE, M. HAUSSLER, S. LIAO, E. MILGROM, G. MUELLER, and F. SARKAR

Chapter 8

Characterization of the Nuclear Binding Sites (Acceptor Sites) for a Steroid Receptor

T. SPELSBERG, A. GOLDBERGER, J. HORA, M. HORTON, AND B. LITTLEFIELD

Introduction

The nuclear binding sites (acceptor sites) for steroid receptors are important because they represent the first nuclear event in the steroid alteration of gene transcription (Thrall et al., 1978; Spelsberg et al., 1983). Although steroids can affect the rate of processing and stability of mRNAs (Moore et al., 1984) as well as membrane transport, one of the major sites of action appears to be directed at the transcription of the genes. For steroids to regulate gene expression, an interaction between steroid hormone receptors and the nuclear "acceptor sites" in the genome is required. This laboratory has taken the approach of using cell-free nuclear-binding assays with intact chromatin, the native state of DNA in all living cells, to analyze these nuclear acceptor sites.

For clarification's sake, a brief description of the mechanism of steroid-induced regulation of gene transcription is presented. This is outlined in Fig. 1. Steroid hormones circulate in the blood and are taken up by target cells via complexes with intranuclear-binding proteins termed *receptors*. These receptors are hormone and tissue specific (Jensen and DeSombre, 1972). Each receptor binds its specific steroid with very high affinity, having an equilibrium dissociation constant (K_d) in the range of 10^{-9} to 10^{-10} M. Once bound by their specific steroid hormones, the steroid receptors undergo a conformational change termed "activation" that allows them to bind with high affinity to specific sites (acceptor sites) on the chromatin (Thrall et al., 1978; Jensen and DeSombre, 1972; Spelsberg et al., 1979; Spelsberg, 1982). There are estimated to be 5,000–10,000 of these sites expressed with an even larger number that are not expressed, i.e., that are "masked," in intact chromatin (Thrall et al., 1978). The result of the binding to nuclear acceptor sites is the alteration of gene transcription (Spelsberg, 1982; O'Malley and Means, 1974; Breathnatch and Chambon, 1981) or, in some cases, alteration of posttranscriptional steps (Moore et al., 1984) as measured by the changing levels of specific RNAs and proteins in that target tissue. Each steroid regulates specific effect on the RNA

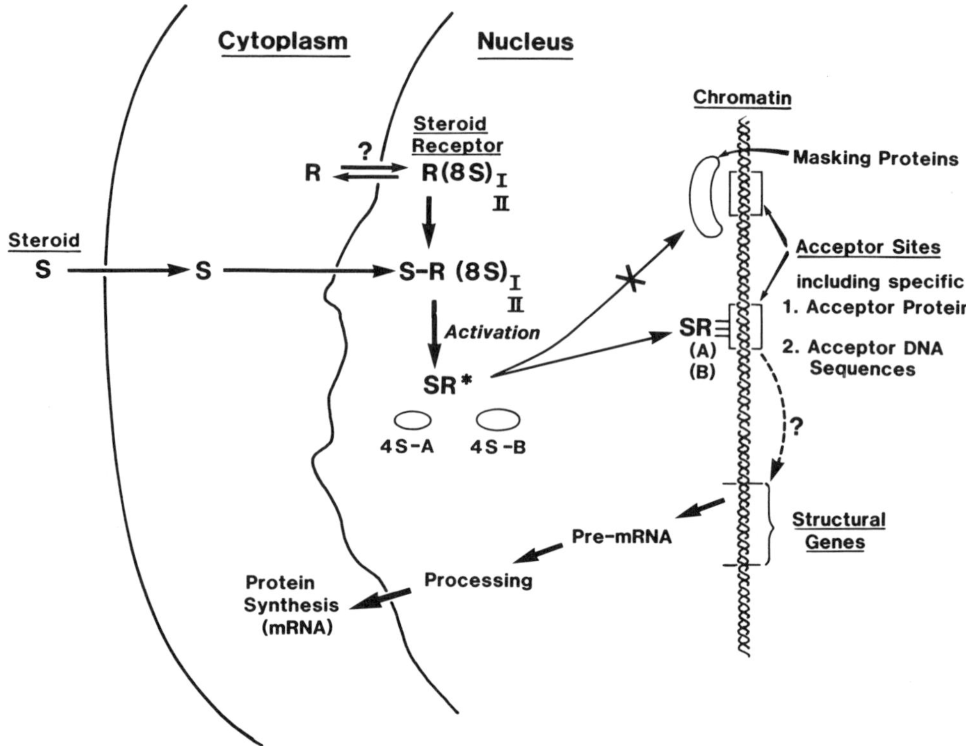

Fig. 1. Model describing the potential mechanism of action of steroid hormones in target cells. This model has been modified to fit data pertaining to the avian PRov including the two species of 8S form of the receptor (Puri et al., 1982; Dougherty et al., 1984), the A and B subunits of the 4S form of the receptor (Schrader et al., 1975; Buller et al., 1976; Schrader and O'Malley, 1972), as well as the properties of the nuclear-binding sites including masking (Spelsberg et al., 1983; Martin-Dani and Spelsberg, 1985) and the acceptor protein - acceptor DNA sequence composition of the nuclear-binding sites (Thrall et al., 1978; Spelsberg et al., 1983, 1984a; Toyoda et al., 1985).

and protein profiles. Since there is a controversy regarding the chemical composition of the specific nuclear-binding sites (acceptor sites) for steroid receptors, the investigators present a brief discussion of both sides of the controversy.

Studies of Specific DNA Sequences Neighboring Structural Genes as Acceptor Sites

There have been a series of papers reporting the enhanced cell-free binding of steroid receptors to specific DNA sequences in or immediately upstream from steroid regulated genes. Two types of experiments suggest that at least some of these DNA sequences are involved in mediating steroid regulation of gene transcription. First, plasmids containing glucocorticoid receptor-binding sequences become regulated by the glucocorticoids (Lee et al., 1981; Payvar et al., 1981). Second, subfragments of long terminal repeat DNA have been fused to herpes simplex viral thymidine kinase gene and a *cis*-acting glucocorticoid response of this kinase gene was established (Chandler et al., 1983). These sequence insertion experiments together with other deletion experiments support an important role of these domains in steroid hormone action on gene transcription.

The Cell-Free Nuclear Binding of Steroid Receptors to Pure DNA Sequences: Observations and Problems

Many laboratories have reported a twofold enhancement of binding of a variety of steroid receptor complexes to these 5′ flanking regions of structural genes (Payvar et al., 1981; Chandler et al., 1983; Compton et al., 1982; Dean et al., 1983; Payvar et al., 1983; Bailley et al., 1983; Mulvihill et al., 1982; Miller et al., 1984; de Ahe et al., 1985). DNA footprinting analyses have shown that steroid receptors do have a preference for some base sequences in these 5′ gene flanking regions of structural genes, suggesting some specific interactions between these two entities. Unfortunately, none of these papers address the binding of these same receptors to intact nuclei or chromatin, which is the more natural state of the DNA in cells. There are several other dilemmas that are of concern when considering the DNA sequences alone as the acceptor sites. *First,* in most instances the binding to the sites 5′ to gene-coding regions remains nonsaturable even at high ratios of receptor to DNA (Payvar et al., 1981; Chandler et al., 1983; Compton et al., 1982; Dean et al., 1983; Payvar et al., 1983; Bailley et al., 1983; Mulvihill et al., 1982; Miller et al., 1984; de Ahe et al., 1985). Thus, no affinity or specific binding can be ascertained. *Second,* only a two- or threefold enhanced binding to these sequences over that of calf thymus or bacterial DNA can be measured. *Third,* some of the ‘‘specific’’ binding sequences for sex steroid receptors are A-T rich (Compton et al., 1982; Dean et al., 1983; Bailley et al., 1983; Mulvihill et al., 1982). This is of concern since this laboratory has recently found that A-T rich synthetic DNAs also display an enhanced (twofold) binding of avian oviduct progesterone receptor compared with binding to whole hen DNA (Toyoda et al., 1985). *Fourth,* the exact sequences that have been reported to be the binding sites upstream from the mouse mammary tumor viral genome in rat liver for the glucocorticoid receptor have differed among six to eight different laboratories (Miller et al., 1984). *Lastly,* recent studies on the steroid receptor bindings to specific DNA sequences have shown that 5′ gene flanking regions lack steroid receptor specificity for these nuclear binding sites (de Ahe et al., 1985). Such a specificity has been reported between avian oviduct estrogen and progesterone receptors (ERov and PRov, respectively) (Kon and Spelsberg, 1982).

Conclusions and Alternative Model

These facts have led some investigators who analyzed the steroid receptor binding directly to these DNA sequences to conclude that these particular DNA sequences, although necessary for the process of steroid regulation of gene expression, still may not represent the direct binding sites (Miller et al., 1984; Jost et al., 1984). These investigators further concluded that "other factors," such as chromosomal acceptor proteins, must participate to generate specific acceptor sites (Miller et al., 1984; Jost et al., 1984). *It is our opinion that the required sequence domains upstream from structural genes do not necessarily have to serve as receptor-binding sites.* These regions could just as well serve as binding domains for important intermediate "regulatory" macromolecules (RNA or protein) coded by regulatory genes, which in turn are directly regulated by steroid receptors (Spelsberg et al., 1983). Alternatively, it has been proposed that, at least for the chick oviduct PRov, two classes of sites may naturally exist (Schrader et al., 1975; Buller et al., 1976; Schrader and O'Malley, 1972). Two subunits for the PRov have been reported, as depicted in Fig. 1. One subunit, the B-receptor species of PRov (as a dimer with the A species), is proposed to bind to one class of nuclear acceptor sites composed of acceptor protein-DNA complexes. This binding would be followed by the dissociation of the A species from the dimer and the binding of this A species to a second but different class of acceptor sites, which are composed of pure DNA sequences flanking structural genes (Schrader et al., 1975; Buller et al., 1976; Schrader and O'Malley, 1972). Which of these proposed mechanisms of steroid hormone action exists in vivo remains unknown.

Studies of Acceptor Protein-DNA Complexes as Acceptor Sites

This laboratory has taken the approach that the analyses of the nuclear acceptor sites initially should utilize chromatin, the native structure of DNA. Studies in this laboratory using the avian PRov have indicated that intact chromatin, as well as partially deproteinized chromatin (i.e., native complexes of nonhistone proteins and DNA), contains saturable, high-affinity binding sites not found with pure DNA (Thrall et al., 1978; Spelsberg et al., 1983, 1979; Spelsberg, 1982). The model of such nuclear acceptor sites is shown in Fig. 1, whereby the steroid receptors bind to acceptor sites composed of specific acceptor proteins bound to specific DNA sequences. Evidence for such acceptor sites is described in the following sections. These sites are also steroid receptor specific (Kon and Spelsberg, 1982). Similar acceptor sites as found for the PRov have been identified for the avian estrogen receptor (ERov) (Ruh and Spelsberg, 1982) for the estrogen receptors from rabbit, cows, and other animal uteri (Ruh and Spelsberg, 1982; Ruh et al., 1981; Ross and Ruh, 1984; Singh et al., 1984; Perry and Lopez, 1978); for guinea pig uterine progesterone receptor (Cobb and Leavitt, 1985); for sheep brain estrogen and progesterone receptor (Perry and Lopez, 1978); for rat testicular androgen receptors (Klyzsejko-Stefanowicz et al., 1976); and, finally, for the glucocorticoid receptors in rat liver (Hamana and Iwai, 1978) and in a human leukemia cell line (Ruh et al., 1985). Some specifics of these nuclear-binding sites are described in the following sections.

Early Work in This Laboratory on the Nuclear Acceptor Sites for PRov and ERov

Many studies have been performed in this laboratory to assess the proper conditions necessary to achieve in vivo-like cell-free binding of the PRov and ERov to intact chick oviduct chromatin (Thrall et al., 1978; Spelsberg et al., 1983, 1979; Spelsberg, 1982; Ruh and Spelsberg, 1982). These conditions include minimizing protease and nuclease activity (Spelsberg et al., 1976, 1984; Webster et al., 1976; Pikler et al., 1976; Webster and Spelsberg, 1979; Martin-Dani and Spelsberg, 1985), achieving the proper ionic strength (Thrall et al., 1978; Webster et al., 1976; Pikler et al., 1976; Spelsberg et al., 1976a), pH (Thrall, et al., 1978; Spelsberg et al., 1976b; Pikler et al., 1976; Webster and Spelsberg, 1979), and time of incubation to obtain an optimal, specific binding. Further, it is important that cell-free binding assays contain intact, fully functional steroid receptors (Boyd and Spelsberg, 1979; Spelsberg and Halberg, 1980; Boyd-Leinen et al., 1982, 1984). These optimal conditions have been summarized by Thrall et al. (1978). The avoidance of artifacts in the progesterone receptor (PR) binding assay, including those generated by damaged DNA (Thrall and Spelsberg, 1980), was recently summarized by Littlefield and Spelsberg, 1985. These data serve to emphasize the need to fully characterize the receptors and the binding assay conditions, as well as the acceptor sites.

Figure 2 briefly reviews the past work in this laboratory on the cell-free binding of PRov to nuclear acceptor sites using the above described conditions. The PRov binding to intact chromatin or to partially deproteinized chromatin, i.e., the nucleoacidic protein (NAP), representing the nonhistone "acceptor" proteins tightly bound to DNA, was shown to have the following properties:

1. The binding is saturable with very high affinity (panel A) (Spelsberg et al., 1983; Ross and Ruh, 1984; Spelsberg et al., 1984a; Webster et al., 1976; Pikler et al., 1976; Spelsberg et al., 1976a; Boyd and Spelsberg, 1979; Littlefield and Spelsberg, 1984)
2. The binding displays a steroid receptor specificity, i.e., unlabeled ERov will compete with the [^{3}H]ERov for nuclear-binding sites in vivo or in

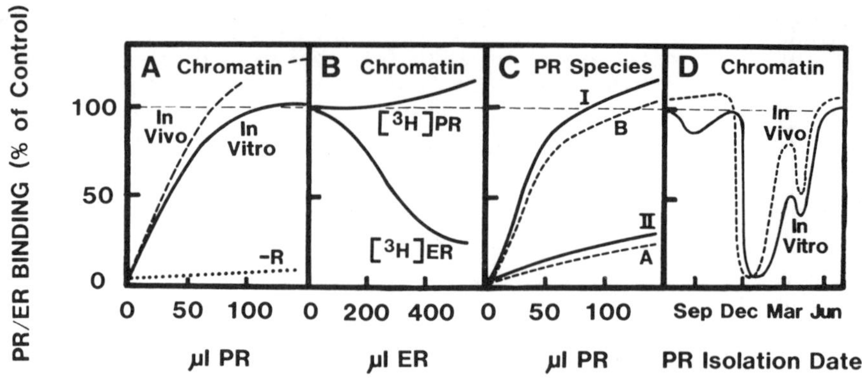

Fig. 2. Outline of the properties of the cell-free nuclear binding of the avian PRov. This figure outlines the basic properties previously published regarding the cell-free nuclear-binding assay of the avian PRov and ERov. The following represent the sources for the data presented in each of the panels. *Panel A* (Thrall et al., 1978; Spelsberg, 1976); *panel B* (Kon and Spelsberg, 1982); *panel C* (data in preparation; *panel D* (Boyd and Spelsburg, 1979; Spelsberg and Halberg, 1980).

the cell-free binding assay, but unlabeled PR will not compete (24) (panel B)

3. The saturable high-affinity binding was found using highly purified PR (panel C)
4. The cell-free nuclear binding also mimics the binding in vivo with regard to functional and nonfunctional receptors (panel D) (Boyd and Spelsberg, 1979; Boyd-Leinen et al., 1984)

It should be mentioned that the removal of the tightly bound acceptor proteins from the DNA destroys these specific sites (Spelsberg et al., 1983, 1976b; Pikler et al. 1976). Therefore, pure DNA displays a nonsaturable binding that is not competable with unlabeled PR.

Specific Acceptor Proteins Involved in the PR Acceptor Sites in the Avian Oviduct

General Approach to Characterizing the PRov Acceptor Sites

Specific acceptor sites have been reconstituted by reannealing the acceptor protein to pure hen genomic DNA (Spelsberg et al., 1984a). These methods have allowed us to isolate specific acceptor proteins and to identify and enrich specific DNA sequences essential for these acceptor sites as described below (Spelsberg et al., 1976b; Webster and Spelsberg, 1979; Littlefield and Spelsberg, 1984). It should be mentioned that a masking of many of the PR and estrogen receptor (ER) acceptor sites in oviduct chro-

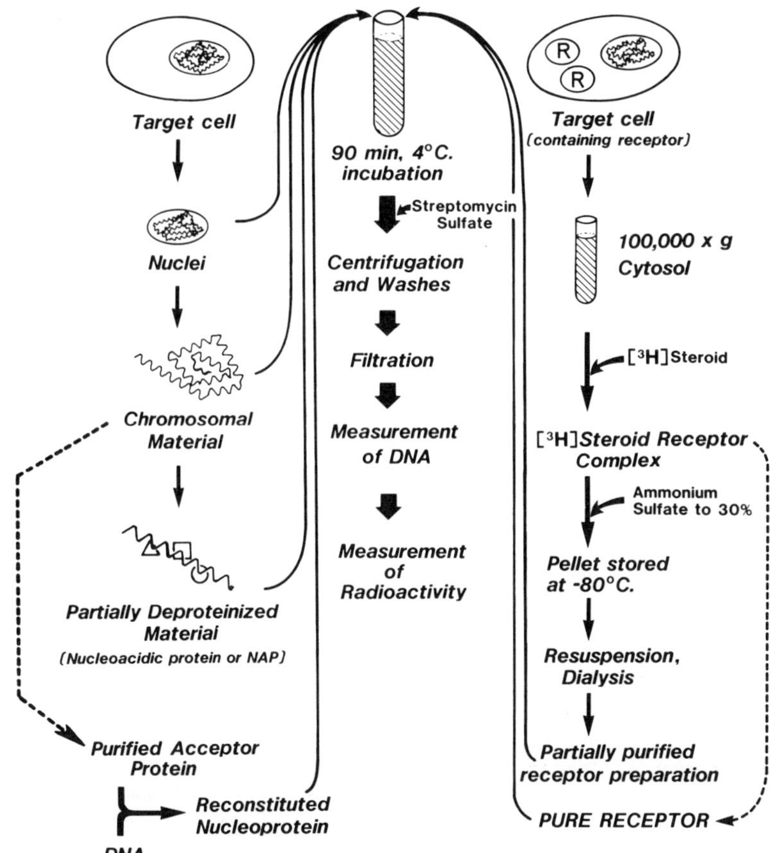

Fig. 3. Outline of the isolation of the nuclear acceptor sites and the PR from the avian oviduct and the cell-free interactions between these two entities. This figure outlines the basic approach taken by the laboratory for the isolation of the nuclear acceptor sites as well as the PR from the avian oviduct (Thrall, et al., 1978; Spelsberg et al., 1983, 1979; Spelsberg, 1982). Included is an outline of the cell-free nuclear-binding assay using the streptomycin sulfate method that was previously reported (Spelsberg, 1983). (Reprinted with permission from Spelsberg et al., 1983, Recent Prog Hormone Res 39:463–517)

matin by other chromatin proteins was identified.* This masking is described in more detail elsewhere (Martin-Dani and Spelsberg, 1985).

A general approach taken for the initial analyses of the acceptor sites for PRov is outlined in Fig. 3. First, the chromosomal material (chromatin) is isolated from the estrogen-treated chick oviducts as well as from other

*Thrall et al., 1978; Spelsberg et al., 1983; Ruh and Spelsberg, 1982; Ruh et al., 1981; Ross and Ruh, 1984; Singh et al., 1984; Perry and Lopez, 1978; Cobb and Leavitt, 0000, Klyzsejko-Stefanowicz, et al., 1976; Hamana and Iwai, 1978; Ruh et al., 1985; Spelsberg et al., 1976b, 1984a; Webster et al., 1976; Martin-Dani and Spelsberg, 1985.

tissues from the chicken, or other sources. In some cases the chromatin is dissociated into protein and DNA, followed by enrichment of the acceptor proteins and reconstitution of the acceptor sites. The oviducts of estrogen-treated chicks also serve as a source of the PRov (Boyd and Spelsberg, 1979; Boyd-Leinen et al., 1982). The PRov is obtained in the partially purified or highly purified state (Puri et al., 1982; Dougherty et al., 1984) for use in the cell-free binding assays by the method described previously (Spelsberg et al., 1984b) and outlined in Fig. 3.

Evidence for the Existence of Acceptor Proteins and Masking of Sites

Figure 4 outlines the results of early studies in the selective removal of proteins from chromatin resulting in the unmasking of the PR acceptor

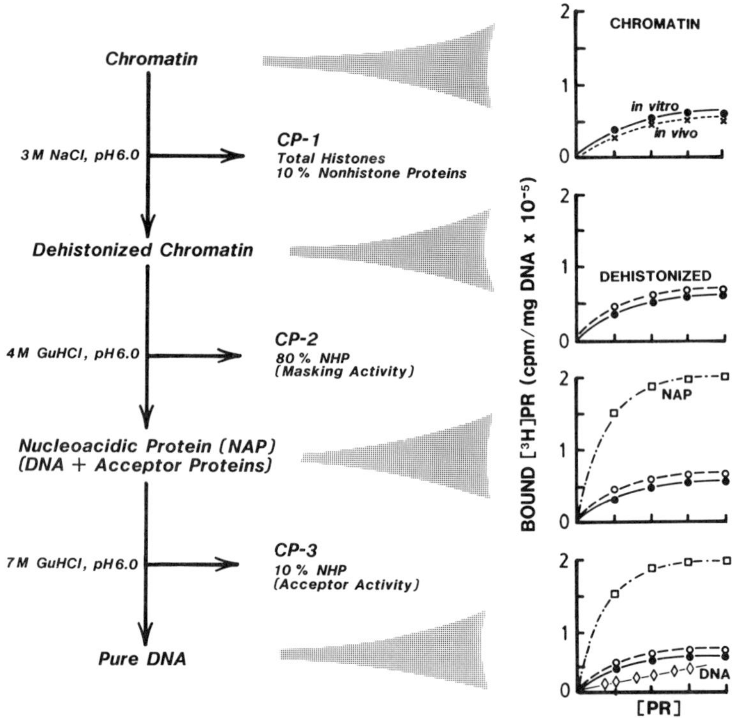

Fig. 4. Outline of the fractionation of the chromatin proteins. This figure outlines past work on the chemical dissociation of proteins from the chromatin with the concomitant effects on PR binding using the cell-free nuclear-binding assay. An unmasking of sites is achieved by the removal of CP-2 fraction of proteins. The binding to the residual NAP displays all the properties of native PR binding as does intact chromatin (Thrall, et al., 1978; Spelsberg et al., 1983, 1979; Spelsberg, 1982). The removal of the CP-3 fraction of proteins from the NAP to obtain pure DNA results in a loss of specific binding.

sites followed secondly with the removal of all the protein by the loss of the specific binding (acceptor sites) on the DNA. The selective removal of chromosomal proteins using various reagents can be achieved using differential centrifugation or absorbing chromatin to hydroxylapatite resin. As shown in Fig. 4, the removal of total histones, termed chromosomal protein fraction 1 (CP-1), results in a minimal effect on the PR binding levels. However, the subsequent removal of fraction CP-2 from the DNA, representing 80% of the total nonhistone chromosomal proteins, yields a nucleoacidic protein (NAP) which displays a marked increase in PR binding. This has been described as an unmasking of binding sites. The PRov (Spelsberg et al., 1979; Webster et al., 1976; Martin-Dani and Spelsberg, 1985) binding to these unmasked sites on the NAP are shown to be identical to those on native chromatin with regard to (1) the saturable high-affinity binding which mimics that observed in vivo, and (2) the binding by the functional/nonfunctional receptors (Spelsberg et al., 1983, 1979; Webster et al., 1976; Martin-Dani and Spelsberg, 1985; Boyd and Spelsberg, 1979; Spelsberg and Halberg, 1980). Figure 4 shows that the removal of the remaining chromosomal proteins from the DNA using 7 *M* guanidine hydrochloride, termed CP-3, results in a loss of the specific binding sites (Spelsberg et al., 1983, 1979, 1984a). The binding to residual pure DNA is markedly reduced and nonsaturable and does not reflect in vivo-like patterns of binding using the nonfunctional receptors (Toyoda et al., 1985; Spelsberg et al., 1984a; Boyd and Spelsberg, 1979; Spelsberg and Halberg, 1980). Thus, it is the CP-3 fraction of the chromosomal proteins that apparently plays a role in the specific binding of PR to the chromatin.

Reconstitution of the Sites

Since removal of the CP-3 fraction results in loss of the PRov acceptor sites, and since proteolytic activity, but not ribonuclease or DNase activities, destroys the binding to the NAP, it was speculated that specific "acceptor proteins" in fraction CP-3 may be involved (Spelsberg et al., 1983). To further purify and characterize these "acceptor proteins," it was found that the proteins must be reannealed back to the DNA in order to generate the specific binding (Thall, et al., 1978; Spelsberg et al., 1983, 1979, 1984a). After finding that the PRov failed to bind to the isolated CP-3 proteins, whether free or absorbed to cellulose, methods were developed to reconstitute the nuclear acceptor sites (Spelsberg et al., 1984a). Using these methods, the CP-3 fraction is reannealed to pure DNA using a regressing gradient of guanidine hydrochloride in specially designed mixing chambers (Spelsberg et al., 1984a). These reconstituted protein-DNA complexes are termed "reconstituted NAP." As shown in Fig. 5, panel A, the reannealing of the CP-3 fraction, but not the CP-2 fraction, to the DNA results in a marked PR binding. This binding is titratable with the amount of CP-3 protein bound to the DNA, and the resulting reconstituted protein-DNA complex displays a high-affinity binding of the PRov sat-

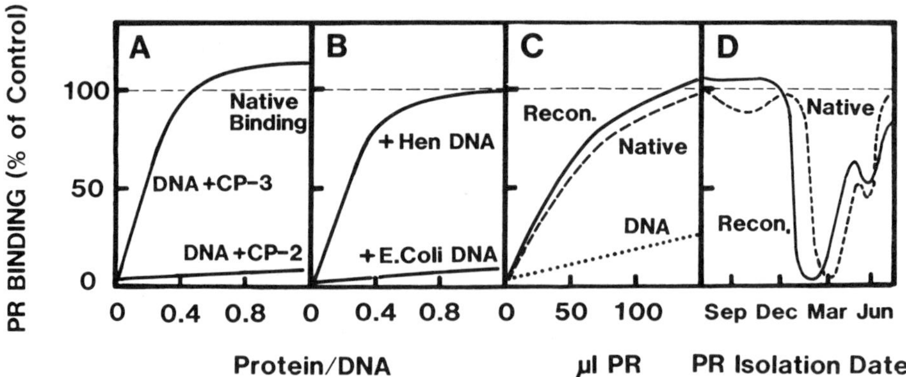

Fig. 5. Properties of reconstituted acceptor sites for the avian PRov. This figure outlines the various properties of the cell-free PR binding to reconstituted acceptor sites as described elsewhere (Thrall et al., 1978; Spelsberg et al., 1983, 1984a). *Panel A* shows that the reconstitution of CP-3 protein but not CP-2 fraction of the chromosomal proteins results in a regeneration of specific PR-binding sites. *Panel B* shows that the substitution of bacterial DNA for hen DNA in these reconstitution assays also fails to regenerate these sites. *Panel C* demonstrates that the reconstituted NAP displays similar levels of saturable binding as does the native NAP. Panel D shows that the reconstituted NAP displays this same pattern of seasonal binding as does the native NAP when using receptors isolated at various periods of the year. (Data taken from Spelsberg et al., 1984, Biochemistry 23:5103)

urable. Thus, the protein fraction that, when removed from the DNA, results in a loss of the specific PR binding when reannealed to the DNA causes a regeneration of these sites. Interestingly, as shown in panel A, Fig. 5, there are a limited number of PR binding sites that are regenerated by the CP-3 even when high ratios of CP-3 proteins are reconstituted to DNA. Panel B of Fig. 5 shows that substitution of *Escherichia coli* DNA for hen DNA in the reconstitution assays results in a DNA protein complex that fails to specifically bind the PR. The limited number of sites that can be regenerated on hen DNA with excess of CP-3 protein and the failure of bacterial DNA to regenerate these sites support the existence of specific DNA sequences in the acceptor sites. Panel C of Fig. 5 shows that reconstituted NAP contains roughly the same number of saturable PR binding sites as the native NAP. Further, panel D of Fig. 5 shows that the reconstituted acceptor sites, i.e., reconstituted NAP, display similar seasonal variation in the patterns of binding, as do the native acceptor sites (NAP) (Spelsberg and Halberg, 1980). Pure DNA fails to display these sites. Therefore, NAP, reconstituted with proteins other than the CP-3 fraction or with *E. coli* DNA, fails to show specific PRov binding with the functional receptors and shows no difference between the functional and nonfunctional PRov. In contrast, the reconstituted native-like NAPs containing CP-3 and hen DNA display a specific binding with functional PR but not with nonfunctional PR.

Characterization of Acceptor Proteins

Using previously described reconstitution methods, the CP-3 proteins were further fractionated to enrich for the acceptor activity. As shown in Fig.

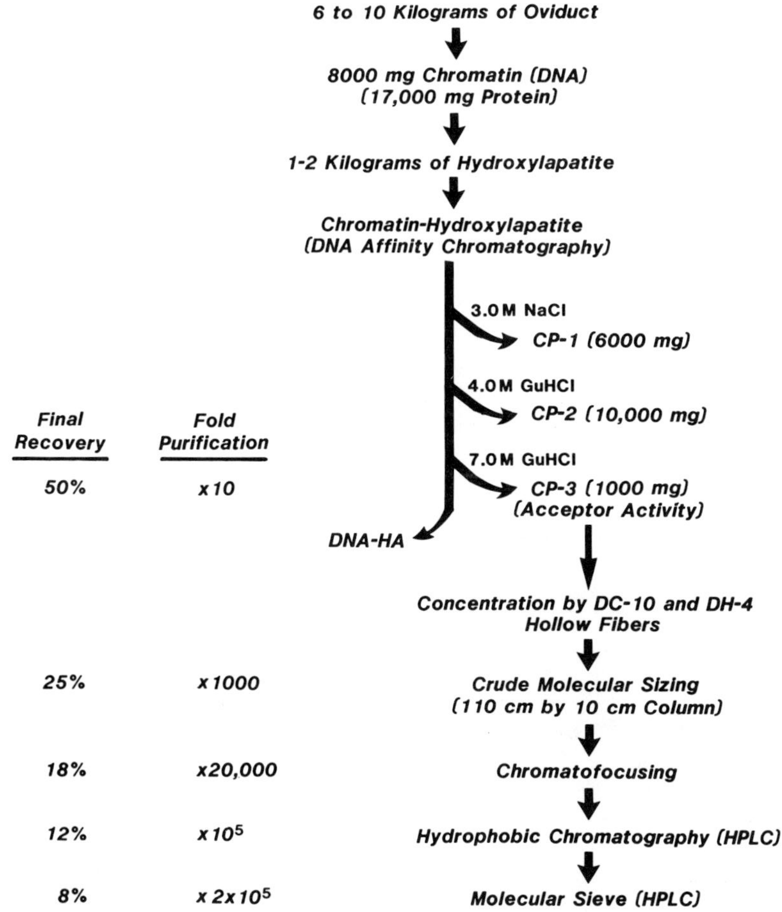

Fig. 6. Schematic for purifying bulk quantities of the acceptor protein for the avian PRov. This figure outlines the method of isolating the acceptor protein from hen oviduct chromatin. Large quantities of the hen oviduct chromatin stored frozen at $-80°$ are homogenized with a special, low-shear continuous homogenizer (Yamato Scientific, Chicago, Illinois). The chromatin is isolated from the nuclei and attached to hydroxylapatite, which is then extracted with various chemical reagents. The CP-3 protein fraction extracted with 7 *M* guanidine is then concentrated by hollow fiber filters and further fractionated using molecular sieve chromatography, chromatofocusing, and hydrophobic chromatography. The figure shows the recovery of the acceptor activity as well as the fold purification of this activity. The purification was calculated by the amount of protein required to regenerate the saturation of acceptor sites on hen DNA as described elsewhere (Spelsberg et al, 1983, 1984a).

6, large quantities of mature avian hen oviducts are homogenized using a low shear continuous homogenizer (Yamato Scientific, Chicago, Illinois) to isolate large quantities of nuclei (Spelsberg et al., 1984b). These preparations of nuclei are then extracted to yield chromatin, as described elsewhere (Thrall et al., 1978; Pikler et al., 1976). This chromatin is adsorbed to hydroxylapatite and the various chromosomal fractions, CP-1, CP-2, and CP-3, are selectively dissociated. The CP-3 fraction is then concentrated by hollow fiber filters as shown in Fig. 6 and subjected to molecular sieve chromatography using CL-Sepharose 6B. When the low molecular weight fraction in the range of 12,000–20,000 daltons is subjected to chromatofocusing (Pharmacia Fine Chemicals, Upsula, Sweden), two peaks of binding activity are eluted at isoelectric points of around 6.5–7.0 and 5.5–6.0. Upon further fractionation by high pressure liquid chromatography using hydrophobic and molecular sieve resins, the activity in each of these subfractions was identified by reannealing of the proteins to hen DNA as described elsewhere (Spelsberg et al., 1984a). In each subsequent step of purification, significantly less protein was needed to regenerate the full complement of specific PR acceptor sites. Table 1 summarizes the properties of the acceptor activity, suggesting that it represents a nonconju-

Table 1. General Properties of Acceptor Proteins for Progesterone Receptor in Avian Oviduct

Property	Value	Evidence
Proteinaceous	Yes	1. Activity destroyed by proteases but not RNase, DNase, or phosphatase
Simple or conjugated	If so, only minor	2. Purified activity contains no RNA or DNA 3. Isopycnic centrifugation
Side chain modification	Unknown	
Molecular weight	13,000–18,000	1. Molecular sieve 2. SDS-PAGE
Isoelectric point	pH6; pH7.0 possibly more	1. Isoelectric focusing 2. Chromatofocusing
Density	1.24 g/cc	Isopycnic centrifugation
Molecular species	One major, one minor, possibly more	1. Isoelectric focusing/ chromatofocusing 2. Hydrophobic chromatography
Steroid receptor specificity	Yes	Does not bind estrogen receptor
DNA sequence specificity	Yes	Activity reconstituted with most animal DNA; not bacteria, viral, fish, or plant DNA
Tissue specificity	Tentative no	Activity masked and protein present in nontarget tissue chromatins
Species specificity	Unknown	

gated, hydrophobic, small molecular weight protein that is slightly acidic, with minimal heterogeneity.

Evidence for Conservation of the Specific DNA Sequences in the PRov Acceptor Sites

Since *E. coli* DNA reconstituted with the chromatin acceptor proteins (CP-3) from hen oviduct failed to regenerate the specific PR binding, it was of interest to examine the DNA from other sources as well. As shown in Fig. 7, the binding of PR to the pure DNA alone from various sources was found to yield a similar level of nonsaturable (nonspecific) PR binding. However, when the hen oviduct acceptor protein (CP-3 fraction) is reannealed to these DNAs, a specific binding of the PR is observed (Toyoda et al., 1985) (Fig. 8, panel A). This specific binding is defined as one that displays a saturable PR binding with minimal binding by nonfunctional receptors (Spelsberg et al., 1984a; Boyd and Spelsberg, 1979; Spelsberg and Halberg, 1980; Boyd-Leinen et al., 1982, 1984), as well as a limited number of acceptor sites per mass of DNA (Toyoda et al., 1985). As shown in Fig. 8 (panel A), human DNA generated even a greater level of PR acceptor sites then did hen DNA. The DNA from cow, dog, salmon, and frog could also substitute for hen DNA in generating of the specific PR-

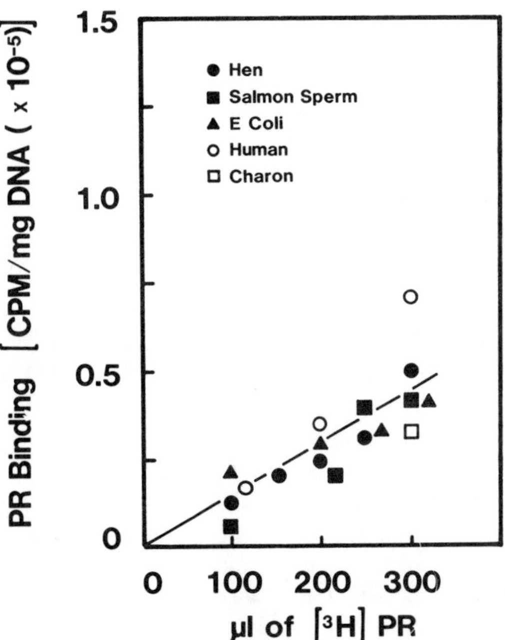

Fig. 7. Binding of the PRov to pure genomic DNAs of various sources. The DNA from hen (●), salmon sperm (■), *E. coli* (▲), human (○), and charon virus (□) were isolated and subjected to the cell-free nuclear-binding assay as described previously (Toyoda et al., 1985; Boyd and Spelsberg, 1979). The values represent the mean of three replicate analyses at each level of receptor for each of the DNAs. (Data reproduced with permission from Toyoda et al., 1985, Proc Natl Acad Sci USA 82:4722)

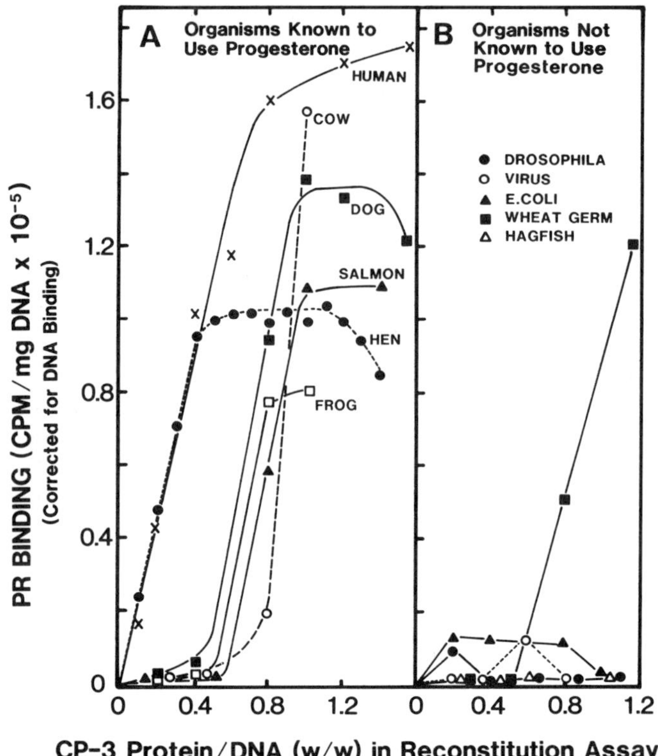

Fig. 8. Reconstitution of the avian oviduct CP-3 protein to the DNA from various sources. The DNAs from a variety of sources were purified as described previously (Spelsberg et al., 1984a). These DNAs are represented in *panel A* by hen (●), human (x), cow (○), dog (■), frog (□), salmon (▲). In *panel B*, the DNAs represent Drosophila (●), virus (○), *E. coli* (▲), wheat germ (■), and hagfish (△). These DNAs were reconstituted with varying ratios of CP-3 protein to DNA and subjected to the cell-free nuclear-binding assay as described elsewhere (Toyoda et al., 1985; Boyd and Spelsberg, 1979). The values represent the mean of triplicate analysis for each ratio of CP-3 protein to DNA in the binding assays. (Data reproduced with permission from Toyoda et al., 1985, Proc Natl Acad Sci USA 82:4722)

binding sites. However, a larger quantity of the hen oviduct CP-3 protein had to be reannealed to these DNAs to achieve a saturation of acceptor sites. In all instances, a limited number of acceptor sites could be generated on these DNAs using an excess of the oviduct acceptor protein.

In contrast, as shown in panel B of Fig. 8, the DNA from Drosophila, viruses, bacteria, and hagfish were found to be largely incapable of regenerating specific PR-binding sites when reannealed with the hen oviduct CP-3 fraction. For unknown reasons, the DNA from wheat germ reconstituted with CP-3 was found to generate some PR binding; however, a saturation of the number of acceptor sites per mass DNA was not achieved even at high ratios of CP-3 protein to DNA.

These data indicate the presence, within many animal genomes, of specific DNA sequences that are capable of generating PR acceptor sites when reannealed with hen oviduct CP-3 protein. This is supported by the limited number of sites that can be regenerated on these DNAs even in excess of the hen oviduct acceptor protein. These acceptor site DNA sequences do not appear to reside in more evolutionary distant organisms including procaryote DNA. The acceptor site DNA sequences do appear to be conserved over evolution in that the genomes of human, cow, dog, fish, and frog can substitute for hen DNA. This is not surprising since the steroids and their receptors are also highly conserved over evolution (Toyoda et al., 1985; Greene et al., 1977). It can be speculated that when organisms adapted the use of steroid hormones and their receptors, the genomes of these organisms also adapted the capacity to bind and respond to these steroid receptor complexes.

PRov Acceptor Sites Do Not Appear to Reside in or Near the Ovalbumin Gene

To ascertain whether or not the acceptor sites reside in or near a structural gene regulated by progesterone, reconstitution studies were performed using the hen oviduct CP-3 protein with the cloned ovalbumin gene (Lai

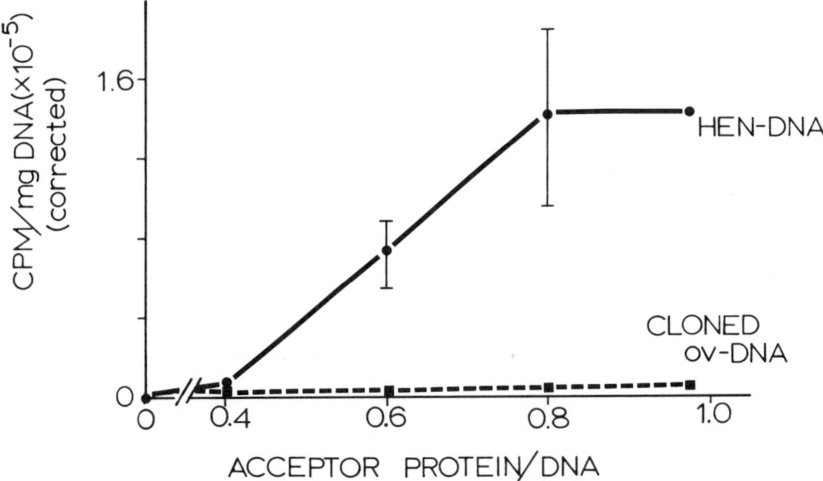

Fig. 9. Reconstitution of acceptor sites using hen genomic DNA as well as the genomic clone of the ovalbumin gene pOV12. Whole genomic hen DNA (●) as well as cloned ovalbumin gene (■) (kindly supplied by Dr. Bert O'Malley, Baylor College of Medicine) (Dean et al., 1983) were reannealed with avian oviduct CP-3 protein as described elsewhere (Spelsberg et al., 1984a). The reconstituted samples were subjected to the cell-free nuclear-binding assay using the partially purified PRov. The values represent the mean of triplicate analysis at each ratio of acceptor protein DNA.

et al., 1980) (kindly supplied by Dr. Bert O'Malley, Baylor College of Medicine). As shown in Fig. 9, while the reconstitution of the CP-3 protein to hen DNA did regenerate a saturable level of PR-binding sites with increasing ratios of acceptor protein to DNA, the substitution of the pOV-12 genomic clone of the ovalbumin gene failed to generate these binding sites. Since this clone contains the genomic sequences for the ovalbumin gene including introns and 3,000 base flanking regions of this gene (Lai et al., 1980), it appears that the putative specific acceptor site DNA sequences for the oviduct PR probably do not lie within or near this steroid regulated gene.

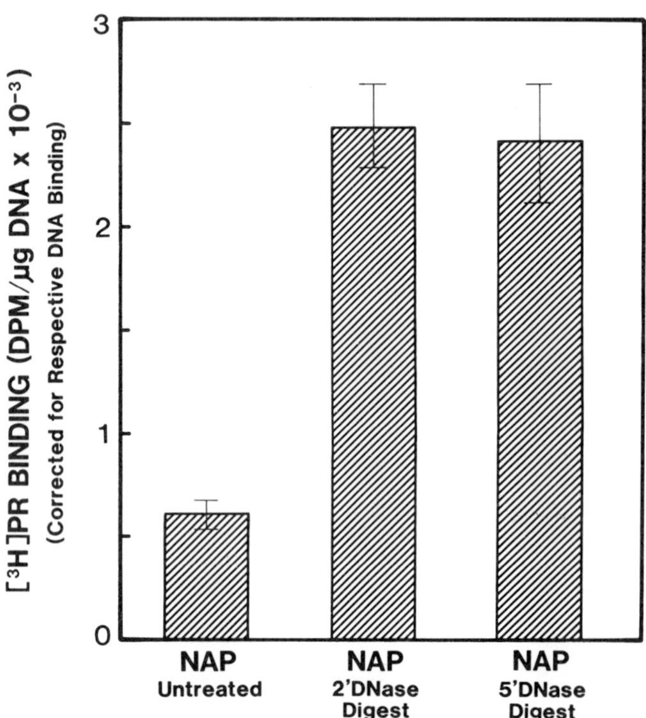

Fig. 10. Enrichment of the native acceptor sites for PRov: PR binding to DNase-treated NAPs (corrected for respective DNA binding). Native NAP from hen oviduct was treated with DNase for 2- and 5-min periods using a ratio of DNase to NAP (DNA) of 300 units of DNase I per mg DNA. Incubations were carried out at 4° in buffer at pH 7.4. The NAPs were centrifuged at 100,000 × *g* average for 1h and then the pelleted NAP_r were subjected to the cell-free nuclear-binding assay with PRov. The data represent the mean and standard deviation for replicate analysis of a typical binding assay (Hora et al., 1986).

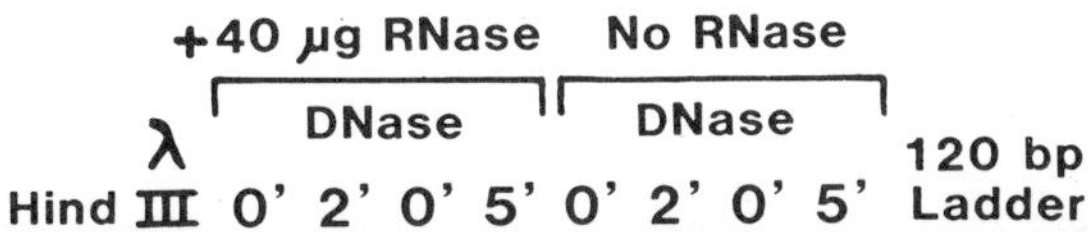

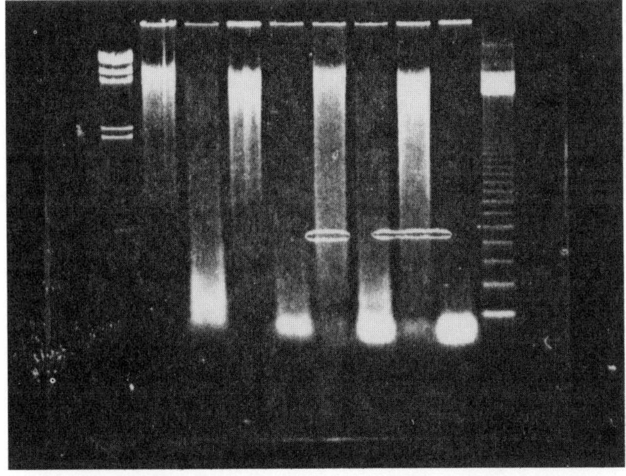

Fig. 11. Agarose gel electrophoresis of the DNA from 0-, 2-, and 5-min digestions with DNase I of the NAP. The products were phenol extracted, ethanol precipitated and electrophoresed through 1.4% agarose gels. Samples were treated with 100 μg/ml of ribonuclease A before analysis. The standards represent the Hind III digested λ DNA as well as 123 bp ladder from Bethesda Research Laboratories. Each of the gel columns are labeled appropriately as to the source or treatment of the DNA (Hora et al., 1986).

Enrichment of the Native Acceptor Sites for PRov

It is desirable to enrich the specific acceptor site DNA sequences involved in the PR nuclear acceptor sites. Interestingly, it was found that the PRov acceptor sites on NAP were resistant to nuclease digestion (DNase I). The DNase digestion of NAP, therefore, generated fragments termed NAP_f, which displayed an enrichment of these sites. Figure 10 shows that with increasing nuclease treatment of the NAP, there is an increase in the specific PR binding per mass DNA. An excess of DNase activity eventually causes a loss in the specific PR binding. Figure 11 shows the agarose gel electrophoresis of the DNA from the digested NAP. It was found that the reduction of the length of the DNA can reach 150 to 250 bp while still maintaining the specific PR binding to the fragments (i.e., acceptor sites). The generation of smaller fragments results in a loss of this binding. The specific PR binding to these fragments (NAP_f) is saturable both when titrating with radioactive PR as well as when competing with unlabeled PR (Hora et al., 1985). Thus, a saturable PR binding is maintained in these fragments. Other studies in this laboratory have suggested that the DNA isolated from these enriched fragments contain largely repetitive sequences of DNA (Toyoda and Spelsberg, 1981). The fact that the nuclear acceptor sites for the avian oviduct PR are resistant to DNase activity suggests that they differ from the DNase susceptible, transcriptionally active regions of the genome (Igo-Kemenes et al., 1982). However, this nuclease resistance does resemble the reports of nuclear acceptor sites residing in the nuclear matrix (Barrack and Coffey, 1980, 1982). Whether or not these sequences lie in the active or inactive regions of the chicken genome is unknown.

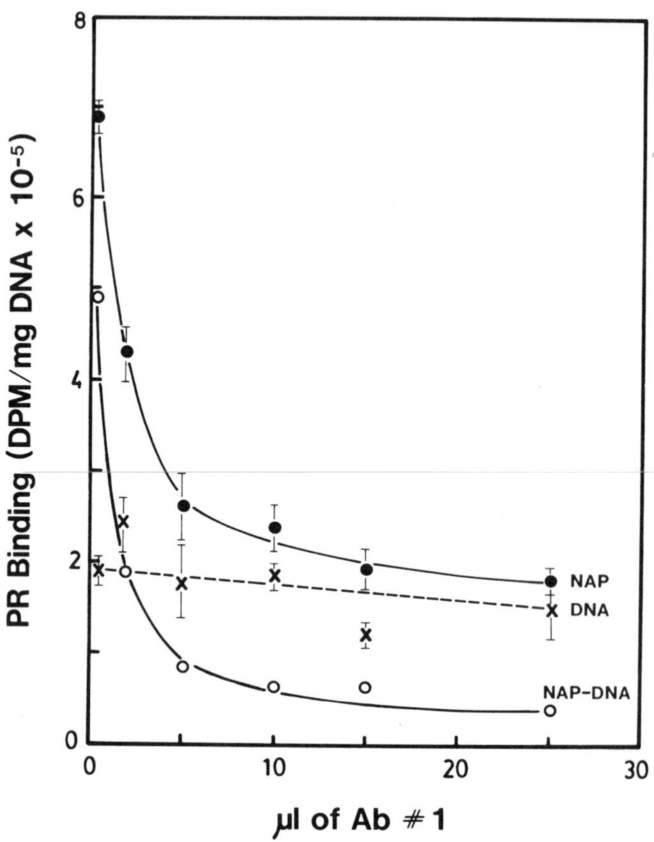

Fig. 12. The effects of ascites fluids from hybridoma 77-1 on the PR binding to NAP and DNA. Monoclonal antibodies prepared against reconstituted PRov acceptor sites containing partially purified acceptor protein; whole genomic hen DNA were added in increasing amounts to a typical cell-free nuclear-binding assay using isolated hen oviduct NAP, purified DNA, and PRov. The antibodies were added 15 min prior to the addition of the receptor. (●) represents PR binding to NAP, (x) represents DNA binding, and (○) represents PR binding to NAP corrected for DNA binding. The values represent the mean and standard deviation of triplicate analysis within a typical binding assay (Goldberger et al., 1986).

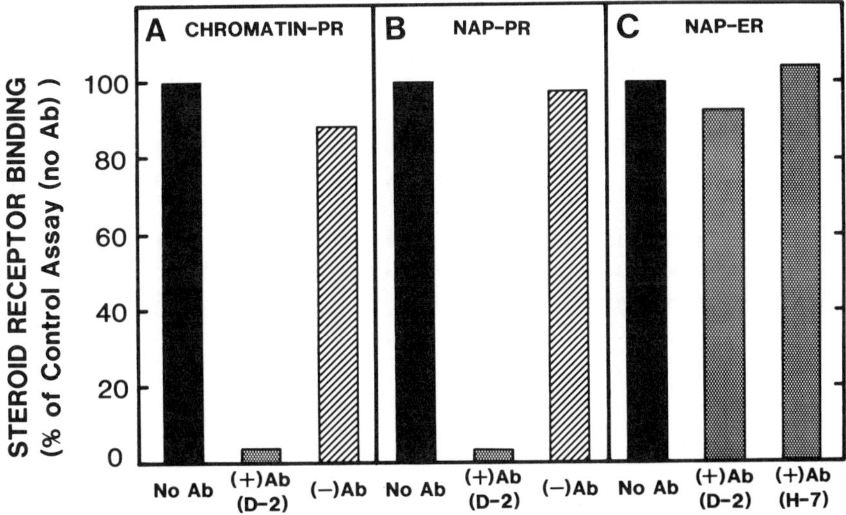

Fig. 13. Effects of anti-PRov acceptor site monoclonal antibodies on the PRov and ERov binding to nuclear acceptor sites. *Panel A* represents the effects of the positive and negative monoclonal antibodies on the PR binding to isolated hen oviduct chromatin. *Panel B* represents the effects of positive and negative antibodies on the PR binding to hen oviduct NAP. *Panel C* represents the effects of different clones of anti-PRov acceptor site monoclonal antibodies on the binding of the hen oviduct ER to hen oviduct NAP. The values represent the mean of triplicate analysis within one binding assay and are plotted as the percent of control PR binding (i.e., that with no antibody).

Novel Monoclonal Antibodies (MAbs) Against the Nuclear Acceptor Sites for the Avian Oviduct PR

Recently this laboratory has been able to generate monoclonal antibodies (MAb) against the nuclear-binding sites (acceptor sites for the avian oviduct PR) (Littlefield et al., 1985; Goldberger et al., 1985). To obtain the antigen, a partially purified acceptor protein was reannealed to whole genomic hen DNA and the reconstituted acceptor sites subjected to DNase I action to remove the excess of DNA. These protein DNA complexes were injected into mice, and the spleen cells were fused with myeloma cells to obtain hybridomas. These hybridomas were then screened for the production of antibodies that inhibit the PR binding to the acceptor sites. As shown in Fig. 12, clones were obtained which block the binding of PR to the NAP but not to the pure DNA. The specific binding to the NAP was shown to be almost totally inhibited with 10 μl of the ascites fluids. Panel A, Fig.

13, shows that the antibodies will inhibit the PR binding to intact native chromatin whereas the control antibodies fail to do so. In panel B, Fig. 13, the positive antisera are shown to inhibit the PR binding to NAP whereas control antisera do not. Panel C, Fig. 13, shows that the MAbs fail to inhibit the estrogen receptor binding to NAP. These studies show that the monoclonal antibodies prepared against a reconstituted PRov acceptor site consisting of a partly purified acceptor protein reannealed to hen DNA do recognize these specific PR-binding sites in native NAP and chromatin. Under the same conditions control antisera prepared against other types of antigens fail to inhibit this binding. Further, the studies indicate that the MAbs display a specificity for acceptor sites of different

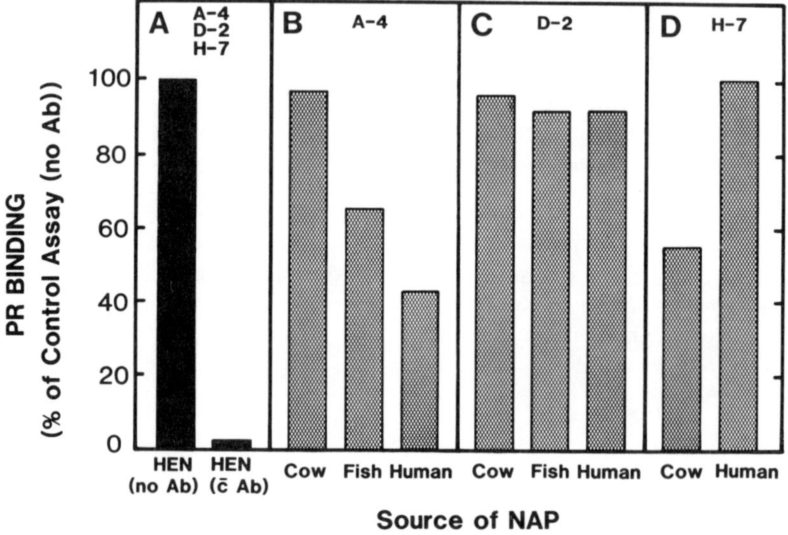

Fig. 14. Animal species specificity of the anti-PR acceptor site monoclonal antibodies: Effects on PR binding to nuclear acceptor sites from various animals. The NAPs from cow and human uteri and hagfish liver were isolated as described elsewhere. These NAPs were subjected to the cell-free nuclear binding assay using the PRov. 20 μl of each of the antiacceptor site clones were added 15 min before the addition of the PRov. *Panel A* represents the pattens of the effects of each of the cloned antibodies on the PR binding to hen oviduct NAP. *Panel B* represents the effects of clone A-4 antiacceptor antibodies on the PRov binding to cow, fish, and human NAP. *Panel C* represents the effects of the D-2 antiacceptor site antibodies on the PR binding to cow, fish, and human NAP. *Panel D* represents the effects of H-7 antiacceptor site antibody on the PR binding to cow and human NAP. The values represent the mean of triplicate analysis within a typical assay and are plotted as a percent of control assays as representing no antibodies, i.e., the control represents the PRov binding to NAP without antibody (Goldberger et al., in press).

steroid receptors. Thus, the reconstituted acceptor sites for PRov appear to be antigenically similar to the native acceptor sites.

Examination of the nuclear acceptor sites for PRov from other organisms indicates a partial recognition by the anti-PRov acceptor site antibodies in these tissues. As shown in Fig. 14, the NAPs from human and cow uterus and fish liver display partial inhibition of PR binding with some cloned MAbs compared with the hen oviduct NAP. Thus, the PR acceptor sites in mammals such as human and cow, and even fish, appear to be at least partially similar (antigenically) to those of the hen oviduct. This suggests that not only are the steroids and their receptors highly conserved over evolution (Goldberger et al., 1985), but also that the nuclear acceptor sites for these steroid receptors are conserved. These data, combined with the reconstitution of heterologous NAPs described above, support the conservation of these nuclear acceptor sites for the progesterone receptor.

Models of the Acceptor Sites of the Avian Oviduct PR

If the nuclear acceptor sites for the avian oviduct PR do not lie in or near the structural genes they regulate, one may speculate that these acceptor sites may be initial "docking sites" to concentrate the receptors. This

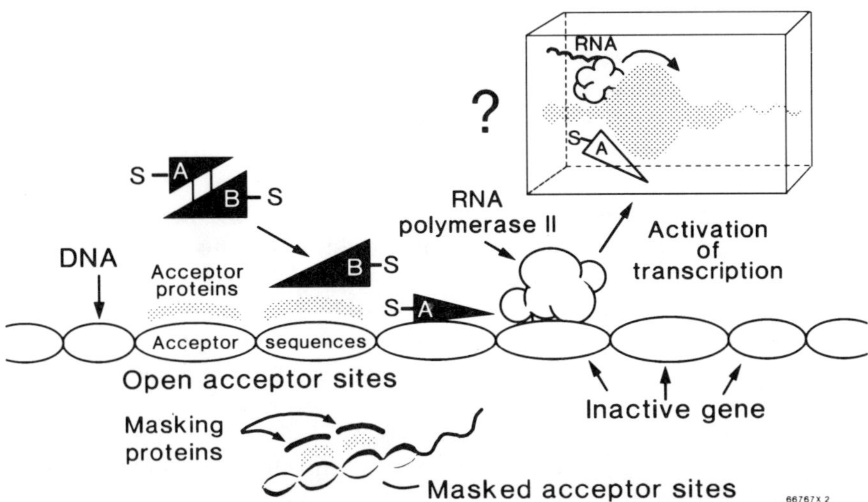

Fig. 15. Model of the possible mechanism of action of steroid action at the level of structural genes. This model was taken from previous reports primarily involving the PR chick oviduct (Spelsberg et al., 1983: Jensen and DeSonbre, 1972). The PR represented by the A and B dimer bind to the nuclear acceptor sites near structural genes via the B subunit. The A subunit then is dissociated and binds to the pure DNA sequences in a 5′ flanking region of structural genes. This results in an alteration of gene transcription. The lower part of the figure displayed a model of the masked acceptor sites which are incapable of binding the PRov.

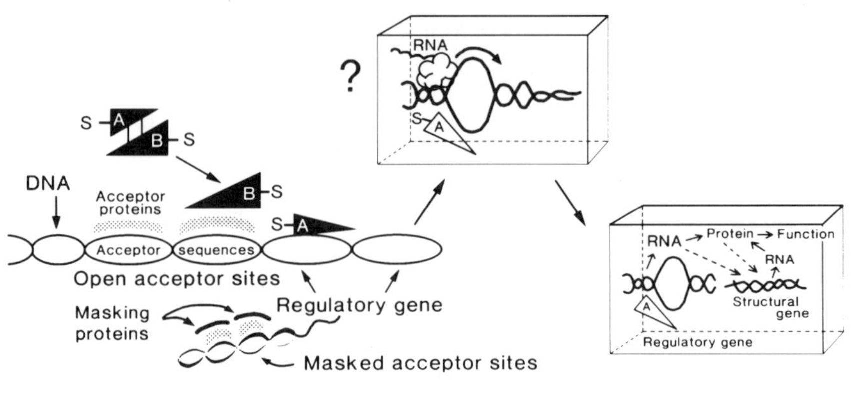

Fig. 16. Model for possible mechanism of the action of steroid receptors on regulatory genes. This model follows closely that described in fig. 15 with the exception that the steroid receptor will bind to nuclear acceptor sites adjacent to regulatory genes. The activation of these genes provides trans-acting factors, either RNA or protein, which migrate long distances to regulate structural gene expression by binding to the 5′ flanking regions of these genes.

process would be followed by the receptor shifting to secondary sites near the steroid regulated genes. Figure 15 represents such a model (model I) as proposed by Schrader and O'Malley (O'Malley and Means, 1974; Jost et al., 1984; Schrader et al., 1975; Buller et al., 1976). Here the binding of the PRov to the nuclear acceptor sites occurs when the receptor is a dimer. The B subunit of this receptor dimer would bind to the nuclear acceptor sites composed of specific acceptor proteins and specific acceptor DNA sequences. Immediately after the binding of the steroid receptor dimer via the B subunit, the A subunit would dissociate and bind near a structural gene to regulate its expression. Below this basic scheme in Fig. 15 are similar acceptor sites that are covered by masking proteins, making them incapable of binding the steroid receptor complex. Alternatively, in model 2 these acceptor sites may lie in enhancer-type regions further upstream from this gene and regulate gene expression by altering the activity of these regions. In the case of the ovalbumin gene, these sites might be 5′ to the two pseudogenes of the true gene and thus at many thousands of base pairs from this true gene. A third model, however, proposes that the PR acceptor sites might lie at still greater distance from the structural genes. In Fig. 16, model 3 is proposed wherein the steroid receptor might bind near regulatory genes that code for trans-acting molecules that migrate long distances to bind to the 5′ flanking regions of structural genes to regulate the transcription of the latter. Thus, these trans-acting molecules would serve as intermediates in the regulation of the structural gene transcription by the steroids. The regulatory genes might code for the synthesis of regulatory RNA (antisense) or regulatory protein. Either of these classes of molecules would then migrate and bind near structural genes and affect their expression within minutes after the steroid receptor binding. Such regulatory proteins might be transcription factors or might be proteins acting subsequently to transcription, e.g., stabilizing mRNA. If activation of the structural genes involves binding of the regulatory substances to specific sequences in their 5′ flanking regions, a requirement for these 5′ flanking regions of structural genes in steroid regulation would then be established. This could explain the essential 5′ regions of glucocorticoid regulated genes discovered through transfection experiments (Lee et al., 1981; Payvar et al., 1981, 1983; Chandler et al., 1983). Lastly, a fourth model should be mentioned whereby glucocorticoids might act in a fundamentally different manner from sex steroids.

Conclusions

Specific nuclear-binding sites (acceptor sites) for the avian oviduct PRov have been identified. Cell-free studies have demonstrated that these sites are saturable with high affinities (K_d to $10^{-9}M$) and require an intact, activated receptor. These nuclear sites resemble the nuclear binding in vivo in terms of the numbers of nuclear-binding sites per cell, the receptor specificity, and the degree of binding of functional versus nonfunctional steroid receptors. These nuclear sites appear to contain specific chromosomal proteins and DNA sequences. The substitution of other chromosomal proteins or genomic DNAs from evolutionarily distant organisms results in a loss of the specific nuclear binding. The nuclear acceptor sites appear to be resistant to the DNase activity which is not characteristic of transcriptionally active gene domains of the genome. Further studies using the ovalbumin gene sequences from genomic clones also indicate that none of the sequences within this domain appear to contain the specific acceptor sequences. These observations have led to development of a model suggesting that the steroid receptors bind to acceptor sites near regulatory genes which code for regulatory substances (as secondary messengers), which in turn regulate the structural gene expression. This model might better fit the sex steroids that require 1 to 2 h to measurably alter gene transcription (Palmiter et al., 1976), as opposed to the glucocorticoids, which more rapidly alter gene expression (Payvar et al., 1981, 1983; Chandler et al., 1983). As described earlier in this chapter, many of the properties of the PRov acceptor sites recently have been reported to be concordant with those of nuclear acceptor sites of many mammalian steroid target tissues (Ruh et al., 1981, 1985; Ross et al., 1984; Singh et al., 1984; Perry and Lopez, 1978; Cobb and Leavitt, 1985; Klyzsejko-Stefanowicz, 1976; Hamana and Kwai, 1978).

Acknowledgments. This work was supported by National Institutes of Health grants HD9140-P1 and HD16705 and the Mayo Foundation. J. Hora and B. Littlefield were supported by grant HD7108; A. Goldberger and M. Horton were supported by training grant CA90441.

References

Bailley A, Atger M, Atger P, Cerbon MA, Alizon M, Vu Hai MT, Logeat F, Milgrom E (1983) J Biol Chem 258:10384–10389
Barrack ER, Coffey DS (1980) J Biol Chem 255:7265–7275
Barrack ER, Coffey DS (1982) Recent Prog Horm Res 38:133–195
Boyd PA, Spelsberg TC (1979) Biochemistry 18:3679–3685
Boyd-Leinen PA, Fournier D, Spelsberg TC (1982) Endocrinology 111:30–36
Boyd-Leinen P, Gosse B, Martin-Dani G, Spelsberg TC (1984) J Biol Chem 259:2411–2421
Breathnach R, Chambon P (1981) Ann Rev Biochem 50:349–383
Buller RE, Schwartz RJ, Schrader WT, O'Malley BW (1976) J Biol Chem 251:5178–5186
Chandler VL, Maler BA, Yamamoto KR (1983) Cell 33:489–499
Cobb A, Leavitt WW (abstract) (1985) 67th Annual Meeting of the Endocrine Society June 19–21 (Baltimore, MD), pp 83, No 330
Compton JG, Schrader WT, O'Malley BW (1982) Biochem Biophys Res Commun 105:96–104
deAhe D, Janich S, Scheidert C, Renkawitz R, Schutz G, Beato M (1985) Nature 313:706–709
Dean DC, Knoll BJ, Riser ME, O'Malley BW (1983) Nature 305:551–554
Dougherty JJ, Puri RK, Toft DO (1984) J Biol Chem 259:8004–8009
Goldberger A, Littlefield BA, Katzmann J, Spelsberg TC (1986) Endocrinology 118:2235–2261
Goldberger A, Horton M, Katzmann J, Spelsberg TC (1986) Biochemistry (in press)
Greene GL, Closs LE, Fleming H, DeSombre ER, Jensen EV (1977) Proc Natl Acad Sci USA 74:3681–3685
Hamana K, Iwai K (1978) J Biochem (Tokyo) 83:279–286
Hora J, Horton MJ, Toft DO, Spelsberg TC (1986) Proc Natl Acad Sci USA (in press)
Igo-Kemenes T, Horz W, Zachau HG (1982) Ann Rev Biochem 51:89–121
Jensen EV, DeSombre ER (1972) Ann Rev Biochem 41:203–230
Jost J-P, Seldram M, Geiser M (1984) Proc Natl Acad Sci USA 81:429–433
Klyzsejko-Stefanowicz L, Chui JF, Tsai YH, Hnilica LS (1976) Proc Natl Acad Sci USA 73:1954–1958
Kon OL, Spelsberg TC (1982) Endocrinology 111:1925–1936
Lai EC, Woo SLC, Bordelon-Riser ME, Fraser TH, O'Malley BW (1980) Proc Natl Acad Sci USA 72:244–248
Lee F, Mulligan R, Berg P, Ringold G (1981) Nature 294:228–232
Littlefield BA, Spelsberg TC (1984) Endocrinology 117:412–414
Martin-Dani G, Spelsberg TC (1985) Biochemistry 24:6988–6997
Miller PA, Ostrowski MC, Hager GL, Simons SS Jr (1984) Biochemistry 23:6883–6889
Moore JT, Norvitch MD, Wieben ED, Veneziale CM (1984) J Biol Chem 259:14750–14756

Mulvihill ER, LePennec JP, Chambon P (1982) Cell 28:621–632
O'Malley BW, Means AR (1974) Science 183:610–620
Palmiter RD, Moore PB, Mulvihill ER, Emtage S (1976) Cell 8:557–572
Payvar F, DeFranco D, Firestone GL, Edgar B, Wrange O, Okret S, Gustafsson JA, Yamamoto KR (1983) Cell 35:381–392
Payvar F, Wrange O, Carlstedt-Duke J, Okret S, Gustafsson JA, Yamamato K (1981) Proc Natl Acad Sci USA 78:6628–6632
Perry BN, Lopez A (1978) Biochem J 176:873–883
Pikler GM, Webster RA, Spelsberg TC (1976) Biochem J 156:399–408
Puri RK, Grandics P, Dougherty JJ, Toft DO (1982) J Biol Chem 257:10831–10837
Ross P Jr, Ruh TS (1984) Biochim Biophys Acta 782:18–25
Ruh TS, Ross P Jr, Wood DM, Keene JL (1981) Biochem J 200:133–142
Ruh MF, Singh RK, Bellone CJ, Ruh TS (1985) Biochim Biophys Acta 844:24–33
Ruh TS, Spelsberg TC (1982) Biochem J 210:905–912
Schrader WT, Heuer SS, O'Malley BW (1975) Biol Reprod 12:134–142
Schrader WT, O'Malley B (1972) J Biol Chem 247:51–59
Singh RK, Ruh MF, Ruh TS (1984) Biochim Biophys Acta 800:33–40
Spelsberg TC (1976) Biochem J 156:391–398
Spelsberg TC (1982) In: Litwack G (ed) Biochemical actions of hormones, Vol. 9. Academic Press, pp 141–204
Spelsberg TC (1983) Biochemistry 22:13–21
Spelsberg TC, Gosse B, Littlefield B, Toyoda H, Seelke R (1984a) Biochemistry 23:5103–5112
Spelsberg TC, Halberg F (1980) Endocrinology 107:1234–1244
Spelsberg TC, Knowler J, Boyd PA, Thrall CL, Martin-Dani G (1979) J Steroid Biochem 11:373–388
Spelsberg TC, Littlefield BA, Seelke R, Martin-Dani G, Toyoda H, Boyd-Leinen P, Thrall C, Kon OL (1983) Recent Prog Horm Res 39:463–517
Spelsberg TC, Pikler GM, Webster RA (1976a) Science 194:197–199
Spelsberg TC, Reinhart GD, Barham S (1984b) Analytical Biochem 143:237–248
Spelsberg TC, Webster RA, Pikler GM (1976b) Nature 262:65–67
Thrall TC, Spelsberg TC (1980) Biochemistry 19:4130–4138
Thrall C, Webster RA, Spelsberg TC (1978) In: Busch H (ed) The cell nucleus Vol. VI, Part 1. Academic Press, New York, pp 461–529
Toyoda H, Seelke R, Littlefield BA, Spelsberg TC (1985) Proc Natl Acad Sci USA 82:4722–4726
Toyoda H, Spelsberg TC (1981) Abstract of the 63rd Annual Meeting for the Endocrine Society, Cincinnati, Ohio, No 857, p 297
Webster RA, Pikler GM, Spelsberg TC (1976) Biochem J 156:409–419
Webster RA, Spelsberg TC (1979) J Steroid Biochem 10:343–351

Discussion of the Paper Presented by T. Spelsberg

CLARK: Do I understand correctly that the only evidence you have is that the binding region is far away from the structural gene, that you didn't observe anything with O'Malley's gene?

SPELSBERG: Right now that's all we have done, Jim [Clark]. That's about 3 kilobases upstream from the ovalbumin gene.

CLARK: The other question has to do with the lack of functionality on the NAP. In about the third or fourth slide that you showed of estrogen-withdrawn chickens, everything looked very low and depressed. We all know that estrogen-withdrawn chickens are highly responsive to progesterone and to estrogen, right?
SPELSBERG: They are responsive to estrogen, but not so much to progesterone. The ovalbumin gene is turned on by progesterone and other steroids. However, we see about one hundreth the radioactivity of ^{3}H progesterone getting into the nucleus compared with the nonwithdrawn chick. We found that the receptor in our hands is almost 95%–98% inactive. We see that a marked decrease of about 90%–95% of the progesterone action on RNA synthesis. Lastly we see a total loss of the induction of avidin by progesterone in the estrogen-withdrawn bird.
MUELLER: Can you tell us something about the molecular size of these acceptor proteins?
SPELSBERG: I have a lot of graphs and tables, but just to summarize, we're seeing three of four species. Their low molecular weight is between 18 and 10,000. They are very hydrophobic. We know this by their solubility and by their binding to a variety of hydrophobic resins, i.e., how tight they are bound and how they elute. They're banding in isopycnic centrifugation indicates simple proteins that are not conjugated to any large prosthetic groups. We've treated these acceptor proteins with ribonucleases, lipases, and glycosidases and we haven't seen any effect. If you are not careful these proteins are very susceptible to protease, even in chromatin, in NAP, or in the free state, and even in sodium dodecyl sulfate (SDS). We don't know the roles of these different acceptor protein species. They have individual activity, so obviously they function alone. When we purify them, we separate one activity from another so they must work independently. Now, whether they are bound to different DNA sequences, or what, we have no idea.
MUELLER: Do the progesterone receptors interact with them in any physical way without having the DNA there?
SPELSBERG: No. We put the acceptor protein on cellulose or absorbed them on glass beads, or to straight cellulose, and we didn't see any specific binding at all. In fact, you can see that with the *E. coli* DNA, even though they're bound to this DNA, we see no PR binding.
MUELLER: It is clear that your acceptor sites are related to receptor localization in the chromatin, as you have shown. But what is the argument basically that they are functionally important?
SPELSBERG: We can't answer that. We don't know whether we are looking at a collecting depot or a functional site. We feel that we are measuring the real acceptor sites, which we classify as the initial interaction in the chromatin. I think it was B. O'Malley that stated we should get a transcription induction going with our system. We think that its important to do these kind of studies. I would like to add that a lot of people jumped into the "get a gene, get a sequence and bind it." It's very important for some researchers to do what we have done, and that is to take a look at whole chromatin, which is the native state of the DNA.

Discussants: J. CLARK, G. MUELLER, and T. SPELSBERG

Chapter 9

Antibodies to Estrogen, Progesterone, Glucocorticoid, Vitamin D Receptors and Autoantibodies to Androgen Receptor

S. LIAO, D. WITTE, C.V.R. MURTY, AND A.K. ROY

Introduction

Steroid hormones regulate cell function by interacting with specific receptor proteins (Gorski et al., 1968; Jensen and DeSombre, 1972; O'Malley and Means, 1974; Liao, 1975; Muldoon, 1980; Schrader, et al., 1981; Schmidt and Litwack, 1982; Jensen et al., 1982; Ringold, 1985). Specific receptors for a number of steroid hormones have been purified (O'Malley and Schrader, 1976; Govindan, 1979; Kuhn et al., 1975; Pike et al., 1982; Jensen et al., 1982). Most of our present knowledge of steroid receptors and their action in target cells has been derived on the basis of receptor binding to radiolabeled steroid. The labeled ligand technique has certain inherent limitations: (1) The labeled steroid can only bind to unoccupied binding sites on the receptor protein; (2) it cannot interact with inactive or nonfunctional receptor proteins that may exist under certain physiological and pathological conditions; (3) other steroid binding proteins present in the target cell interfere with the ligand-binding assay. These difficulties have led to the search for a better alternative for identification and quantification of steroid receptors such as the immunochemical approach. Preparation of monospecific antibody generally requires purified antigen, and steroid receptors constitute only a minor component of cellular proteins. Therefore, purification of steroid receptors to homogeneity has always been fraught with insurmountable difficulties. Although, initially, highly purified estrogen receptor from calf uterus was used for the successful production of polyclonal antibodies (Greene et al., 1977, 1979), recently the purification hurdles have largely been circumvented by the hybridoma procedure for the production of monoclonal antibody with only partially purified starting material. Since "epitopes" of such monoclonal antibodies are directed against different segments of the receptor proteins, monoclonal antibodies have also provided important tools for molecular analysis of the structural and functional domains of the receptor molecule. Several groups of investigators have raised antibodies to different receptor proteins (Govindan, 1979; Coffer et al., 1980; Radanyi et al., 1979; Raam et al., 1981). These antibodies have been used for detection, purification, localization, and measurement of receptor proteins in normal and neoplastic tissues and in cultured cells as well as to analyze receptor structure.

Antibodies to Estrogen Receptor

Estrogen receptors have been purified to near homogeneity from calf uterus (Jensen et al., 1982) and MCF-7 human breast cancer cells (Greene et al., 1980c). The molecular weight based on sodium dodecyl sulfate (SDS)-polyacrylamide gel electrophoresis has been estimated to be 65,000 daltons (Merrill et al., 1981). Fox (1978) found that nonspecific stimulation of immune system can result in serum factors that form complexes with estrogen receptor. These factors that copurify with IgG fraction react with 5S "activated" form of the receptor complex and show no species specificity. Naturally occurring IgG which reacts with estrogen receptor antibody was detected in the sera of a population of male and female humans ranging in age from 1 to 85 years (Mudarris and Peck, 1985). However, the levels of this antibody significantly differed with sex and age of the individual. At present the etiology of this antibody is not clear.

Greene and co-workers prepared antibodies against calf uterine estrogen receptor in the rabbit, goat, and rat (Jensen et al., 1982; Greene et al., 1980a) which recognized all mammalian and nonmamalian estrogen receptors tested so far (Jensen and Greene, 1980). When the labeled estradiol-receptor complex is incubated with immunoglobulin obtained from immune and preimmune sera and analysed on a sucrose gradient containing 10 m*M* KCl, a sedimentation profile shown in Fig. 1 is obtained. The interaction of excess immunoglobulin from immunized animal with estrogen receptor produces a shift in the receptor peak from 4.5S to greater than 9S.

Cross-reacting antibodies have also been produced in animals immunized with estrogen receptor from rat mammary tumor (Al-Nuami et al., 1979), human myometrium (Coffer et al., 1980), and human breast cancer (Raam et al., 1981). This indicates that one or more antigenic determinants (thus amino acid sequences) are conserved during evolution. These polyclonal antibodies have proved useful for structure and function analysis (Jensen et al., 1982; Greene et al., 1977) and for the purification and immunoassay of receptor.

With the help of hybridoma technology, monoclonal antibodies have been raised against calf uterine and MCF-7 human breast cancer estrogen receptors (Greene et al., 1980a, 1980b, 1980c). These antibodies react with both 4S and 5S forms of the receptor and do not interfere with the ability of the receptor to bind steroid (Greene et al., 1980b). Hence, these antibodies make excellent probes for detecting intact receptor and receptor fragments or in any situation where receptor is either occupied with unlabeled steroid or otherwise unable to bind estrogen (Greene and Jensen,

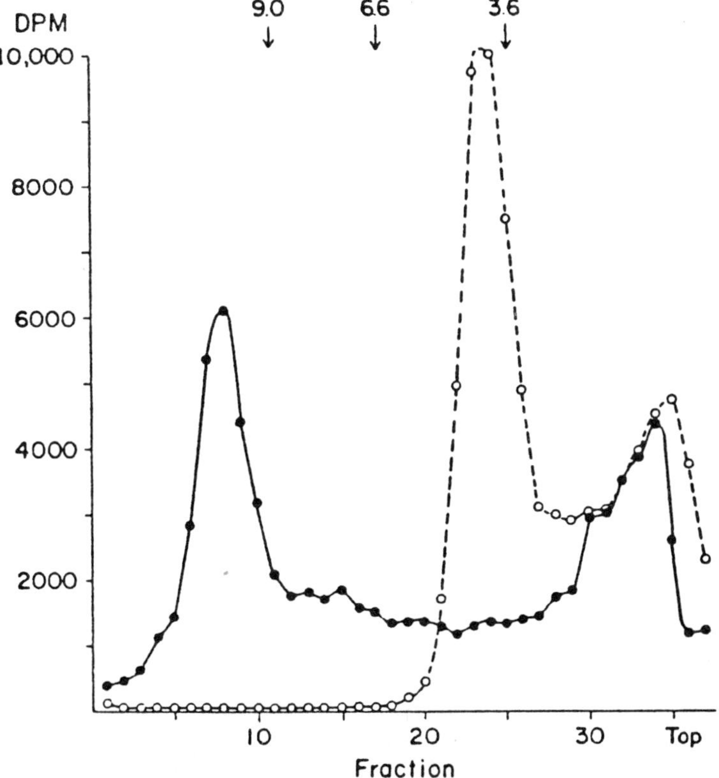

Fig. 1. Sedimentation profile for purified estradiol-receptor complex from calf uterine nuclei in 10%–30% sucrose gradients containing 10 m*M* KCl, after incubation with immunoglobulin from an immunized (●) or control (○) rabbit (Greene et al., 1979).

1982; Greene, 1983). Immunocytochemical assays have been developed for estrogen receptor to determine intracellular distributions of receptors in tissue sections and in dispersed cells (King and Greene, 1984; Press and Greene, 1984; Shimada et al., 1985; McCarty et al., 1985; Pertschuk et al., 1985). Using these cytochemical techniques, several investigators (King and Greene, 1984; McClellan et al., 1984) concluded that estrogen receptor resides in target-cell nuclei even in the absence of ligand and that the presence of cytosolic receptors after tissue homogenization does not represent cytoplasmic localization. Immunocytochemical analysis at the electron-microsopic level has led to localization of the estrogen receptor in the euchromatin structure but not in the marginated heterochromatin or nucleoli of epithelial and stromal nuclei of human endometrium (Press et al., 1985). Furthermore, monoclonal antibodies against estrogen receptor have been employed to determine whether breast and ovarian carcinomas contain high levels of estrogen receptors and to evaluate whether the hormonal therapy should be employed for cancer patients (Jensen et al., 1982; Lorincz et al., 1985; Holt et al., 1985).

Monoclonal antibodies have also been used to probe various molecular forms of the receptor (Moncharmont et al., 1984). An IgM class of monoclonal antibody raised against calf uterine estrogen receptor was used to probe the structure of estrogen receptor bound to either estradiol or to 4-hydroxytamoxifen (Borgna et al., 1984). This antibody interacts more strongly with the activated receptor than with the nonactivated complex and also with nuclear forms of receptor bound to both estradiol and antiestrogen and inhibits the binding of receptor to DNA. Further studies (Borgna et al., 1984) suggested that the conformational changes of receptor induced by estradiol or antiestrogen differ at the hormone-binding site, the DNA-binding domain, and the antibody-binding domain. Thus, these antibodies can be used to discriminate different forms of receptor and to characterize some of their binding domains.

Recently, Abbott laboratories developed an enzyme immunoassay kit for the quantification of estrogen receptor. This assay has been tested under different conditions, and the results are comparable with those obtained by ligand-binding assay (Goussard et al., 1985). By using monoclonal antibodies, Walter et al. (1985) identified cDNA clones corresponding to the estrogen receptor mRNA. The nucleotide sequence of the cDNA has also been established (Greene et al., 1986). All of these developments attest to the analytical power of monoclonal antibodies for exploration of receptor function.

Antibodies to Progesterone Receptor

Considerable amount of data are currently available on progesterone receptors derived from chick oviduct (Vedeckis et al., 1978) and rabbit uterus (Loosfelt et al., 1984). The chicken receptor is thought to be made up of two hormone-binding subunits, A and B (M_r 79,000 and 106,000, respectively) which have kinetically identical progesterone-binding characteristics (Vedeckis et al., 1978; Schrader and O'Malley, 1972). The B subunit seems to have a binding preference to chromatin structure whereas the A subunit has a higher affinity for the DNA (Schrader et al., 1981). Thus these two subunits may have different roles in the mechanism of progesterone action. Antibodies against the subunits will, therefore, be highly useful for studies on structure-function relationship of receptor subunits. Weigel et al. (1981) found a spontaneous sheep antibody of the IgG type to chick progesterone receptor. It is specific to chicken progesterone receptor and binds neither to other steroid hormone receptors nor to progesterone receptors derived from other species. This spontaneous antibody reacts with almost equal affinity to the A and B subunits, indicating that at least portions of these two subunits are quite similar. Rabbit and goat antibodies against chicken progesterone receptor have been prepared (Renoir et al., 1982; Tuohimaa et al., 1984). The goat antiserum cross-reacts not only with the mammalian progesterone receptor, but more surprisingly with chicken glucocorticoid receptor (Renior et al., 1982). It recognizes the nontransformed 8S receptor as well as the transformed receptor. The rabbit antiserum to the B subunit also reacts with the A subunit (Touhimaa et al., 1984). Immunohistochemical studies show that the progesterone receptor is present in cell nucleus and hormone administration does not significantly modify the intracellular distribution of the receptor, providing additional support to the revised model for the steroid hormone action (Gasc et al., 1984). Edwards et al. (1984) produced monoclonal antibodies against the B subunit of chicken progesterone receptor in the rat. These antibodies failed to react with the subunit A, and this tends to support the earlier observation that A and B subunits are separate gene products (Birnbaumer et al., 1983). These antibodies were also used to map antigenic sites in relationship to the hormone-binding domain. Immunoblotting of proteolytic peptides reveals that antigenic sites for both monoclonal antibodies are localized within a 3,000-dalton region on the C-terminal side of the 31,000-dalton fragment, the smallest fragment containing hormone-binding site (Edwards et al., 1984). Monoclonal antibodies against 8–9S molybdate stabilized chicken oviduct progesterone receptor have also been produced (Radanyi et al., 1983). The antibody activity to this receptor resides in the IgG fraction, and the antibody binding does not prevent hormone binding to the receptor. These investigators have also reported an epitope of monoclonal antibody that is specific for a nonhormone binding 90,000-dalton protein component of progesterone receptor (Radanyi et al., 1983). Using this antibody, it was shown that the 90,000-dalton protein is present in the untransformed molybdate stabilized 8S forms of androgen, estrogen, and glucocorticoid receptors (Joab et al., 1984). Sullivan et al. (1985) generated four IgG secreting hybridoma cell lines against 90,000-dalton nonhormone-binding component of the progesterone receptor and used it to show the presence of this protein in the brain, liver, and skeletal muscle.

Milgrom's group has been working with progesterone receptor from the rabbit uterus, which apparently consists of one steroid-binding subunit of 110,000 daltons. The antiserum prepared against the cytosolic progesterone receptor also reacts with the nuclear form of the receptor (Logeat et al., 1981). Cross-reaction has been observed with progesterone receptors from other mammalian tissues, especially the human receptor (Logeat et al., 1981); but the chicken oviduct progesterone receptor does not cross-react with this antibody. Logeat et al. (1983) produced five monoclonal antibodies against the rabbit uterine cytosolic progesterone receptor. All five antibodies cross-reacted with the rabbit nuclear receptor, human cytosolic receptor, and other mammalian but not avian progesterone receptors (Logeat et al., 1983). These monoclonal antibodies were used for the im-

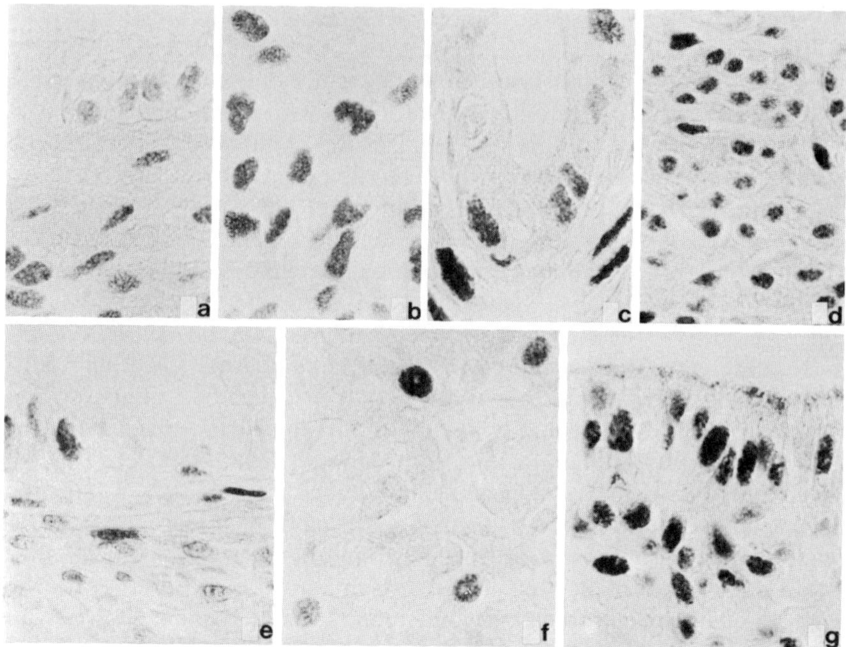

Fig. 2. Progesterone receptor (PR) is present in nuclei even in the absence of its ligand. Immunostaining of PR is shown in different target tissues from estradiol-primed prepubertal rabbits (**a–f**) or estradiol-primed ovariectomized guinea pigs (**g**). Nuclear straining was observed in epithelial and stromal cells of rabbit (**a**) and guinea pig (**g**) endometrium, in stromal cells (**b**) and glandular cells (**c**) of rabbit endometrium, in myometrial cells (**d**), in stromal cells of rabbit oviduct (**e**), and in cells of the pituitary gland (**f**). Paraffin sections were treated according to an indirect immunoperoxidase technique using Mil-2 (**b, c**), Mi60-10 (**d, g**), or a mixture of monoclonal antibodies (**a, e, f**) at 6 g Ig/ml (**a–e**) or 12 μg Ig/ml (**f, g**) and peroxidase-labeled antibodies to mouse IgG. Magnification, × 1,000 (Applanat et al., 1985).

munocytochemical detection of progesterone receptors in various mammalian tissues (Applanat et al., 1985). Immunohistochemical analysis (Fig. 2) shows that the rabbit uterine progesterone receptor is exclusively localized in the nucleus even in the absence of the hormone, confirming earlier reports on nuclear localization of the estrogen receptor (King and Greene, 1984; McClellan et al., 1984; Gorski et al., 1984; Welshons et al., 1984). No progesterone receptor immunoreactivity is observed in diaphragm, spleen, or intestine, which is in conformity with previous biochemical observations and does not support the concept that nonsteroid-binding forms of the receptor protein might be present in such tissues (Applanat et al., 1985). The availability of monoclonal antibodies against the progesterone receptor made it possible to prepare receptors in the native forms, which are able to bind their ligands and interact with DNA, for the purification of unoccupied receptors (Logeat et al., 1985). Logeat et al. claim that the existence of two (or more) subunits of the progesterone receptor in other mammalian species is due to proteolytic degradation of the native receptor during experimental procedures. Such a contention is supported by the results of immunoblot studies of receptors from guinea pigs, rats, hamsters, and humans (Logeat et al., 1985). Monoclonal antibodies against nonactivated rabbit uterine progesterone receptors raised by Nakao et al. (1985) are specific to progesterone receptor from rabbit uterus and MCF-7 breast cancer cells and did not bind to rat uterine, guinea pig uterine, or chick oviduct progesterone receptors, showing clear-cut structural differences in the receptors of various species.

Antibodies to Glucocorticoid Receptor

The native mouse liver glucocorticoid receptor monomer has a molecular weight of 81,000 daltons and is composed of three functional domains: hormone-binding domain, DNA-binding domain, and a domain that may be involved in specifying the site of interaction of the receptor in the nucleus (Vedeckis, 1983). Antiserum raised against the rat liver glucocorticoid receptor reacts specifically with the 78,000-dalton hormone-binding receptor, but not with transcortin (Eisen, 1980). These antibodies are used to establish immunochemical differences (Stevens et al., 1981) between glucocorticoid receptors from corticoid-sensitive and corticoid-resistant malignant lymphocytes. Stevens et al. (1981) immobilized immune IgG to Sepharose and examined the ability of the affinity gel to recognize cytosolic ^{3}H-triamcinolone acetonide receptor complexes from corticoid-sensitive and corticoid-resistant strains of mouse lymphoma P1798. The immunoaffinity column retained 70%–84% of 58–62 A° (Stokes radius) receptor complex characteristic of the glucocorticoid-sensitive mouse and human lymphocytes, but failed to recognize the 27–28 A° complex present in corticosteroid-resistant mouse lymphoma cells (Stevens et al., 1981). Harmon et al. (1984) raised antibodies against human lymphoid glucocorticoid receptors from IM-9 cells. These antibodies recognize 90,000-dalton protein from both IM-9 and CEM-C7 cells as the steroid-binding component, and the concentration of this protein is markedly reduced in mutants containing diminished quantities of the glucocorticoid-binding activity (Harmon et al., 1984). Polyclonal antibodies against the rat liver receptor has been highly useful in separating functional domains of the glucocorticoid receptor (Carltedt-Duke et al., 1982; Govindan and Sekeris, 1978). Based on immunochemical analyses, it has been suggested that loss of a part of the glucocorticoid receptor equal to domain C might result in glucocorticoid resistance in P1798 lymphoma cells. Thus, it appears that this domain may be required to retain biological activity of the receptor.

Okret et al. (1981) have established the immunochemical similarity between glucocorticoid receptors in different tissues as well as in different species and have found that antibodies to the glucocorticoid receptor do not cross-react with other steroid receptors. Immunofluorescence studies made possible direct analysis of glucocorticoid receptors in neurons and adrenal medullary cells in culture (Okret et al., 1981).

Weinberger et al. (1985) used the polyclonal antiserum raised against human lymphoid glucocorticoid receptor to isolate glucocorticoid receptor cDNA clones. Screening a λgtll cDNA library with the antiserum resulted in the identification of four positive clones. The fusion proteins from these clones were immobilized and incubated with the antiserum to select epitope-specific antibodies that were subsequently eluted and assayed for their ability to bind the receptor in IM-9B cellular extracts. Three of the four clones exhibited cross-hybridization with the nick translated probe prepared from one of the clones, showing that they have a common nucleic acid sequence that might encode an immunogenic domain that was identified in the glucocorticoid receptor using many antibody preparations (Okret et al., 1981; Westpahl et al., 1982).

Antibodies to Vitamin D Receptor

Consistent with the general mechanism of steroid hormone action, 1,25 dihydroxy vitamin D_3 is also known to exert its biological responses through the involvement of specific cytoplasmic receptor proteins (Pike and Haussler, 1979). In this case not much information is available regarding the characteristics of the receptor protein. Recently Pike et al. (1983) partially purified a 64,000-dalton protein and tentatively concluded that this protein represents the chick receptor for 1,25 dihydroxy vitamin D_3. This partially pure preparation was used to raise monoclonal rat antibodies. Four monoclonal antibodies specific for the receptor have been characterized, and they react with both occupied and unoccupied chicken intestinal receptor. These antibodies also cross-react with vitamin D_3 receptors from a variety of tissue and cell types (Pike et al., 1982). Since these antibodies are able to specifically recognize the 1,25 dihydroxy vitamin D_3 receptor protein, they can be used as probes for the elucidation of the chemistry and function of the vitamin D_3 receptor.

Antibodies to Androgen Receptor

Although polyclonal and monoclonal antibodies to estrogen (Greene et al., 1977; Greene and Jensen, 1982), glucocorticoid (Eisen, 1980; Grandics et al., 1982; Okret et al., 1984) and progestin (Logeat et al., 1981; Feil, 1983; Logeat et al., 1983) receptors have provided new methods for identification, localization, and characterization of steroid receptors, an antibody against androgen receptors has yet to be produced. This is mainly due to the difficulty in obtaining sufficiently pure androgen receptors in an amount that is necessary for efficient immunization of experimental animals. Recently, Liao and Witte (1985) found that high-titer antibodies to androgen receptor are present in the blood sera of some male patients. It may therefore be feasible to use lymphocytes from individuals with high antibody titers to produce monoclonal antibodies against androgen receptors. Future studies may also reveal whether distinct clinical syndromes are associated with the presence of autoantibodies to steroid receptors.

Immunointeraction of Prostatic Androgen Receptors with Human Serum Antibodies

In low-salt media, steroid-receptor complexes sediment either as 7 to 12 S units or as heavier aggregates, depending on the pH and temperature

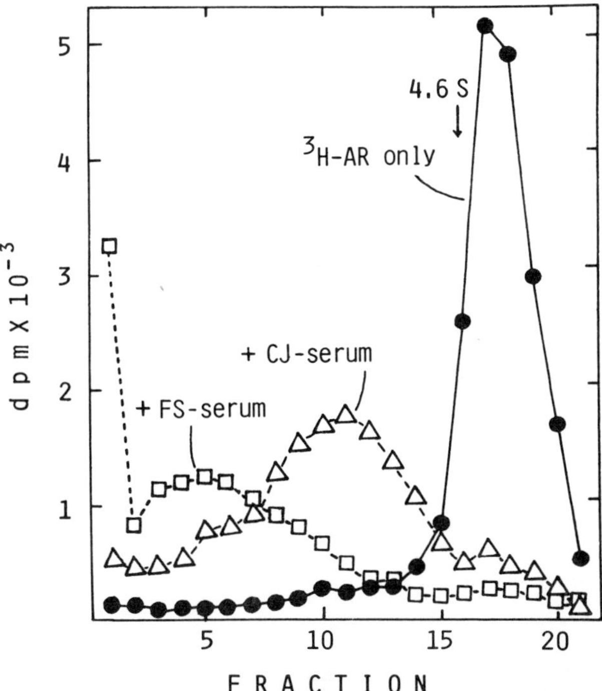

Fig. 3. Sedimentation profiles of androgen-receptor complexes in the presence of human sera containing autoimmune antibodies. Salt precipitated [^{3}H] DMNT-R (20,000 dpm) alone (●) or with 40 μl of serum from male patient. CJ (△) or FS (□) were incubated at 4°C for 4 h. The incubated mixtures (100 | 1) were layered on the top of sucrose gradients (5%–20%) containing 1.5 m*M* EDTA, 2 m*M* dithiothreitol, 0.4 *M*KCl, and 20 m*M* Tris-HCl, pH 7.5. Centrifugation was performed at 34,000 rpm for 16 h at 2°C using a Beckman SW 60 rotor. The gradient fractions (0.2 ml each) were collected and numbered from the bottom of each tube. Human IgG (7S), bovine albumin (4.6S), and [^{3}H] DMNT-R alone (3.8S) sedimented, respectively, at the vicinity of fractions 13, 16, and 17 (Liao and Witte, 1985).

used in treatment of the receptor preparations and on the presence of other cellular components, such as nucleic acids or nucleoproteins (Liao et al., 1973, 1980), metal ions and thiols (Liao et al. 1975, 1984), and proteases (Wilson and French 1979). In our study we carried out gradient centrifugation and immunoprecipitation in medium containing 0.4 *M* KCl, in which the receptor complexes are dissociated from aggregated materials and sediment rather uniformly at 3.8 ± 0.3 S (Liao et al., 1975). As the radioactive androgen, we employed [^{3}H] DMNT(7α, 17α-dimethyl-19-nortestosterone or mibolerone), a synthetic androgen that is highly specific for androgen receptor. By using this androgenic ligand (Schilling and Liao, 1984), we minimized the difficulties that may be caused by the nonspecific interaction of blood or prostate proteins with natural androgens such as 5α-dihydrotestosterone (DHT).

As shown in Fig. 3, the radioactive androgen-receptor complexes had a sedimentation coefficient of about 4S. In the presence of a serum sample from patient CJ, however, the complexes sedimented as 9–12S units. When a serum sample from patient FS was present, the major radioactive peak was at 14–19S, and some material sedimented at the bottom of the tube. These results suggest that the serum samples of these patients contained antibodies to androgen receptor. The changes in the sedimentation patterns suggest that different groups of immunoglobulins may be involved. IgG and IgM may be, respectively, the predominant antibodies in the serum samples obtained from patients CJ and FS. When immunoprecipitation was carried out with serum from patient CJ or FS, the radioactive receptor complexes precipitated only in the presence of antihuman immunoglobulins (Table 1). In the absence of the human serum, the second antibodies did not precipitate the receptor complexes or the radioactive androgens.

Table 1. Antihuman Immunoglobulin-Dependent Precipitation of the Androgen-Receptor Complexes in the Presence of Human Serum

Addition	Immunoprecipitable receptor complexes (dpm)
[^{3}H] DMNT-R	210
+ FS-serum	53
+ Sheep antihuman-Ig	90
+ FS-serum + sheep antihuman-Ig	19,283

Note: Immunoprecipitation of the [^{3}H] DMNT-R (50 μl prostate cytosol, containing about 20,000 dpm hydroxylapative assayable complexes) was carried out in the absence or presence of 10 μl of a human serum sample from male patient. FS and/or 200 μl of sheep antihuman immunoglobulins as the second antibodies. Similar results were obtained with affinity purified goat or sheep antihuman IgG that are H and L chains specific or affinity purified rabbit antihuman IgG, but not with preimmune rabbit sera.
Source: Liao and Witte, 1985.

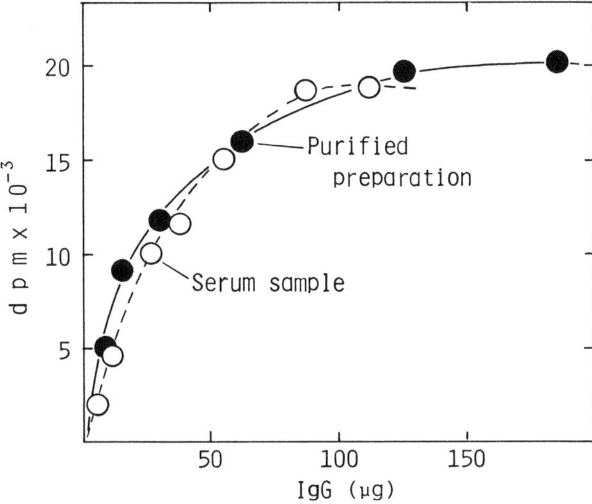

Fig. 4. Immunoprecipitation of [^{3}H] DMNT-R. Prostate cytosol (50 μl) containing about 20,000 dpm hydroxylapatite-filter assayable complexes was mixed with various amounts of IgG in serum or purified IgG shown on abscissa. Virtually all the radioactive receptor complexes were precipitated by 100 μg of IgG present in the serum or in the purified preparation.

Characterization of IgG Antibodies

The antibodies to androgen receptor in the serum of patient CJ were apparently due to IgG, because essentially all the antibody activity in the whole serum could be recovered (Fig. 4) in an IgG fraction (95% γ-globulin on electrophoresis) purified from his serum by ammonium sulfate precipitation and diethylaminoethyl (DEAE)-cellulose chromatography (Livingston, 1974). The dissociation constant for the antibody-receptor complex appeared to be well below 0.1 n*M* so that under the assay conditions with excess antibodies, there was complete precipitation of the receptor complexes.

Receptor Specificities

The antibodies from patient CJ immunoprecipitated DHT- or DMNT-receptor complexes from different sources such as the rat ventral prostate, human prostate, and human breast tumor MCF-7 cells. With several serum samples, no immunoreactivity was found with glucocorticoid receptors of rat liver, estrogen receptors of rat uterus, and MCF-7 cells, or progestin receptors of human breast tumor T-47D. The receptor selectivity was not due to cellular components that promoted or inhibited immunoprecipitation of specific steroid receptors, since mixing of cytosol fractions from dif-

Table 2. Specificity of Human Antireceptor Antibodies

Source of cytosol or TEBG	^{3}H-steroid (40 Ci/mmol)	Immunoprecipitable radioactivity (dpm)
Prostate (0.5 ml)	DMNT	11,051
Prostate (0.5 ml)	17 β-Estradiol	127
Prostate (0.5 ml)	Dexamethasone	166
Uterus (0.5 ml)	17 β-Estradiol	183
Liver (0.5 ml)	Dexamethasone	554
Uterus/prostate (0.25 ml each)	17 β-Estradiol	101
Uterus/prostate (0.25 ml each)	DMNT	6,610
Liver/prostate (0.25 ml each)	Dexamethasone	311
Liver/prostate (0.25 ml each)	DMNT	6,640
TEBG	DHT	56
TEBG	DMNT	126
TEBG	17 β-Estradiol	259
TEBG/0.4 *M* KCl	DHT	67
TEBG/0.4 *M* KCl	DMNT	193
TEBG/0.4 *M* KCl	17 β-Estradiol	147
No cytosol or TEBG	DHT	64
No cytosol or TEBG	DMNT	91

Note: Cytosol fractions from ventral prostate, liver, or uterus of rats deprived of steroid hormones (by castration, adrenalectomy, or ovariectomy) or purified human TEBG (25 μg) were incubated with radioactive DMNT(10 n*M*), 17 β-estradiol (10 n*M*), DHT (20 n*M*), or dexamethasone (30 n*M*) for 60 min at 0°C. Immunoprecipitation of the radioactive steroid complexes was carried out in the presence of 10 μl of serum from male patient. CJ and a rabbit antihuman IgG serum. The amounts of hydroxylapatite-filter assayable receptors in the cytosol preparations of prostate (for DMNT), uterus (for 17 β-estradiol), and liver (for dexamethasone) were about 70 fmol (12,000 dpm)/0.5 ml cytosol.
Source: Liao and Witte, 1985.

ferent sources did not significantly alter the amount of immunoprecipitable radioactivity (Table 2). The antisera also did not immunoprecipitate radioactive steroids bound to other nonreceptor steroid-binding proteins, such as human testosterone/estradiol-binding globulin (TEBG) (Rosner and Smith, 1975), rat epididymal androgen-binding protein (Hanson et al., 1976), or rat ventral prostate α-protein (Chen et al., 1982).

It has been shown that steroid binding, heating, and salt precipitation promote the transformation (also called activation) of various steroid receptors from forms that do not bind to DNA into the DNA-binding forms (Liao, 1975; Schmidt and Litwack, 1982). The experiments illustrated in Table 3 indicate that the human antibodies recognized both DNA-binding and nonbinding forms of [^{3}H] DMNT-receptor complexes, although about 25%–30% of the heat- or salt-activated DNA-binding form resisted immunoprecipitation. This may be attributable to partial proteolysis of the receptor proteins. Both the cytosol- and nucleus-bound receptor complexes extracted from minced prostates that had previously been incubated with radioactive androgen were immunoprecipitated. Because the antibodies recognized a significant portion of the nucleus-bound receptor complexes

Table 3. Immunoprecipitation of Different Forms of the Androgen-Receptor Complexes

	[^{3}H] DMNT-R,dpm HAP-filter assayable	Immuno-precipitable
Naked receptor incubated with [^{3}H] DMNT		
Before heat activation	23,274	21,193
Heated and DNA-bound	20,274	15,383
Ammonium sulfate precipitated [^{3}H] DMNT-R		
Unfractionated	19,940	16,130
DNA-bound	14,260	9,887
DNA-excluded	10,420	11,833
Minced prostate incubated with [^{3}H] DMNT		
Cytosol	25,917	25,413
Nucleus-bound	19,440	26,765

Note: Various radioactive receptor fractions were prepared and [^{3}H] DMNT-R immunoprecipitated with 10 μl of a human serum sample from male patient. CJ and 200 μl of goat antihuman immunoglobulins. The cytosol fraction prepared from rat ventral prostate of castrated rats was passed through a double-layer column containing carboxymethylsephadex and DNA-Sepharose. Receptors passed through the column were used as the "naked receptors." The receptors were labeled with 10 n*M* [^{3}H] DMNT for 30 min at 0°C. For the heat transformation (or activation) of these complexes to the DNA-binding forms, DNA-Sepharose was added and the mixture was incubated for 45 min at 20°C. The heated mixture was packed into a column and the DNA-bound [^{3}H] DMNT-R eluted. The [^{3}H] DMNT-R was also precipitated from the cytosol fraction by ammonium sulfate (40% saturation). The precipitated complexes were desalted and separated into DNA-excluded and DNA-bound fractions. Ventral prostates were minced and incubated with 10 vol of Dulbecco's modified Eagles' medium and 20 n*M* [^{3}H] DMNT for 45 min at 37°C in a shaking water bath. The incubated prostates were homogenized and the cytosol and nucleus-bound [^{3}H] DMNT-R isolated. HAP = hydroxylapatite.
Source: Liao and Witte, 1985.

that were not measured in the hydroxylapatite-filter assay, various forms of androgen receptors may be present in the nuclear extracts. In contrast, all cytosol [^{3}H] DMNT-receptor complexes that were measurable by the hydroxylapatite-filter assay that we employed could be immunoprecipitated.

Sex and Age Difference

Antibodies to androgen receptor are clearly more prevalent in males than in females. About 36.9% (124 of 336) of male and 2.6% (7 of 268) of female serum samples that we screened had significant antibody titers (i.e., they immunoprecipitated 1 fmol or more of [^{3}H] DMNT-receptor complexes per 10 μl of serum) (Table 4). This may reflect the difference in the levels of autoantigens for androgen receptors in the two sexes. Although the major sources of the androgen receptors may be the prostate, it is possible that a significant amount of antigenic androgen receptors is supplied by organs other than the prostate. The tissue concentrations of androgen receptors in organs such as liver, kidney, skin, and muscle are low, but these organs, in total, contain more receptors than does the prostate in an individual.

Table 4. The Activities of Human Serum Antibodies to Androgen Receptors

[^{3}H] DMNT-R immunoprecipitated (fmol/10 μl serum)	Number of serum samples in 3 age groups					
	Male			Female		
	2–45	46–65	66–90	2–45	46–65	66–90
<1[a]	79	43	90	132	66	63
1–3	11	34	55	2	0	2
3–5	1	6	6	1	0	0
5–10	1	0	2	0	1	0
10–15	1	0	1	0	1	0
15–20	1	0	1	0	0	0
>20	0	1	3	0	0	0
Total	95	85	158	135	68	65

[a]Since the variation in the duplicate assays was about 0.5 fmol, this value was considered insignificant.
Source: Liao and Witte, 1985.

Males less than 45 years old (16.0%) are less likely than those older than 45 years (45%) to have significant amounts of antibody. The probability of finding serum with an antibody titer above 20 fmol/10 μl of serum is better in males above 66 years of age than in younger males or females at all ages.

Clinical Disorders

Recent studies (Doar and Fabre, 1981; Stefansson et al., 1985) suggest that autoimmune antibodies may appear not only in the presence of pathologic conditions, but also in healthy individuals, although normally at low titers. In the study shown in Table 4, samples of female serum that showed insignificant antibody titers included those from 57 patients with hirsutism. The individual with the highest antibody titer (14 fmol/10 μl of serum) among females was a heart transplant (female donor) recipient (aged 49). Among young males, the highest titer (16 fmol/10 μl of serum) was that of a kidney transplant patient (aged 36). Six other patients with significant antibody titers were medically immunosuppressed. It is intriguing to suggest that organ transplants, medication, and other clinical conditions that can cause suppression of the immune responses to external antigens are related to the shift in the immune responses and to the increased emergence of antibodies to certain autoantigens.

The clinical characteristics of 22 patients having moderate (3–20 fmol/10 μl) or high (>20 fmol/10 μl) autoantibodies to androgen receptor in serum were as follows (Vogelzang et al., 1986):

1. Their median age was 70 years
2. Eight of the patients had prostate cancer

3. Six patients were immunosuppressed; among them, five had other malignancies, and two had benign prostatic hypertrophy
4. The highest titer (163 and 137 fmol/10 μl serum) was associated with hormone-resistant prostatic cancer.

An analysis of 41 prostatic cancer patients showed no apparent relationship between the antibody levels and stage of the disease. In addition, 22 patients with stage D_2 disease exhibited no apparent relationship between antibody levels and hormone dependence or exposure to chemotherapeutic agents. Other patients who had elevated antibody levels included a pregnant woman, three of six patients with germ cell tumors, and one cardiac and one renal transplantation patient. Further evaluation is required to establish whether any distinct clinical syndromes are associated with the presence of high-titer antibodies to androgen receptor in serum samples.

Acknowledgments. Studies in the laboratories of S. Liao and A.K. Roy are supported by National Institutes of Health grants AM-09461, DK-37694(SL), and AM-14744(AKR).

References

Al-Nuami N, Davies P, Griffiths K (1979) Cancer Treat Rep 63:1147
Applanat MP, Logeat F, Picard MTG, Milgrom E (1985) Endocrinology 116:1473–1484
Birnbaumer ME, Schrader WT, O'Malley BW (1983) J Biol Chem 258:7331–7337
Borgna JL, Fauque J, Rochefort H (1984) Biochemistry 23:2162–2168
Carltedt-Duke J, Okret J, Okret S, Wrange O, Gustafsson JA (1982) Proc Natl Acad Sci USA 79:4260–4264
Chen C, Schilling K, Haiipakka RA, Huang IY, Liao S, (1982) J Biol Chem 257:116–121
Coffer AI, King RJB, Brockas AK (1980) Biochem Intl x1:126–132
Doar AK, Fabre JW (1981) Clin Exp Immunol 45:37–47
Edwards DP, Weigel NL, Schrader WT, O'Malley BW, McGuire WL (1984) Biochemistry 23:4427–4435
Eisen HJ (1980) Proc Natl Acad Sci USA 77:3893–3897
Feil PD (1983) Endocrinology 112:396–398
Fox TO (1978) Proc Natl Acad Sci USA 75:2664–2668
Gasc JM, Renoir JH, Radanyi C, Joab I, Tuohimaa P, Baulieu EE (1984) J Cell Biol 99:1193–1201
Gorski J, Toft D, Shyamala G, Smith D, Notides A (1968) Recent Prog Horm Res 24:45–80
Gorski J, Welshons W, Sakai D (1984) Mol Cell Endocrinol 36:11–15
Goussard J, Leghevrel C, Martin PM, Roussel G (1985) Bull Cancer 72:168–182
Govindan MV (1979) J Steroid Biochem 11:323–332
Govindan MV, Sekeris CE (1978) Eur J Biochem 89:95–104
Grandics P, Gasser DL, Litwack G (1982) endocrinology 111:1731–1733

Greene GL (1983) In: Roy AK, Clark JH (eds) Gene Regulation by Steroid Hormones II. Springer-Verlag, New pp 191–199
Greene GL, Closs LE, DeSombre ER, Jensen EV (1977) Proc Natl Acad Sci USA 74:3681–3685
Greene GL, Closs LE, DeSombre ER, Jensen EV (1979) J Steroid Biochem 11:333–341
Greene GL, Closs LE, DeSombre ER, Jensen EV (1980a) J Steroid Biochem 12:159–167
Greene GL, Fitch FW, Jensen EV (1980b) Proc Natl Acad Sci USA 77:157–161
Greene GL, Nolan C, Engler JP, Jensen EV (1980c) Proc Natl Acad Sci USA 77:5115–5119
Greene GL, Jensen EV (1982) J Steroid Biochem 16:353–359
Greene GL, Gilna P, Waterfield M, Baker A, Hort Y, Shine J (1986) Science 231:1150–1154
Hanson V, Weddington SC, French FS, McLean W, Smith A, Nayfeh SN, Ritzen EM, Hagenas L (1976) J Reprod Fertil Suppl 24:17–33
Harmon JM, Eisen HJ, Brower ST, Simons SS, Langley CL, Thomason EB (1984) Cancer Res 44:4540–4547
Holt JA, Lorincz MA, Greene GL, Herbst AL (1985) Gynecol Concol 20:270
Jensen EV, DeSombre ER (1972) Annu Rev Biochem 41:203–230
Jensen EV, Greene GL (1980) In: Mahesh VB, Muldoon TG (eds) hormone receptors in reproduction. Elsevier Science, New York, pp 317–333
Jensen EV, Greene GL, Closs LE, DeSombre ER, Nadri M (1982) Recent Prog Horm Res 38:1–40
Joab I, Radanyi C, Renoir JM, Buchow T, Catelli MG, Binart N, Mester J, Baulein EE (1984) Nature (London) 308:850–853
King WJ, Greene GL (1984) Nature (London) 307:745–747
Kuhn RW, Schrader WT, Smith RG, O'Malley BW (1975) J Biol Chem 250:4220–4228
Liao S (1975) Int Rev Cytol 41:87–172
Liao S, Liang T, Tymoczko JL (1973) Nature (London) 41:211–213
Liao S, Smythe S, Tymoczko JL, Rossini GP, Chen C, Hiipakka RA (1980) J Biol Chem 255:5545–5551
Liao S, Tymoczko JL, Castaneda E, Liang T (1975) Vitam Horm NY 33:297–317
Liao S, Witte D (1985) Proc Natl Acad Sci USA 82:8345–8348
Liao S, Witte D, Schilling K, Chang C (1984) J Steroid Biochem 20:11–17
Livingston DM (1974) Methods Enzymol 34:723–731
Logeat F, Pamphile R, Loorefelt H, Jolivet A, Fournier A, Milgrom E (1985) Biochemistry 24:1029–1035
Logeat F, VuHai MT, Milgrom E (1981) Proc Natl Acad Sci USA 78:1426–1430
Logeat F, VuHai MT, Fournier A, Legrain P, Buttin G, Milgrom E (1983) Proc Natl Acad Sci USA 80:6456–6459
Loosefelt H, Logeat F, VuHai MT, Milgrom E (1984) J Biol Chem 259:144196–14202
Lorincz MA, Holt JA, Greene GL (1985) J Clin Endocrinol Metab 61:412–417
McCarty KS, Miller LS, Cox EB, Kourath J (1985) Arch Pathol Lab Med 109:716–721
McClellan MC, West NB, Tacha DE, Greene GL, Brenner RM (1984) Endocrinology 114:2002–2014

Merrill CR, Goldman D, Sedman SA, Ebert MH (1981) Science 211:1437–1438
Moncharmont B, Anderson WL, Rosenberg B, Parikh I (1984) Biochemistry 23:3907–3912
Mudarris A, Peck EJ (1985) Clin Chem 31:1009
Muldoon TG (1980) Endocrine Rev 1:339–364
Nakao K, Myers JE, Faber LE (1985) Can J Biochem Cell Biol 63:33–40
Okret S, Duke J, Wrange O, Carlson K, Gustafsson AJ (1981) Biochem Biophys Acta 677:205–219
Okret S, Wilstrom AC, Wrange O, Anderson B, Gustafsson JA (1984) Proc Natl Acad Sci USA 81:1609–1613
O'Malley BW, Means AR (1974) Science 183:610–620
O'Malley BW, Schrader WT (1976) Sci Am 234:32–43
Pertschuk LP, Eisenberg KB, Carter AG, Feldman JG (1985) Cancer (Phila) 55:1513–1518
Pike JW, Donaldson CA, Mariow SL, Haussler MR (1982) Proc Natl Acad Sci USA 79:7719–7723
Pike JW, Haussler MR (1979) Proc Natl Acad Sci USA 76:5485–5489
Pike JW, Mariow SL, Donaldson CA, Haussler MR (1983) J Biol Chem 258:1289–1296
Press MF, Greene GL (1984) Lab Invest 50:480–486
Press MF, Goebl NAN, Greene GL (1985) J Histochem Cytochem 33:915–924
Raam S, Meneth E, Tamura H, O'Brian DS, Cohen JL (1981) Mol Immunol 18:143–156
Radanyi C, Joab I, Renoir JM, Foy HR, Baulein EE (1983) Proc Natl Acad Sci USA 80:2854–2858
Radanyi C, Redemith G, Eigenmann E, Lebeau MC, Massol N, Secco C, Balieu EE, Richard-Foy H (1979) CR Acad Sci Paris Series D. 288:255–258
Renoir JM, Radanyi C, Yang CR, Baulein EE (1982) Eur J Biochem 127:81–86
Ringold GM (1985) Ann Rev Pharmacol Toxicol 25:529
Rosner W, Smith RN (1975) Biochemistry 14:4813–4820
Schilling K, Liao S (1984) Prostate 5:581–588
Schmidt TJ, Litwack G (1982) Physiol Rev 62:1131–1192
Schrader WT, Birnbaumer ME, Hughes MR, Weigel NL, Gordy WW, O'Malley BW (1981) Recent Prog Horm Res 37:583–633
Schrader WT, O'Malley BW (1972) J Biol Chem 247:2401–2407
Shimada A, Kimura S, Abe K, Nagasaki K, Adachi I, Yamaguchi K, Suzuki M, Nakazima T, Miller LS (1985) Proc Natl Acad Sci USA 82:4803–4807
Stefansson K, Marton LS, Dieperink ME, Molnar GK, Schlaepfer WW, Helgason CM (1985) Science 228:1117–1119
Stevens J, Eisen HJ, Stevens YW, Haubenstock H, Rosenthal RL, Artishevsky A (1981) Cancer Res 41:134–137
Sullivan WP, Vroman BT, Bauer VJ, Puri RK, Riehl RH, Pearson GR, Toft DO (1985) Biochemistry 24:4214–4222
Tuohimaa P, Renoir JM, Radyani C, Hester J, Joab I, Buchou T, Baulien EE (1984) Biochem Biophys Res Commun 119:433–439
Vedeckis WV (1983) Biochemistry 22:1975–1983
Vedeckis WV, Schrader WT, O'Malley BW (1978) Biochem Action Horm 5:321–372
Vogelzang N, Chodak G, Witte D, Liao S (1986) Am Soc Clin Oncol Proc Abstract

Walter P, Greene S, Greene GL, Krust A, Bornert JA, Jeltsch JM, Stanb A, Jensen EV, Scrace G, Waterfield M, Chambon P (1985) Proc Natl Acad Sci USA 82:7889–7893
Weigel NL, Pousette A, Schrader WT, O'Malley BW (1981) Biochemistry 20:6798–6803
Weinberger C, Hollenberg SM, Ong ES, Harmon JM, Brower ST, Cidlowski J, Thompson EB, Rosenfeld MG, Evans RM (1985) Science 228:740–742
Welshons WV, Liberman ME, Gorski J (1984) Nature 307:747–749
Westpahl HM, Moldenhaver G, Beaton M (1982) EMBO J 1:1467–1471
Wilson EM, French FS (1979) J Biol Chem 254:6310–6319

Disscussion of the Paper Presented by S. Liao

SCHRADER: In the case of human sera, or those individuals suffering from various forms of likely autoimmune disease, have you looked to see whether or not there is a presence of these spontaneous antibodies that might correlate with that?
LIAO: We haven't but we will pay attention to any relationship between the autoimmune antiandrogen receptor antibodies and diseases.
SCHRADER: Are you trying to get monoclonal antibodies by such a route.
LIAO: Yes. That is what we're interested in.
SCHRADER: One likely class of patients that might be worth screening would be those suffering from idiopathic thrombocytopenic purpura syndrome because one of the treatments for that is a spleenectomy.
ROY: Are you willing to provide some more information about those antibodies? What kind of immunoglobulins are they, and do they cross-react with any other receptors?
LIAO: The one we studied most was a serum sample from a prostate cancer patient. By electrophoresis and purification methods, we found the antibody in the IgG fraction. The serum from another prostate cancer patient, however, had the antiandrogen receptor antibodies in the IgM fraction.
ROY: It couldn't react with any other steroid receptors?
LIAO: No, not with estrogen or progestin receptor of rat uterus or cultured human cells, nor with glucocorticoid receptors of rat liver.

Discussants: S. LIAO, A.K. ROY, AND W. SCHRADER

Chapter 10

Isolation and Characterization of cDNA Probes for Human CBG and Rat ABP

G.L. HAMMOND, J. REVENTOS, N.A. MUSTO, G.L. GUNSALUS, AND C.W. BARDIN

Human plasma corticosteroid-binding globulin (CBG) and rat testicular androgen-binding protein (ABP) belong to a group of high-affinity steroid-binding proteins that are traditionally thought of as "extracellular" steroid transport proteins. Both proteins have been purified to apparent homogeneity, specific antibodies have been raised against them, and their physicochemical characteristics and physiological functions have been the subject of numerous articles.*

In human blood, CBG is the major transport protein for cortisol. It has a single steroid-binding site that also binds corticosterone and progesterone with relatively high affinity (Rosner, 1972). When analyzed either in its native state by gel filtration or under denaturing conditions by sodium dodecyl-sulfate-polyacrylamide gel electrophoresis (SDS-PAGE), CBG exhibits a relative molecular mass (M_r) of ~60K (Robinson et al., 1985). However, this value is probably influenced by its high carbohydrate content (~23%) and the M_r of the peptide is thought to be closer to 42K (Robinson et al., 1985). Information about the amino acid sequence of CBG is limited to the first eight residues from its amino terminus (Rosner, 1972; Fernlund and Laurell, 1981), and information about the molecular composition of its steroid-binding domain is limited (Defaye et al., 1980).

Immunochemical analyses of human CBG (hCBG) and CBG from a variety of primates and other vertebrate species have indicated that antibodies raised against hCBG cross-react only with CBG in the blood of Old World apes and monkeys (Robinson et al., 1985). Failure to detect any cross-reactivity with CBG from New World primates may be due to either a total lack of CBG in these species or to very low serum levels of a protein that must be immunochemically different to CBG in Old World primates.

---

*Rosner, 1972; Rosner et al., 1983; Robinson et al., 1985; Fernlund and Laurell, 1981; Defaye et al., 1980; Musto et al., 1980; Gunsalus et al., 1978; Brien, 1981; Tindal and Means, 1980.

The liver is thought to be the site of CBG synthesis. This originates from early evidence that a partial hepatectomy decreases plasma CBG-binding capacity in rats (Gala and Westphal, 1966), and from the observation that low plasma concentrations of CBG are found in patients with liver diseases (Doe et al., 1964). More direct evidence that CBG is produced by the liver has since been obtained from experiments in which a translation product of mRNA extracts from rat and guinea-pig livers can be immunoprecipitated with the relevant anti-CBG antiserum. These immunoreactive products exhibit electrophoretic mobilities similar to the CBGs isolated from the blood of these species (Wolf et al., 1981; Perrot-Applanat and Milgrom, 1979). In addition, a hepatoma cell-line (HepG2) has also been found to secrete a protein that binds cortisol and shares the immunochemical characteristics of CBG in human plasma (Khan et al., 1984). There is, however, no evidence that the liver is the only site of CBG synthesis in the human or any other species.

Estrogens increase the plasma concentrations of CBG, and their potency in this regard appears to be proportional to the affinity of their interaction with the estrogen receptor (Moore et al., 1978). Although this suggests that CBG synthesis is increased by estrogen-receptor-mediated events at the genomic level, an alteration in the clearance of the protein cannot be excluded as an explanation for an increase in plasma CBG concentrations. Apart from estrogens, the only other compound known to influence plasma CBG concentrations in humans is vitamin B_{12} (Doe et al., 1982). Adequate reserves of cobalamin appear to be necessary for the maintenance of normal plasma concentrations of CBG, but the mechanisms involved remain, as in the case of estrogens, at present speculative. The only way these questions can be answered directly is to quantify the hormonal regulation of CBG mRNA production and/or translational capacity.

It is generally assumed that CBG functions exclusively as a plasma steroid transport protein that reduces the metabolic clearance rate of a number of steroid hormones and thereby prolongs their biological half-life (Siiteri et al., 1982). This is particularly important in the case of cortisol because its binding affinity to albumin is very low and, consequently, more than 90% of the cortisol in blood plasma is bound to CBG (Siiteri et al., 1982). The concept that CBG-bound steroids are "biologically inactive" in plasma stems from the demonstration that CBG suppresses cortisol-induced liver glycogen maintenance in adrenalectomized mice (Slaunwhite et al., 1962). However, recent evidence suggests that CBG may interact with the plasma membranes of a number of tissues (Hryb et al., 1985) and has even been localized in specific cell types that are known to be sites of glucocorticoid action (Siiteri et al., 1982; Perrot-Applanat et al., 1984; Werthamer et al., 1973). It has therefore been proposed that the protein may actively promote the cellular accumulation of specific steroids or modulate their action in target cells in some way (Siiteri et al., 1982). The testis and epididymis of the rat and many other species, including man, contain an ABP that is distinct from the androgen receptor (Cheng et al., 1983; Cheng and Musto, 1982; Taylor et al., 1980; Hsu and Troln, 1978). In all species studied, ABP appears to be a glycoprotein composed of a heterogeneous mixture of two protomers that may be separated on the basis of their M_r when analysed by SDS-PAGE, and these are referred to as the heavy (50K) and light (48K) protomers (Musto et al., 1980; Cheng et al., 1983; Cheng and Musto, 1982). The native dimeric form of ABP also exhibits charge heterogeneity when analyzed by isoelectrofocusing (Cheng et al., 1983), and it may be separated into two forms on the basis of Concanavalin A binding ability (Cheng et al., 1983; Cheng and Musto, 1982; Hsu and Troen, 1978). The amino acid sequence of testicular ABP in the rat or any other species has not been determined, and it is not known how the two protomers combine to form a single steroid-binding site. There is also no information about the molecular composition of the steroid-binding site but it may be affinity-labeled with $[^3H]$-Δ^6-testosterone (Taylor et al., 1980).

In man, testicular ABP exists in two forms: one of which is physicochemically similar to testosterone-estradiol-binding globulin (TeBG) and the other is distinct from this serum protein. Both forms are immunologically almost indistinguishable from TeBG (Cheng et al., 1984). This latter protein, like CBG, is assumed to be produced by the liver. In the rat, ABP appears to be produced exclusively by Sertoli cells (Gunsalus et al., 1980), and there is apparently no rat equivalent of the plasma TeBG in adult rats. There is, however, some secretion of ABP from testis into the blood and this is maximal prior to maturation of the blood-testis barrier (Gunsalus et al., 1980). Sertoli cell production and secretion of rat ABP (rABP) into the seminiferous tubules is stimulated by follicle-stimulating hormone (FSH) (Tindal and Means, 1980; Gunsalus et al., 1980), and its function within the male reproductive tract is believed to be associated with maintaining a high local androgenic environment for sperm maturation between the germ cells and the epididymis. There is also evidence that rABP is internalized by the epithelial cells of the caput epididymis (Pelliniemi et al., 1981), and this process may involve binding to specific membrane receptors similar to those recently described for TeBG in other cell types (Hryb et al., 1985; Strel'chyonok et al., 1984). The physiological significance of this is not well understood, but it is possible that ABP, like other "extracellular" steroid-binding proteins, can modulate steroid action or metabolism in target tissues.

Thus, despite this wealth of information, a number of important aspects of the structure, biosynthesis, and function of CBG and ABP remain obscure. This has prompted us to learn more about these molecules using specific cDNAs. In this report, we review our recent results on the preparation and use of these probes.

Screening the λgt_{11} Libraries

Details of the antisera against hCBG (Robinson et al., 1985) and rAPB (Gunsalus et al., 1978) employed to screen the libraries have been published. The protocol used to screen the libraries for plaque production of immunoreactive fusion proteins was essentially the same as that described by Young and Davies (1983), except that peroxidase-labeled protein A was used as the antibody detection system instead of [^{125}I]-labeled protein A.

The human liver and rat testis cDNA libraries were both screened at a density of ~3 × 10^4 pfu (plaque-forming units) per plate. Prior to use, antisera [diluted 1:200 in 10 m*M* Tris/150 m*M* NaCl (TBS) buffer, pH 8, and 20% fetal calf serum] were incubated overnight at room temperature with 100 μl plating bacteria (Y1090), and then centrifuged at 4,000 × *g* for 15 min. This treatment removes most of the anti-*Escherichia coli* antibodies present in the antisera without reducing the titer of the antibodies of interest. A single IPTG (isopropylthio-β-galactoside) impregnated nitrocellulose filter was used to blot the plaques during their growth phase. These were removed, blocked with TBS-containing 20% fetal calf serum, and then incubated with antisera for 2 h. The filters were washed in TBS, in TBS + 1% Nonidet P40, and finally in TBS. Protein A-peroxidase (1 mg/ml TBS), diluted 1:500 in TBS-10% fetal calf serum, was incubated with the filters for 30 min. After washing the filters again, they were stained by incubation with 50 ml of a 1:4 (vol/vol) mixture of 300 mg 4-chloro-1-napthol/100 ml methanol and TBS containing 100 μl hydrogen peroxide (30%) for 15–20 min at room temperature.

In the initial screen of the human liver cDNA library, a mixture of two different anti-hCBG antisera were used, and 51 plaques were found to give positive signals. These were isolated and, when the first 30 were rescreened individually with the two antisera, 10 of them produced a signal with both antisera. Two of these (CBG_1 and CBG_7) gave a more intense signal than the others, out of which a third clone (CBG_3) was selected as an example. This represents a relative abundancy of ~ 0.03–0.1% of the total cDNA population, if it is assumed that only one in six of the recombinant phage contains a cDNA in the correct orientation and reading frame for expression of a fusion protein.

The rat testis cDNA library was initially screened using an antiserum obtained from the National Hormone and Pituitary Program for the radioimmunoassay (RIA) of rat ABP, and this gave 21 positive signals. When the phage producing these signals were recloned and screened with an additional, highly specific antiserum, six clones were isolated on the basis that they gave positive signals with both antisera. Again one of these (ABP_4) gave a more intense signal than any of the others. In terms of relative abundance, it can be calculated that cDNAs for ABP represent ~0.01%–0.04% of the total population of cDNAs in the rat testis library.

Immunochemical Analyses of Fusion Protein Products

The identities of the recombinant λgt_{11} phage, isolated as described above, were confirmed by analysis of β-galactosidase fusion proteins using specific RIAs (Robinson et al., 1985; Gunsalus et al., 1978) or by the epitope selection-Western blot method (Weinberger et al., 1985).

The production of adequate amounts of fusion proteins for analysis by RIA was accomplished as follows: $\sim 1 \times 10^4$ pfu of individual λgt_{11} clones to be tested were used to infect (20 min at 37°C) plating bacteria (Y1090 strain). After the addition of 8 ml NZC top agar® (Gibco) and 10 m*M* IPTG, they were poured onto 150-mm culture plates containing a layer of 1.5% agar in LB, and incubated at 42°C for 3 h followed by an additional 3 h at 37°C. The bacterial lawn by this time was completely lysed, and the top layer of agar was scraped off and centifuged at 25,000 × *g* in a swing-out rotor for 30 min at 4°C. The supernatant containing fusion protein was then removed for analysis.

The β-galactosidase fusion proteins produced by some of these recombinant phage clones were analysed by RIAs for CBG and ABP, and the results are shown in Fig. 1. In the both cases the relative efficiency with which the fusion proteins displaced labeled CBG and ABP in the RIAs mimicked the intensity of the signals seen in the screening assays. The CBG-β-galactosidase fusion proteins produced by CBG_7 displaced more than 75% of the pure CBG used as the labeled ligand in the RIA. This indicates that this fusion protein contains a number of epitopes that are common to native CBG, and the displacement of labeled CBG observed is similar to the level of cross-reaction that has been found to exist between

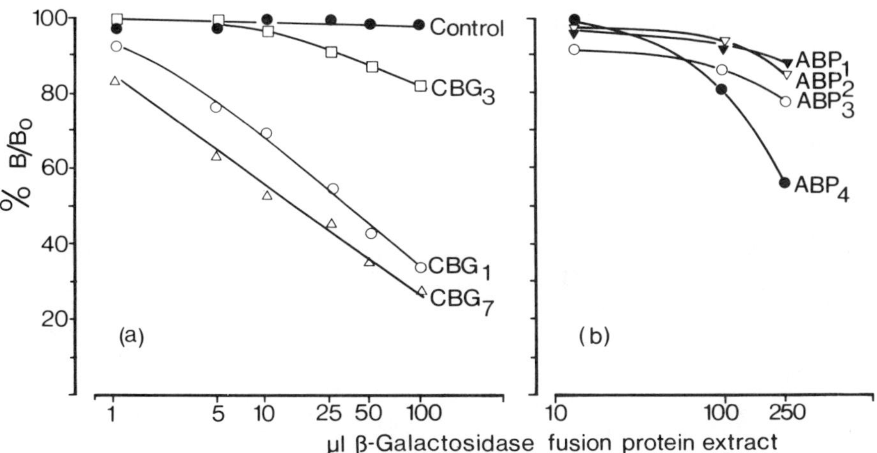

Fig. 1. Analysis of β-galactosidase fusion protein extracts produced by (**a**) human CBG-λgt_{11} and (**b**) rat ABP-λgt_{11} phage by RIA.

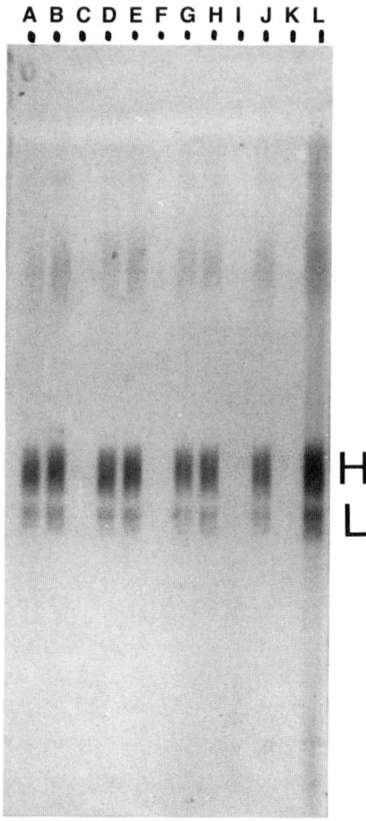

Fig. 2. Demonstration that the six rat ABP-λgt_{11} clones isolated immunochemically contain epitopes that will select antibody populations that recognize the heavy and light protomers of rat ABP on a Western blot of epididymal cytosol proteins. The blank lanes (C, F, I, and K) represent the negative control obtained using λgt_{11} without an ABP cDNA insert in the immunoselection procedure. The far right lane (L) is unselected antiserum, and lanes A and J are duplicates. The positions of the heavy (H) and light (L) protomers are indicated.

rhesus monkey CBG and human CBG (Robinson et al., 1985). The fusion protein produced by CBG_3 reacted relatively weakly in the RIA, despite the fact that the size of its cDNA insert (1.19 kb) is intermediate between that of CBG_1 (1.07 kb) and CBG_7 (1.43 kb).

In the ABP RIA, one particular clone (ABP_4) produced a fusion protein that displaced ~50% of the pure ABP used as the labeled ligand. Fusion proteins produced by the other 5 ABP clones gave little or no displacement of the labeled ABP (Fig. 1b). Therefore, to test further the identity of the ABP clones, each one of them was used to epitope-select antibodies from the antiserum originally used to isolate them from the rat testis cDNA

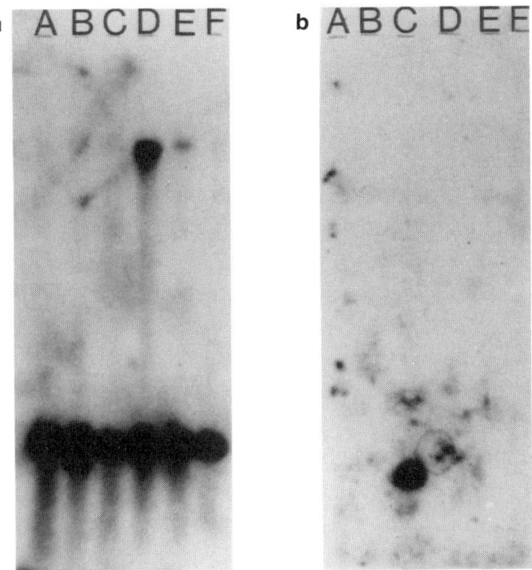

Fig. 3. Southern blot of *Eco*RI digests of the six rat ABP-λgt_{11} clones using the ^{32}P-labeled ~ 0.7 kb (**a**) and ~ 0.5 kb (**b**) fragments of $rABP_4$. The high molecular weight signal in one of the lanes (D) in panel (**a**) represents a partial cut and runs with the λgt_{11} arms at ~ 21 kb. The *Eco*RI digest of $rABP_4$ is in lane C of both panels.

library. When these epitope-selected antibodies were used to probe a Western blot of an epididymal cytosol, all six recombinant ABP phage were found to be capable of producing a fusion protein that selected antibodies against the characteristic heavy and light protomers of rat ABP (Musto et al., 1980), while unmodified λgt_{11} lacks the ability to do this (Fig. 2).

When the cDNA inserts from the ABP clones were excised with *Eco*RI and examined by agarose gel electrophoresis it was found that an insert of ~0.7 kb was common to all of them, but that ABP_4 contained an additional DNA fragment of ~0.5 kb. Both fragments were cloned into the *Eco*RI site of pBR322 and propagated on a large scale. They were then re-excised with *Eco*RI, purified on an 8% PAGE gel, and electroeluted for nick-translation with ^{32}p-labeled nucleotide. These labeled fragments were used to examine a Southern blot of *Eco*RI digests of all six phage-containing ABP inserts (Fig. 3). Only ABP_4 produced a signal with the ~0.5 kb fragment, while all six clones hybridized with the ~0.7 kb fragment. The latter indicates homology between all six ABP clones, and the possibility that the ~0.5 kb fragment of ABP_4 may have arisen as a result of the tandem insertion of an irrelevant cDNA was excluded by isolating a number of other ABP clones that contain inserts which hybridize with both *Eco*RI fragments of ABP_4.

Northern-Blot Analyses

Tissues used for the isolation of RNA were removed, frozen in liquid nitrogen, and stored at −70°C. Frozen pieces of tissue (~2 g) were pulverized in liquid nitrogen, homogenized in 3*M* LiCl/6*M* urea, and the RNA that precipitated was processed (Auffray and Rougeon, 1980) and subjected to two cycles of affinity chromatography on oligo-dT cellulose to obtain poly $(A)^+$ RNA (Aviv and Leder, 1972). Approximately 2 μg of poly$(A)^+$ RNA from each tissue was electrophoresed in a 1% agarose/formaldehyde gel (Rave et al., 1979), and the resolved mRNAs were transferred to GeneScreen Plus™ membranes (NEN) by capillary blotting (Thomas, 1980). The Northern blots were then baked (2 h at 80°C), prehybridized (6 h at 42°C), hybridized with ^{32}P-labeled cDNA in 50% formamide (16 h at 42°C), and washed according to the manufacturer's instructions for the use of GeneScreen Plus™. Autoradiography was performed using Lightening-Plus™ enhancing screens (Dupont) and XAR film (Kodak) at −70°C.

To study the tissue distribution of CBG mRNA, we nick-translated CBG_I cDNA and used it to probe a Northern blot of mRNA extracts of rhesus monkey liver, kidney, testis, and epididymis. It is known that there is some homology between human and rhesus monkey CBGs when they are

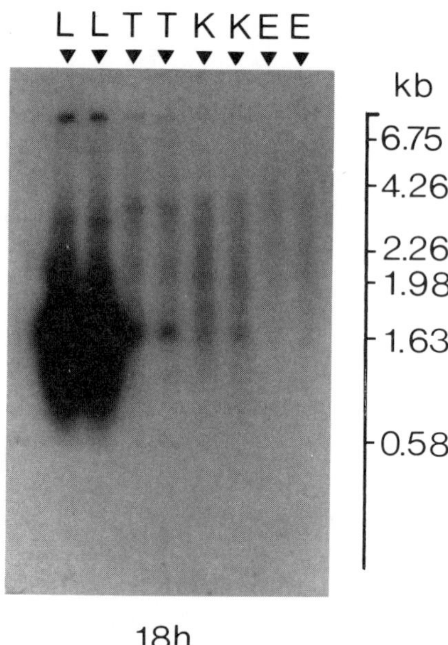

Fig. 4. Northern blot of mRNA extracted from monkey liver (L), kidney (K), testis (T), and epididymis (E) using ^{32}P-labeled CBG_I cDNA as the probe. approximately 2 μg of L, K, and T mRNA and 0.2 μg of E mRNA were loaded into the gel. The exposure time was 18 h at -70°C.

examined immunochemically (Robinson et al., 1985), and as expected, a strong hybridization signal was observed between the human CBG cDNA probe and a 1.65 kb mRNA species in rhesus monkey liver, even after only a 2-h autoradiographic exposure. The size of this mRNA is appropriate for a 42K protein, and it is evident that its relative abundance is far higher in the liver when compared with the other tissues studied. However, when the Northern blot was exposed for 72 h, signals could be detected in the kidney and testis and were in the same position as that associated with CBG mRNA in the liver extract (Fig. 4). This is the first direct evidence that CBG may be made in sites other than the liver, and its function in these tissues is therefore the focus of further investigation.

The tissue distribution and messenger unit size of ABP mRNA were also determined as described above for CBG mRNA. In this case poly $(A)^+$ RNA was prepared from cultured rat Sertoli cells and Northern blots were probed with labeled rat ABP cDNA. The hybridization signal also revealed a single band with a size of 1.65 kb. This is an appropriate size to code for a 45K protein, which is consistent with the subunit mass of ABP. Poly$(A)^+$ RNA was also prepared from cells cultured for 3 days in the presence of tesosterone, FSH, or a combination of these two hormones. Northern blots were hybridized with ^{32}P-labeled rat ABP cDNA, and the relative intensity of the hybridization signal was determined. These data revealed that both testosterone and FSH caused a modest (1.5-fold) increase in the signal but the combination produced a three-fold increase in ABP mRNA concentration. These data suggest that these hormones act to increase specifically the steady-state levels of ABP mRNA.

Acknowledgments. This work was supported in part by the National Institute of Health grant HD13541. The human liver and rat testis cDNA λgt_{11} libraries used in this study were generously provided by Dr. S.L.C. Woo and Dr. A.R. Means, respectively.

References

Auffray C, Rougeon F (1980) Eur J Biochem 107:303–314

Aviv H, Leder P (1972) Proc Natl Acad Sci USA 69:1408–1412

Brien TG (1981) Clin Endocrinol (Oxford) 14:193–212

Cheng CY, Frick J, Gunsalus GL, Musto NA, Bardin CW (1984) Endocrinology 114:1395–1401

Cheng CY, Musto NA, Gunsalus GL, Bardin CW (1983) J Steroid Biochem 19:1379–1389

Cheng S-L, Musto NA (1982) Biochemistry 21:2400–2405

Defaye G, Basset M, Monnier N, Chambaz EM (1980) Biochem Biophys Acta 623:280–294

Doe RP, Fernandez R, Seal US (1964) J Clin Endocrinol Metab 24:1029–1039

Doe RP, Lohrenz F, River G, Dosherholmen A, Roberts R (1982) J Clin Endocrinol Metab 54:381–385
Fernlund P, Laurell C-B (1981) J Steroid Biochem 14:545–552
Gala RR, Westphal U (1966) Endocrinology 79:67–76
Gunsalus GL, Musto NA, Bardin CW (1978) Science 200:65–66
Gunsalus GL, Musto NA, Bardin CW (1980) In: Steinberger A, Steinberger E (eds) Testicular Development, Structure and Function, Raven Press, New York, pp 291–297
Hryb DJ, Khan, MS, Rosner W (1985) Biochem Biophys Res Comm 128:432–440
Hryb DJ, Khan MS, Romas N, Rosner W (1985) Presented at the 67th Annual Meeting of the Endocrine Society, Baltimore, Abstract 244
Hsu A-F, Troen P (1978) J Clin Invest 61:1611–1619
Khan MS, Aden D, Rosner W (1984) J Steroid Biochem 20:677–678
Moore DE, Kawagoe S, Davajan V, Mishell DR, Nakamura RM (1978) Am J Obstet 130:475–481
Musto NA, Gunsalus GL, Bardin CW (1980) Biochemistry 19:2853–2860
Pelliniemi LJ, Dym M, Gunsalus GL, Musto NA, Bardin CW, Fawcett DW (1981) Endocrinology 108:925–931
Perrot-Applanat M, Milgrom E (1979) Biochemistry 18:5732–5737
Perrot-Applanat M, Racadot O, Milgrom E (1984) Endocrinology 115:559–569
Rave N, Crkvenjakov R, Boedtker H (1979) Nucl Acids Res 6:3559–3567
Robinson PA, Hawkey C, Hammond GL (1985) J Endocrinol 104:251–257
Robinson PA, Langley MS, Hammond GL (1985) J Endocrinol 104:259–267
Rosner W (1972) J Steroid Biochem 3:531–542
Rosner W, Polimeni S, Khan MS (1983) Clin Chem 29:1389–1391
Siiteri PK, Murai JT, Hammond GL, Nisker JA, Raymoure WJ, Kuhn RW (1982) Recent Prog Horm Res 38:457–510
Slaunwhite WR, Lockie GN, Back N, Sandberg AA (1962) Science 135:1062–1063
Strel'chyonok OA, Avvakumov GV, Survilo LI (1984) Biochim Biophys Acta 802:459–466
Taylor Jr CA, Smith HE, Danzo BJ (1980) Proc Natl Acad Sci USA 77:234–238
Thomas PS (1980) Proc Natl Acad Sci USA 77:5201–5205
Tindal DJ, Means AR (1980) In: Thomas JA, Singhal RL (eds) Advances in Sex Hormone Research, Vol 4, Urban & Schwarzenberg, Baltimore, pp 295–327
Weinberger C, Hollenberg SM, Ong ES, Harmon JM, Brower ST, Cidlowski J, Thomson EB, Rosenfeld MG, Evans RM (1985) Science: 228:740–742
Werthamer S, Samuels AJ, Amaral L (1973)J Biol Chem 248:6398–6407
Wolf G, Armstrong EG, Rosner W (1981) Endocrinology 108:805–811
Young RA, Davis RW (1983) Science 222:778–782

Discussion of the Paper Presented by G.L. Hammond

CLARK: Do you see any relationship between your failure to get CBG in New World monkeys and the steroid resistance state of New World monkeys, the very high levels of steroids in those monkeys?

HAMMOND: That's something that's on the cards. We've got a couple of New World monkey's livers, we need to do some Northern blot testing to see if they are producing any form of CBG mRNA.

CLARK: Of course you would predict the free steroid levels would be higher, because if they had no CBG it wouldn't be bound to CBG.
HAMMOND: That's right, and I think Siiteri has been reporting that for a while.
CLARK: Now you're finding that the CBG message was present in tissues other than the liver. Give us your thoughts on how that relates to Siiteri's and other people's thinking that CBG is an actively transported molecule.
HAMMOND: Well it's certainly something that has to be shown. The question of whether the immunocytochemical localization in tissues is a result of active transport or synthesis within that tissue may be resolved by the availability of this probe.

Discussants: J. CLARK AND G.L. HAMMOND

Chapter 11

Ornithine Decarboxylase mRNAs in Murine Kidney: Structure and Regulation by Androgens

O.A. Jänne, N.J. Hickok, P.J. Seppänen, K.K. Kontula, E. Melanitou, and C.W. Bardin

Introduction

Physiological and synthetic androgens bring about their actions in murine kidney via androgen receptor-mediated mechanisms in a fashion similar to that of other steroid hormones through their respective receptor proteins in a variety of target tissues. Unlike some other steroid-responsive tissues or cell lines, in which large amounts of a few proteins are induced (O'-Malley and Means, 1974; O'Malley et al., 1979; Tata and Smith, 1979), androgen action in mouse kidney involves induction of many proteins that are of low abundancy, with none predominating as a percentage of total protein concentration. For example, two-dimensional gel electrophoretic analysis of soluble renal proteins could not reveal any major differences in their profile in untreated and androgen-induced mice (Swank et al., 1978). These considerations, along with well-established mouse genetics, render the murine kidney an attractive model system to study steroid hormone action and, in particular, to investigate regulation of a number of gene products within a single cell type and by a single steroid hormone.

The best studied markers for androgen action in murine renal cells are β-glucuronidase (Swank et al., 1978; Bardin and Catterall, 1981; Palmer et al., 1983; Watson and Catterall, 1986) and ornithine decarboxylase (Pajunen et al., 1982; Kontula et al., 1984; Jänne et al., 1984). We recently reviewed several comparative aspects in the regulation of these two gene products and demonstrated that their androgenic control is different in terms of kinetics of induction, genetics, and androgen sensitivity (Catterall et al., 1986). In this chapter, we summarize our work on the regulation of ornithine decarboxylase gene(s) in murine kidney.

Characteristics of Ornithine Decarboxylase Protein

Ornithine decarboxylase (OrnDCase) catalyzes conversion of L-ornithine to putrescine and is the first and one of the rate-controlling enzymes in polyamine biosynthesis. It is a ubiquitous enzyme that is induced many fold in a number of tissues within a few hours after exposure to trophic stimuli, such as hormones, drugs, tissue regeneration, and growth factors (Tabor and Tabor, 1984; Jänne et al., 1978; Pegg and McCann, 1982). Under appropriate experimental conditions, these stimuli may elevate OrnDCase activity up to 1,000-fold; however, even when maximally stimulated, the enzyme protein concentration represents only a minute fraction of total cellular protein ranging from 0.0001%–0.04% of the soluble proteins (Tabor and Tabor, 1984; Pegg and McCann, 1982). For reasons not currently understood, murine kidney has a relatively high OrnDCase activity under physiological conditions, and it seems to be the richest eukaryotic source of this enzyme after androgen induction.

OrnDCase protein has been purified to apparent homogeneity from a variety of sources, including the murine kidney. The enzyme protein has a subunit molecular weight of about 50,000 (Fig. 1) and appears to be a dimeric molecule under physiological conditions. However, the individual 50,000-dalton subunits are catalytically active enzymes since they bind the active-site directed inhibitor, 2-difluoromethylornithine (Fig. 1). Independent of the method used for visualization of OrnDCase on two-dimensional gels (protein stain or fluorography after labeling with [^{3}H]2-difluoromethylornithine), the enzyme protein shows charge heterogeneity. Similar heterogeneity has been also observed for the renal enzyme in other mouse strains (Persson et al., 1984; Seely et al., 1985) and for OrnDCase from tissues of other species (Seely et al., 1985; Choi and Scheffler, 1983; McConlogue and Coffino, 1983; Mitchell and Wilson, 1983). The two (or

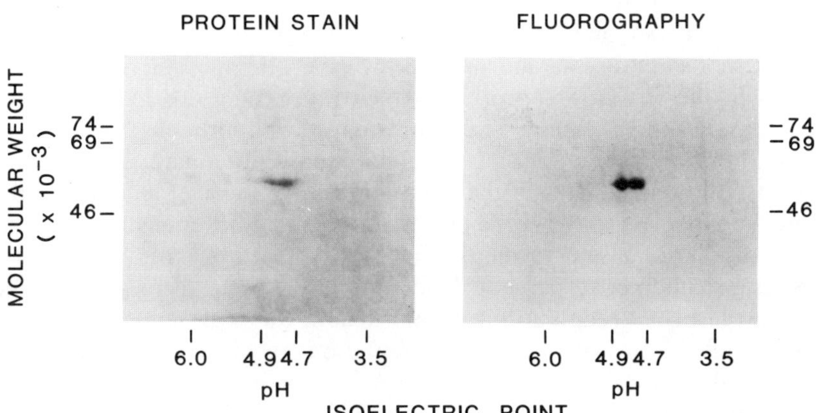

Fig. 1. Two-dimensional gel electrophoretic analysis of OrnDCase purified from kidneys of androgen-treated mice. The enzyme was detected by protein stain (left, 7 μg of enzyme protein) or by fluorography of the [^{3}H]2-difluoromethylornithine-labeled enzyme (right, 25,000 cpm of radioactivity) (Isomaa et al., 1983).

more) forms of the enzyme may represent polypeptides with different degrees of posttranslational modifications such as phosphorylation; however, these putative modifications have to maintain catalytic activity of the subunits since each form is capable of binding [^{3}H]2-difluoromethylornithine (Fig. 1). An alternative possibility is that these forms of OrnDCase are encoded by different mRNAs (Hickok et al., 1986), thus making their charge heterogeneity the consequence of somewhat dissimilar primary amino acid sequences (see below).

Striking characteristics of OrnDCase protein involve its very fast turnover in mammalian tissues and the rapid changes that occur in the amount of this enzyme after exposure to a variety of effectors in vivo (Tabor and Tabor, 1984; Jänne et al., 1978; Pegg and McCann, 1982). The turnover rate of mammalian OrnDCase is faster than that of any other mammalian enzyme; the half-life in mouse kidney is approximately 15 min (Isomaa et al., 1983; Seely and Pegg, 1983). Administration of testosterone at pharmacological doses for 5–7 days changes the turnover rate of OrnDCase; the half-life is prolonged from 10–15 min to 100–150 min. This change in the half-life applies to both the catalytically active enzyme and the immunoreactive OrnDCase protein (Isomaa et al., 1983; Seely and Pegg, 1983). The mechanisms responsible for slowing down the turnover rate by androgens in murine kidney are currently unknown, as is the physiological significance of this phenomenon.

The activity of OrnDCase in various tissues has been proposed to be regulated by a number of posttranslational mechanisms. These include covalent modifications of the enzyme *via* phosphorylation and transglutamination (Atmar and Kuehn, 1981; Russell, 1981) and the presence of macromolecular inhibitors and activators of OrnDCase in certain cells (Heller et al., 1976; Fong et al., 1976; Heller and Canellakis, 1981; Fujita et al., 1982; Kitani and Fujisawa, 1981). When we have studied androgenic control of OrnDCase in murine kidney, the enzyme activity is strictly related to the OrnDCase protein concentration (Fig. 2). In each experimental situation, including acute and long-term androgen exposures and inhibition of androgen action by a concomitant administration of antiandrogens, the changes in OrnDCase activity are always accompanied by parallel changes in the enzyme protein concentrations measured by a specific radioimmunoassay (Isomaa et al., 1983; Kantula et al., 1985). It thus appears that the above posttranslational control mechanisms play a minor, if any, role in the regulation of OrnDCase in the murine kidney.

A salient feature of OrnDCase regulation by virtually all stimuli is a short-lived response that typically peaks within 4–8 h and subsides by 24 h after the inducer treatment (Tabor and Tabor, 1984; Jänne et al., 1978; Pegg and McCann, 1982). This phenomenon may occur owing to several reasons, such as a short half-life of the inducer itself, rapid turnover rate of OrnDCase protein and/or OrnDCase mRNA, and the appearance of

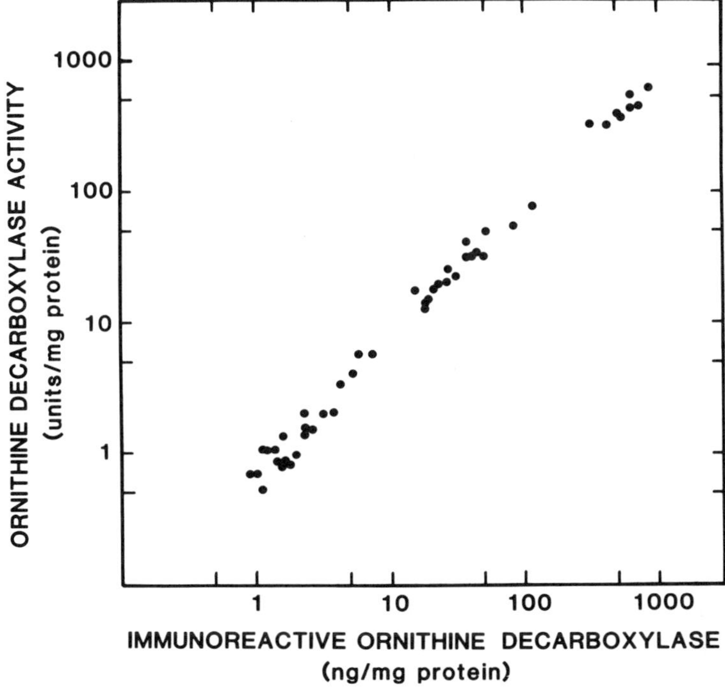

Fig. 2. Correlation between the catalytic activity of OrnDCase and the immunoreactive enzyme protein concentration. Soluble protein fractions from kidneys of intact female mice and of castrated, intact, and androgen-treated male mice were used in these studies. The correlation coefficient (r = 0.986) between the two values is highly significant (P < .001) (Isomma et al., 1983).

molecules that facilitate inactivation of OrnDCase after its maximal induction. However, several pieces of indirect evidence have suggested that polyamines themselves could be responsible for turning off the enzyme activation (see Heby and Jänne, 1981), and, indeed, recent studies by Kahana and Nathans have clearly indicated that putrescine and other polyamines negatively regulate the translation of OrnDCase mRNA, thereby controlling their own synthesis (Kahana and Nathans, 1985a). These latter findings also explain, at least in part, why androgenic induction of OrnDCase activity in murine kidney is much longer-lived than expected; in contrast to most other experimental systems, putrescine and other polyamines are excreted by the renal cells and do not accumulate in this tissue in quantities that reflect the extent of OrnDCase activation (Pajunen et al., 1982).

Structure of OrnDCase mRNAs

Cloning of OrnDCase cDNAs

Despite its 400- to 600-fold increase in response to androgen administration, OrnDCase is still a minor component in murine kidney and represents about 0.03%–0.05% of total soluble protein after maximal induction (Persson, 1981; Seely et al., 1982a; Seely et al., 1982b; Isomaa et al., 1983). To facilitate cloning of DNA sequences complementary to OrnDCase mRNA, we used a modification of the protein A-Sepharose immunoadsorbent technique described by Shapiro and Young (1981) to enrich OrnDCase mRNA sequences in preparations isolated from renal polysomes of androgen-treated mice (Kontula et al., 1984; Jänne et al., 1984). The final purity of the mRNA in different preparations ranged from 5%–20%, as judged by immunoprecipitation of peptides from cell-free translation reactions with monospecific OrnDCase antiserum. Synthesis of double-stranded cDNA (ds-cDNA) from the purified OrnDCase mRNA, insertion of the tailed ds-cDNA into the *Pst* I site of pBR322, and propagation of recombinant plasmids in *Escherichia coli* were performed using standard techniques (Kontula et al., 1984).

The cDNA clones that proved to be complementary to OrnDCase mRNA by a number of criteria (Kontula et al., 1984) fell into two groups with regard to their 3′-ends; one group was about 450 base pairs (bp) longer at the 3′-end than the other one (Jänne et al., 1984). Hybridization probes prepared from overlapping regions of plasmid DNAs in either group recognized two OrnDCase mRNA species on Northern blot analysis, with molecular sizes of 2.2 and 2.7 kilobases (kb). These two mRNAs are constitutively expressed in kidneys of female mice and their accumulation is regulated by androgens (see below).

Size Heterogeneity of OrnDCase mRNAs

Two representative OrnDCase cDNAs with dissimilar 3′-termini (e.g., pODC16 and pODC74) were reasoned to represent cDNAs originating from the two OrnDCase mRNA species. To study this possibility, a number of fragments were isolated from these plasmids carrying OrnDCase cDNA sequences and used as hybridization probes in Northern blot analyses (Hickok et al., 1986). As illustrated in Fig. 3, the most 3′ sequence of pODC74 (*Pst* I/*Pst* I fragment) hybridizes only to the 2.7-kb OrnDCase mRNA, whereas fragments isolated from overlapping regions of pODC16 and pODC74 recognize both 2.2- and 2.7-kb mRNAs. DNA sequencing proved directly that pODC16 and pODC74 correspond to the 2.2- and 2.7-kb OrnDCase mRNAs, respectively. The longer mRNA contains two polyadenylation signals (AAUAAA); the second one is 422 nucleotides downstream from the first one, resulting in a 3′-noncoding region of 748

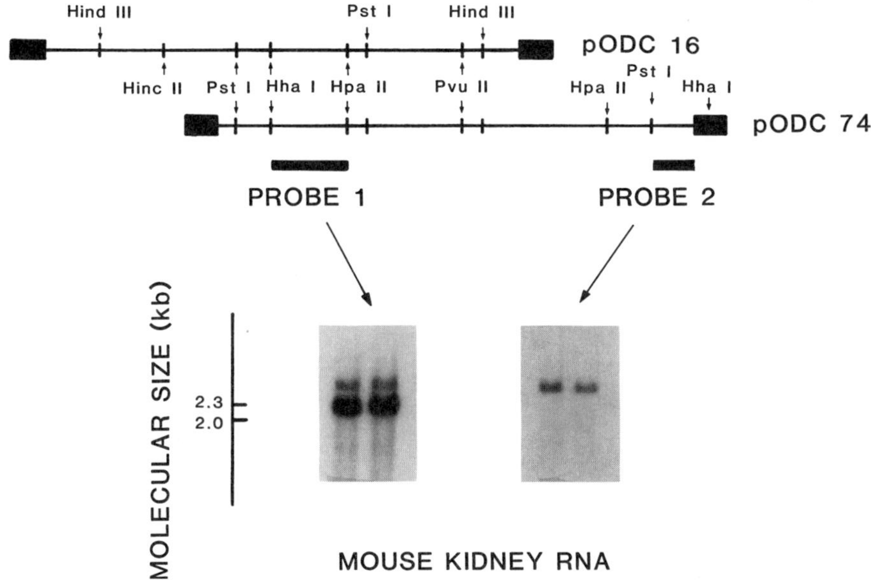

Fig. 3. Northern blot analysis of OrnDCase mRNAs with specific hybridization probes. Poly(A)-containing RNA was isolated from kidneys of testosterone-treated NCS mice, duplicate samples were subjected to electrophoresis on a 1% agarose gel in the presence of 2.2 *M* formaldehyde and transferred to nitrocellulose. Indicated fragments were cleaved from plasmids pODC16 and pODC74 carrying sequences complementary to OrnDCase mRNA, isolated by polyacrylamide gel electrophoresis, nick-translated, and used as hybridization probes (Hickok et al., 1986).

nucleotides long in the 2.7-kb mRNA. The 2.2-kb OrnDCase mRNA has only one polyadenylylation/termination signal that is exactly at the same position as the first AAUAAA in the longer mRNA and results in a 329-nucleotide long 3′-noncoding sequence for the 2.2-kb mRNA (Hickok et al., 1986).

In view of these data, it is obvious that the major reason for the size heterogeneity of the OrnDCase mRNAs in murine kidney is their dissimilar 3′-termini. However, since the cDNA clones used for these studies were not full-length copies of the corresponding mRNAs, we cannot rule out the possibility that some size heterogeneity also exists at the 5′-ends of OrnDCase mRNAs, as would be the case when different promoters are used. The presence of two OrnDCase mRNAs is not restricted to mouse kidney, but these mRNAs are also expressed in several other murine tissues and cell lines (McConlogue et al., 1984; Kahana and Nathans, 1984; Berger et al., 1984; Kahana and Nathans, 1985a; Gupta and Coffino, 1985), and in rat (our unpublished data, 1985) and hamster tissues (Pohjanpelto et al., 1985; Gilmour et al., 1985).

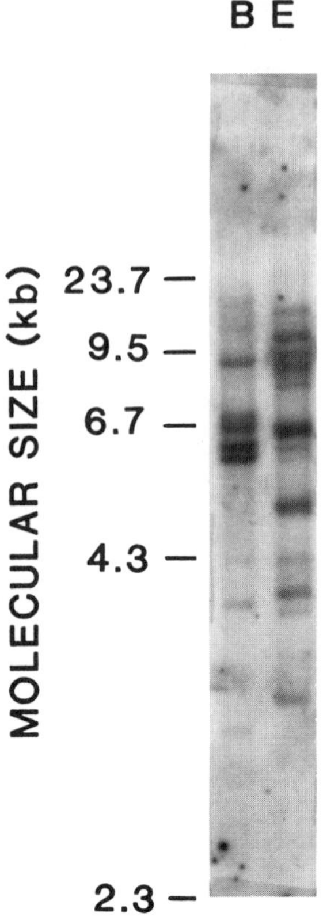

Fig. 4. Southern blot analysis of mouse liver DNA. Aliquots of high-molecular weight DNA (10 μg) were cleaved with restriction endonucleases *Eco*RI (E) or *Bam*HI (B), subjected to agarose gel electrophoresis and transferred to nitrocellulose. OrnDCase gene sequences were visualized by hybridization with nick-translated pODC16 DNA carrying sequences complementary to OrnDCase mRNA.

Although the alternative use of two different polyadenylylation/termination signals within a single OrnDCase gene could explain the presence of the two mRNA species, additional results suggest that more than one gene may be expressed. This is possible, since there are multiple OrnDCase genes in the murine genome (McConlogue et al., 1984; Berger et al., 1984; Kahana and Nathans, 1985b), as illustrated by the Southern blot of mouse DNA in Fig. 4. Comparison of the partial nucleotide sequences of the two cDNAs corresponding to the 2.2- and 2.7-kb OrnDCase mRNAs revealed five mismatches within their 759-bp overlap, suggesting that they originate from two very similar, yet different, OrnDCase genes (Hickok et al., 1986). It remains to be elucidated whether the two mRNAs present in murine kidney have any bearing on the charge heterogeneity observed for the enzyme protein isolated from the renal tissue of mice of the same randomly bred strain (Fig. 1). In addition to charge heterogeneity on two-dimensional gel electrophoresis, two separate forms of OrnDCase have been reported to be present in mouse kidney (Loeb et al., 1984). These enzyme forms differ in their heat sensitivity, half-life in vivo, and androgen-inducibility. It appears, however, that the 2.2- and 2.7-kb OrnDCase mRNAs are not directly related to these latter forms of the enzyme, since the relative amounts of the mRNAs do not correlate with those of the enzyme forms. Furthermore, both OrnDCase mRNA species are induced to about the same extent by androgens, in contrast to only the heat-labible enzyme being androgen-regulated (Loeb et al., 1984).

Regulation of OrnDCase mRNAs

Androgen Induction of the mRNAs

The clear sexual dimorphism in OrnDCase activity, immunoreactive enzyme protein concentration, and OrnDCase mRNA content in renal tissue of most mouse strains indicate that regulation of OrnDCase gene expression is elicited already at physiological androgen concentrations (Pajunen et al., 1982; Isomaa et al., 1983; Kontula et al., 1984; Jänne et al., 1984). As illustrated in Fig. 5, administration of testosterone at pharmacological doses for several days increases the accumulation of the two OrnDCase mRNA species to a similar extent; in this particular inbred strain (129/J), the increase over the intact male values achieved by the higher testosterone dose is about sevenfold for each mRNA species. Other studies have shown that depending on the mouse strain used, OrnDCase mRNA accumulation is increased 10- to 50-fold over the values in intact females or castrated males by a 5- to 7-day androgen treatment. Although the OrnDCase mRNA measurements have been only semiquantitative, it looks as if the increases in the enzyme activity and immunoreactive OrnDCase concentration (400- to 600-fold) are relatively greater than those in the mRNA accumulation. This notion is in agreement with the findings that prolonged androgen treatment stabilizes OrnDCase protein (Isomaa et al., 1983; Seely and Pegg, 1983) and that the rate of enzyme synthesis is increased about 25-fold by androgen administration in murine kidney (Persson et al., 1984).

In contrast to the wild-type animals, androgen-insensitive (Tfm/Y) mice show no change in OrnDCase gene expression during prolonged testosterone treatment, indicating that this regulation is dependent on functional androgen receptors (Kontula et al., 1984; Jänne et al., 1984). It is of interest to note that OrnDCase activities, immunoreactive enzyme concentrations, and OrnDCase mRNA levels are all very similar in kidneys of intact females, castrated males, and Tfm/Y animals (Kontula et al., 1984; Jänne et al., 1984). Therefore, they seem to represent constitutive values rather than levels regulated by low circulating androgen concentrations (females and castrated males) or by a low target cell receptor concentration (Tfm/

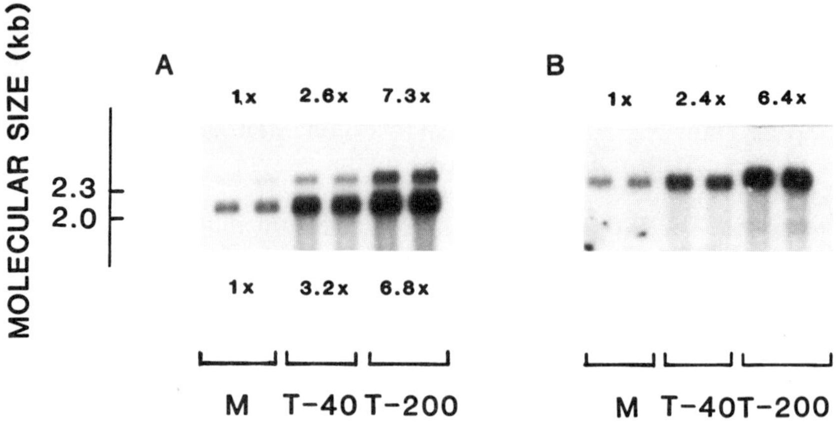

Fig. 5. Androgen-regulation of OrnDCase mRNA species. Poly(A)-containing RNA was isolated from kidneys of intact male mice (M) (129/J strain) and from female animals treated for 7 days with Silastic implants releasing either 40 μg (T-40) or 200 μg (T-200) testosterone per day. Duplicate samples of RNA (4 μg/lane) were fractioned on agarose gels and transferred to nitrocellulose. The most 5′ *Pst* I/*Pst* I fragment from pODC16 (see Fig. 3) was used as the hybridization probe for panel **A**, whereas the most 3′ *Pst* I/*Pst* I fragment from pODC74 (see Fig. 3) was the hybridization probe in panel **B**. The numbers refer to fold increases over the intact male values, as determined by radioactivity measurements. (Adapted from Hickok et al., 1986.)

Y animals). On the basis of available information from other hormone-regulated systems, there is no reason to assume a priori that constitutive and androgen-induced OrnDCase mRNAs would not be products of the same gene(s), although this has not yet been demonstrated directly.

A typical feature of androgenic regulation of OrnDCase activity is the rapid induction kinetics that requires de novo synthesis of the enzyme protein (Pajunen et al., 1982; Isomaa et al., 1983; Seely and Pegg, 1983; Persson et al., 1984). After a single dose of testosterone, the induction of OrnDCase mRNA very closely parallels to that of the enzyme protein (Fig. 6), with the initial increase in the mRNA concentration being detectable between 2 and 6 h after steroid administration (Isomaa et al., 1983; Kontula et al., 1984). These data thus indicate that the androgenic induction of OrnDCase activity results from facilitated accumulation of the mRNAs coding for this protein. The results do not, however, prove that androgens directly stimulate the rate of transcription of the OrnDCase gene(s). The possibility that accumulation of OrnDCase mRNAs would occur, at least, in part, through mechanisms not requiring an increased rate of OrnDCase gene transcription, has been inferred from some recent

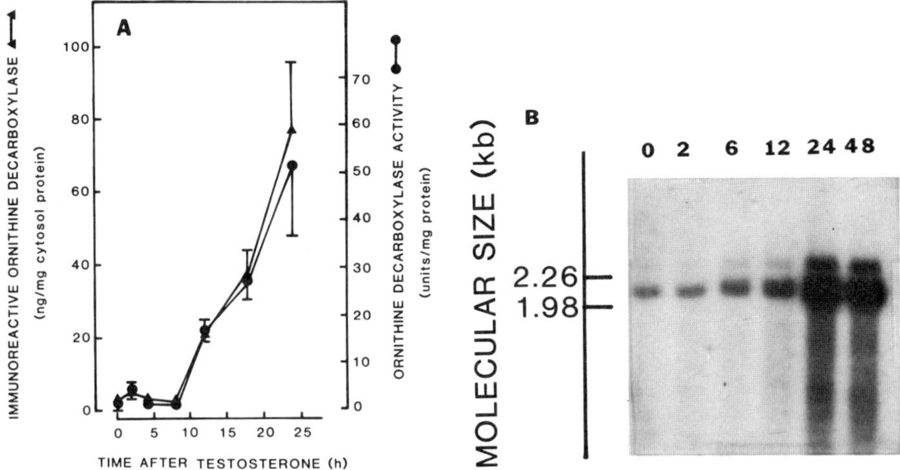

Fig. 6. Induction of OrnDCase activity and the enzyme protein content (**A**) and OrnDCase mRNA acccumulation (**B**) by a single dose of testosterone. Female mice were given intraperitoneally a 10-mg dose of testosterone and killed at indicated time intervals. (Data adapted from Isomaa et al., 1983, and Kontula etal., 1984.)

results. For example, nuclear run-on assays have indicated that there is not more than a two- to threefold increase in the number of RNA polymerase II molecules engaged in transcription of several androgen-regulated genes, including the OrnDCase gene(s), after testosterone administration (Berger et al., 1986, and our unpublished observations, 1985). In addition, concurrent administration of the protein synthesis inhibitor cycloheximide with testosterone decreases, but does not totally abolish, androgen-induced OrnDCase mRNA accumulation in murine kidney during the first 8 h of testosterone exposure (Jänne et al., 1984).

Genetic Variation in Androgen Regulation of OrnDCase Gene Expression

Previous studies have shown that in inbred strains of mice, a number of tissue functions and/or specific gene products are expressed in a genetically-regulated manner. Among the specific, hormone-regulated genes are those coding for renal β-glucuronidase (Palmer et al., 1983; Watson and Catterall, 1986; Catterall et al., 1986) and MK908 mRNAs (Berger et al., 1981; Elliott and Berger, 1983; King et al., 1986), submaxillary gland renin mRNAs (Field and Gross, 1985; Panthier and Rougeon, 1983), and hepatic major urinary protein mRNAs (Kuhn et al., 1984; Bishop et al., 1982;

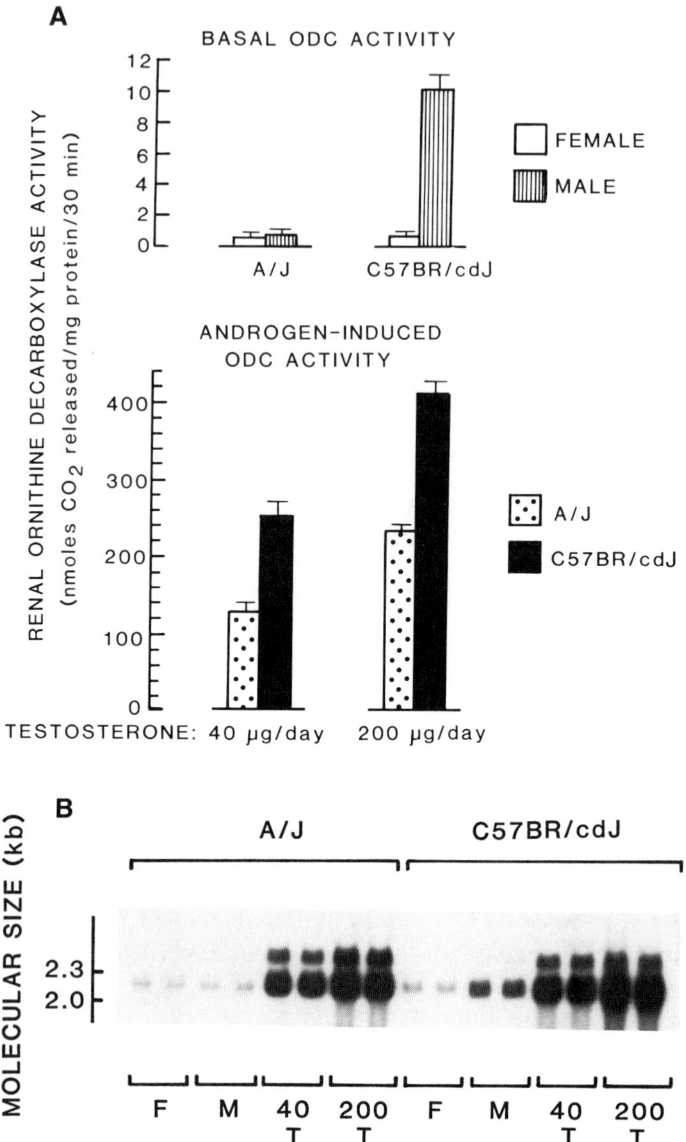

Fig. 7. Androgen regulation of OrnDCase activity (**A**) and OrnDCase mRNA accumulation (**B**) in two inbred strains of mice. The enzyme activity was measured in kidneys of A/J and C57BR/cdJ strains in the indicated groups. Testosterone treatment was accomplished by subcutaneous implants releasing 40 μg and 200 μg steroid/day. For Northern blot analysis, duplicate poly(A)-containing RNA samples (6 μg/lane) from pooled kidneys of 5–10 animals were fractioned on agarose gels and hybridized to nick-translated pODC16 DNA (see legend to Fig. 3). Testosterone (T) treatment was performed at indicated daily doses (in μg) for 7 days. F = female; M = male. (Adapted from Melanitou et al., 1986.)

Clark et al., 1984). Our recent studies have demonstrated that renal OrnDCase gene(s) can be included in this list of genes, the control of which is genetically regulated.

As mentioned above, two OrnDCase mRNA species with molecular sizes of 2.2 and 2.7 kb are constitutively expressed and androgen-regulated in murine kidney. The relative amounts of these mRNAs are not constant but vary among the eight inbred strains of mice studied, with the 2.7-kb mRNA representing 3.3%–19.2% and 2.9%–23.9% of total OrnDCase mRNA content in kidneys of females and males, respectively (Melanitou et al., 1986). This variation is apparently unrelated to the genetic regulation that governs the extent of androgen responsiveness of OrnDCase gene(s). Figure 7 illustrates data on two inbred strains of mice (A/J and C57BR/cdJ) that exemplify this latter genetic control in OrnDCase gene expression. The genetic regulation seems to be confined to the androgenic response, since no strain-dependent variation is detected in the constitutive expression of OrnDCase gene in intact female mice. However, renal enzyme concentrations in untreated male mice exhibited marked strain-dependent variation; for example, the values in C57BR/cdJ strain are 10-fold higher than those in A/J mice (Fig. 7). Similar genetic variation, although of somewhat lesser magnitude, is also seen in the OrnDCase mRNA content in kidneys of intact male mice; C57BR/cdJ males, among other highly responding strains, exhibit much higher mRNA levels than females, while poorly responding strains, such as the A/J, show barely higher male than female values (Fig. 7). In general, results from studies in which exogenous testosterone was administered to mice of different strains corroborated the findings in intact male and female animals (Melanitou et al., 1986).

The factors responsible for this genetic control are not known at present, but they do not seem to be directly related to circulating testosterone levels or concentrations of renal nuclear androgen receptors (Melanitou et al., 1986). We infer from these data that it is the OrnDCase gene(s) whose androgen regulation is under genetic control, rather than genetic regulation of androgen production and nuclear androgen receptor occupancy. The examples of other genes cited above have indicated that several mechanisms may account for the genetic differences, such as polymorphism in the structural gene, differential expression of two nonallelic structural genes, and presence of a *cis*-acting regulatory locus in close proximity to the structural gene. Similar to the major urinary protein gene (Kuhn et al., 1984; Bishop et al., 1982; Clark et al., 1984), the murine OrnDCase gene belongs to a multigene family, and, therefore, expression of different nonallelic structural genes could explain some of the above findings.

Inhibition of Androgen Action by Antiandrogens

Androgen regulation of OrnDCase gene expression in murine kidney appeared to be well suited to studies on the mechanisms by which steroidal and nonsteroidal antiandrogens interfere with androgen action for several reasons: first, induction of OrnDCase in this tissue is specific for androgenic steroids (Pajunen et al., 1982); second, physiological concentrations of androgens are capable of regulating OrnDCase gene activity; and third, induction of OrnDCase synthesis is dependent on functional androgen receptors (Pajunen et al., 1982; Kontula et al., 1984; Jänne et al., 1984). To date, most of our data have been obtained with a representative nonsteroidal antiandrogen, flutamide; however, no significant differences seem to exist between the mode of action of nonsteroidal and steroidal antiandrogens in murine kidney.

Flutamide is capable of increasing the concentration of nuclear androgen receptors in vivo, although to a lesser extent than testosterone (Kontula et al., 1985). This is illustrated by data in Fig. 8; at 1 h after a single dose of flutamide, nuclear androgen receptor concentration in murine kidney

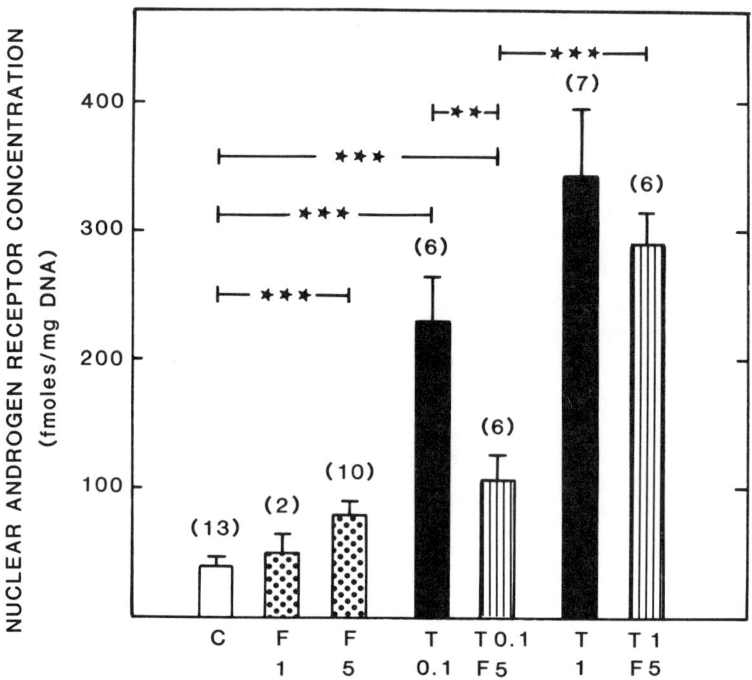

Fig. 8. Nuclear accumulation of androgen receptors in murine kidney after administration of testosterone alone, testosterone and flutamide together, or flutamide alone. The compounds were given intraperitoneally as indicated (C = control animals; T = testosterone; F = flutamide; the numbers refer to doses in milligrams), and the animals killed 1 h thereafter. Nuclear receptor concentration was measured as described by Isomaa et al., 1982. The data represent the mean ± SE for the number of experiments given in parenthesis. *Stars* indicate the statistical significance of the difference between the groups indicated (**$P < .01$; ***$P < .001$). (Kontula et al., © 1985, The Endocrine Society)

is significantly higher than in controls, but not as high as after testosterone administration. The antiandrogenic effect predominates when the flutamide-to-testosterone mass ratio is 50:1, but can be overcome by increasing the dose of testosterone (Fig. 8). The ability of flutamide to increase measurable nuclear receptor levels is not unique to the mouse kidney, as comparable data were achieved using the rat prostate as the androgen target tissue (Kontula et al., 1985). To investigate long-term effects of flutamide on androgen receptor dynamics and OrnDCase gene expression, female mice were implanted for a week with rods releasing a submaximal dose of testosterone (about 40 μg/d) alone, or in combination with flutamide-releasing implants at two different doses (release rates, 150 and 650 μg/day). Under these experimental conditions, flutamide alone is able to bring about a dose-dependent increase in the nuclear androgen receptor content and to almost the same extent as the submaximal dose of testosterone (Fig. 9).

Despite its ability to elevate nuclear androgen receptor concentrations, flutamide is completely devoid of agonistic actions, as it does not affect

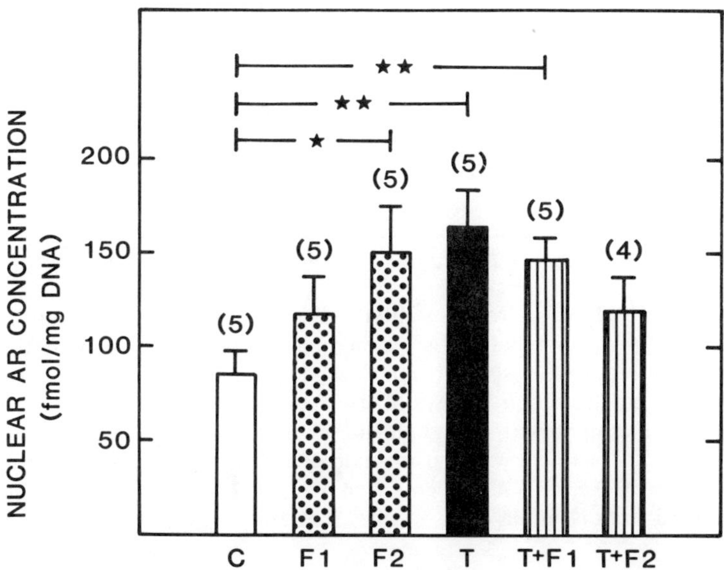

Fig. 9. Effect of long-term flutamide (F) and testosterone (T) administration, alone or in combination, on the nuclear androgen receptor (AR) accumulation in murine kidney. Female mice of NCS strain were implanted with F = releasing Silastic rods (release rates: F1, 150 μg/d; F2, 650 μg/d), T-releasing rods (40 μg/d), or their combinations at these doses (T+F1, T+F2) for 8 days. The data represent the mean ± SE for the number of experiments given in parenthesis. Symbols: *, $P < .05$; **, $P < .01$, as compared with the controls (C). (Kontula et al. © 1985, The Endocrine Society)

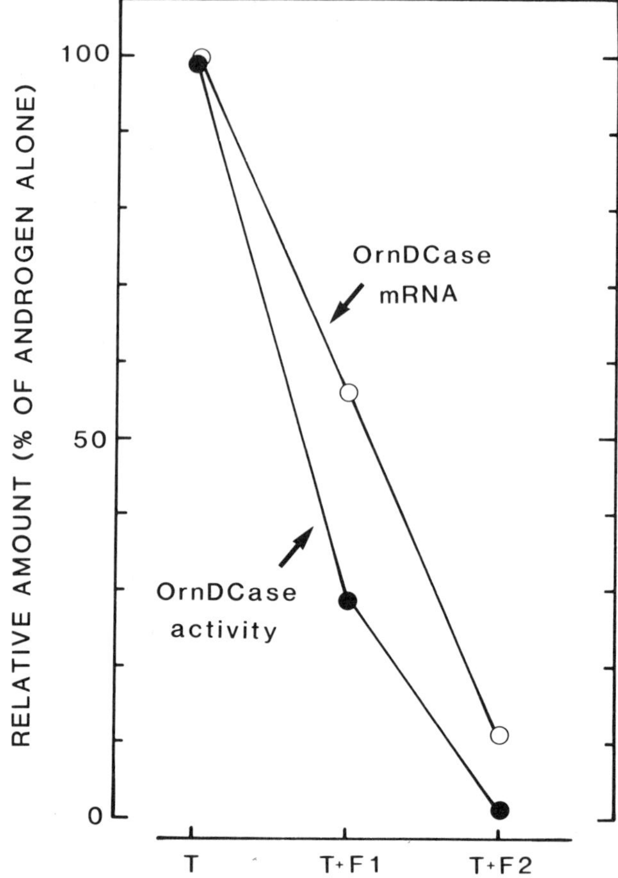

Fig. 10. Inhibition of androgen-elicited increases in OrnDCase activity and OrnDCase mRNA accumulation by a concomitant flutamide administration. Experimental conditions are described in the legend to Fig. 9. The values are expressed relative to those achieved in kidneys of female mice treated with testosterone (T) alone (100%). Flutamide (F) was administered at two different doses (F1, 150 μg/d; F2, 650 μg/d). Treatment with the antiandrogen alone did not increase OrnDCase activity or OrnDCase mRNA concentration over the values in intact female mice. (Data adapted from Kontula et al. © 1985, The Endocrine Society)

renal OrnDCase or OrnDCase mRNA concentrations (Kontula et al., 1985). These data imply that the androgen receptor-antiandrogen complexes are biologically inactive, possibly owing to their inability to interact with appropriate nuclear regulatory sites. Administration of flutamide concurrently with testosterone inhibits androgen-induced elevation in renal OrnDCase activity and OrnDCase mRNA accumulation in a parallel fashion (Fig. 10). Collectively, these results show that inhibition of the testosterone-elicited increase in the enzyme protein concentration by flutamide occurs mainly via decreased accumulation of OrnDCase mRNA, rather than via androgen-induced stabilization of the protein (see above). The mode of action of steroidal antiandrogens, such as cyproterone acetate, seems to be very similar to that of flutamide in terms of changes in nuclear androgen receptor concentrations and the inability of androgen receptor-cyproterone acetate complexes to facilitate OrnDCase activation and OrnDCase mRNA accumulation.

Summary and Conclusions

Regulation of OrnDCase gene expression in murine kidney appears to be a suitable experimental system to study androgen action. In addition to including the strict androgen specificity, the dependency of functional androgen receptors, and the regulation with physiological androgen concentrations, there are several interesting and intriguing features in the expression of this gene. First, OrnDCase belongs to a multigene family that comprises about 10 unique DNA sequences in the murine genome. Their structural characterization has not yet been completed; however, the studies with OrnDCase cDNAs suggest that more than one gene encoding this enzyme could be expressed in the renal tissue. Androgen responsiveness of OrnDCase shows genetic variation among inbred strains of mice, which does not seem to be directly related to circulating testosterone concentrations or occupancy of nuclear androgen receptors in murine kidney.

Two OrnDCase mRNA species are expressed and androgen-regulated in mouse renal cells. These mRNAs are size-heterogeneous at their 3′-termini in that the longer species (2.7 kb) has a 3′-noncoding region of 748 nucleotides as opposed to a 329-nucleotides long 3′-nonconding sequence in the 2.2-kb OrnDCase mRNA. The 2.7-kb mRNA has two polyadenylylation/termination signals and thus the alternative use of AAUAAA signals in a single OrnDCase gene could explain the presence of two mRNAs. However, sequence differences in the cDNAs corresponding to the 2.2- and 2.7-kb mRNAs suggest that they might be encoded by two different, but very similar, OrnDCase genes. Accumulation of OrnDCase mRNAs in response to androgen administration parallels the increase in enzyme activity and OrnDCase protein concentration. It is not known at present, however, the extent to which this accumulation can be ascribed to an increased rate of OrnDCase gene transcription as opposed to mRNA stabilization. Whatever is the mechanism, it is androgen receptor-dependent, as it does not occur in testicular feminized animals and can be abolished by a concomitant antiandrogen treatment.

Androgens regulate OrnDCase activity in a dual fashion: first, by increasing the rate of synthesis of the enzyme protein and second, by prolonging the biological half-life of this rapidly turning over protein. The first event appears to be due to an increase in the mRNA concentration, whereas the mechanism and physiological significance of the latter phenomenon are currently unknown. Posttranslational modifications of OrnDCase protein as regulators of the enzyme activity seem to play a minor role in the androgen-induced control of OrnDCase activity in murine kidney, since changes in the immunoreactive enzyme protein concentration always reflect those in the enzyme activity.

Acknowledgments. This work was supported by the National Institutes of Health (Grant No. HD-13541). The skillful secretarial assistance of Linda McKeiver and Jean Schweis is gratefully acknowledged.

References

Atmar VJ, Kuehn GD (1981) Proc Natl Acad Sci USA 78:5518–5522
Bardin CW, Catterall JF (1981) Science 211:1285–1294
Berger FG, Gross KW, Watson G (1981) J Biol Chem 256:7006–7013
Berger FG, Loose DS, Meisner HM, Watson G (1986) Biochemistry 25:1170–1175
Berger FG, Szymanski P, Read E, Watson G (1984) J Biol Chem 259:7941–7946
Bishop JO, Clark AJ, Clissold PM, Hainey S, Franke V (1982) EMBO J 1:615–620
Catterall JF, Kontula KK, Watson CS, Seppänen PJ, Funkenstein B, Melanitou E, Hickok NJ, Bardin CW, Jänne OA (1986) Recent Prog Horm Res 42:71–109
Choi JR, Scheffler IE (1983) J Biol Chem 258:12601–12608
Clark AJ, Hickman J, Bishop JO (1984) EMBO J 3:2055–2064
Elliott RW, Berger FG (1983) Proc Natl Acad Sci USA 80:501–504
Field LJ, Gross KW (1985) Proc Natl Acad Sci USA 82:6196–6200
Fong WF, Heller JS, Canellakis ES (1976) Biochem Biophys Acta 428:456–465
Fujita K, Murakami Y, Hayashi S (1982) Biochem J 204:647–652
Gilmour SK, Avdalovic N, Madava T, O'Brien TG (1985) J Biol Chem 260:16439–16444
Gupta M, Coffino P (1985) J Biol Chem 260:2941–2944
Heby O, Jänne J (1981): In: Morris DR, Marton LJ (eds) Polyamines in Biology and Medicine. Marcel Dekker Inc, New York, pp 243–310
Heller JS, Fong WF, Canellakis ES (1976) Proc Natl Acad Sci USA 73:1858–1862
Heller JS, Canellakis ES (1981) J Cell Physiol 107:209–217
Hickok NJ, Seppänen PJ, Kontula KK, Jänne PA, Bardin CW, Jänne OA (1986) Proc Natl Acad Sci USA 83:594–598
Isomaa V, Pajunen AEI, Bardin CW, Jänne OA (1982) Endocrinology 111:833–843
Isomaa VV, Pajunen AEI, Bardin CW, Jänne OA (1983) J Biol Chem 258:6735–6740
Jänne OA, Kontula KK, Isomaa VV, Bardin CW (1984) Ann NY Acad Sci 438:72–84
Jänne J, Pösö H, Raina A (1978) Biochim Biophys Acta 473:241–293
Kahana C, Nathans D (1984) Proc Natl Acad Sci USA 81:3645–3649

Kahana C, Nathans D (1985a) J Biol Chem 260:15390–15393
Kahana C, Nathans D (1985b) Proc Natl Acad Sci USA 82:1673–1677
King D, Snider LD, Lingrel JB (1986) Mol Cell Biol 6:209–217
Kitani T, Fujisawa H (1981) J Biol Chem 256:10036–10040
Kontula KK, Seppänen PJ, Van Duyne P, Bardin CW, Jänne OA (1985) Endocrinology 116:226–233
Kontula KK, Torkkeli TK, Bardin CW, Jänne OA (1984) Proc Natl Acad Sci USA 81:731–735
Kuhn NJ, Woodworth-Gutai M, Gross K, Held WA (1984) Nucl Acids Res 12:6073–6090
Loeb D, Houben PW, Bullock LP (1984) Mol Cell Endocrinol 38:67–73
McConlogue L, Gupta M, Wu L, Coffino P (1984) Proc Natl Acad Sci USA 81:540–544
McConlogue L, Coffino P (1983) J Biol Chem 258:8384–8388
Melanitou E, Cohn DA, Bardin CW, Jänne OA (1986) Mol Endocrinol, submitted for publication
Mitchell JLA, Wilson JM (1983) Biochem J 214:345–351
O'Malley BW, Means AR (1974) Science 183:610–620
O'Malley BW, Roop DR, Lai EC, Nordstrom JL, Catterall JF, Swaneck GE, Colbert DA, Tsai M-J, Dugaiczyk A, Woo SLC (1979) Recent Prog Horm Res 35:1–46
Pajunen AEI, Isomaa VV, Jänne OA, Bardin CW (1982) J Biol Chem 257:8190–8198
Palmer R, Gallagher PM, Boyko WL, Ganschow RE (1983) Proc Natl Acad Sci USA 80:7596–7600
Panthier J-J, Rougeon F (1983) EMBO J 2:675–678
Pegg AE, McCann PL (1982) Am J Physiol 243:C212–C221
Persson L (1981) Acta Chem Scand B35:451–459
Persson L, Seely JE, Pegg AE (1984) Biochemistry 23:3777–3783
Pohjanpelto P, Hölttä E, Jänne OA, Knuutila S, Alitalo K (1985) J Biol Chem 260:8532–8537
Russel DH (1981) Biochem Biophys Res Commun 99:1167–1172
Seely JE, Pösö H, Pegg AE (1982a) Biochemistry 21:3394–3399
Seely JE, Pösö H, Pegg AE (1982b) J Biol Chem 257:7549–7553
Seely JE, Pegg AE (1983) J Biol Chem 258:2496–2500
Seely JE, Persson L, Sertich GJ, Pegg AE (1985) Biochem J 226:577–586
Shapiro SZ, Young JR (1981) J Biol Chem 256:1495–1498
Swank RT, Paigen K, Davey R, Chapman V, Labarca C, Watson G, Ganschow R, Brandt EJ, Novak E (1978) Recent Prog Horm Res 34:401–436
Tabor CW, Tabor H (1984) Annu Rev Biochem 53:749–790
Tata JR, Smith DF (1979) Recent Prog Horm Res 35:47–95
Watson CS, Catterall JF (1986) Endocrinology 118:1081–1086

Discussion of the Paper Presented by O. Jänne

Roy: Olli, you showed different strains of mice with different levels of ornithine decarboxylase activity. Could there be similar heterogeneity in the kidney in that some cells are more active than others?

JÄNNE: Do you mean heterogeneity in the induction among individual renal cells?
ROY: In regard to the amount and the degree of induction in particular cells. Are some cells more active than others in those kidneys?
JÄNNE: All renal cells are not androgen responsive; the cells that respond are those in the proximal tubules and Bowman's capsule. We haven't performed in situ hybridizations with the cDNAs to see if all proximal tubule cells express OrnDCase mRNA to the same extent. Some other investigators have done immunocytochemistry on OrnDCase protein, but I don't recall that they have reported any major heterogeneity in the level of expression as measured by immunostaining.
THOMPSON: Frank Berger and Ken Paigen and some others have looked at β-glucuronidase and a number of other androgen-inducible genes in the kidney. I think there has been some work done with alcohol dehydrogenase as well. There seem to be strains of mice that are more sensitive to androgen induction than others. Are there general rules with respect to androgen sensitivity in mice, or does it vary from one gene to another, for example, to your knowledge, is A/J an androgen-resistant mouse?
JÄNNE: No. It depends on the gene product. For example, A/J strain is very sensitive to androgen administration if one measures β-glucuronidase activity or uses β-glucuronidase gene expression as the end point. This is dependent on the presence of a *cis*-acting regulator for β-glucuronidase gene, and A/J happens to be homozygous for the *a* allele that gives high androgen responsiveness to this particular gene. We have also studied some additional gene products such as the kidney androgen-regulated protein (KAP) under the same conditions, and it behaves in a way very similar to that of OrnDCase. For example, OrnDCase and KAP genes are less responsive in the A/J strain than they are in the C57BR strain. So, KAP and OrnDCase seem to follow the same regulation that is distinctly different from that of β-glucuronidase, whose regulation is highly dependent on the genetics of this *cis*-acting regulator.

Discussants: O. JÄNNE, A.K. ROY, and A. THOMPSON

Chapter 12

Glucocorticoid Receptors and the Control of Gene Expression

G.M. Ringold, A.B. Chapman, M. Danielsen, E.S. Klein, D.M. Knight, J.P. Northrop, and J.L. Vannice

Introduction

Steroid hormones mediate physiological responses in target cells via an interaction with specific, high-affinity receptors. Binding of hormone to these receptors results in a structural alteration in the receptor protein, allowing the steroid-receptor complex to bind tightly to acceptor sites within the nucleus (for reviews see Yamamoto and Alberts, 1976; Ringold, 1985). The association of steroid-receptor complexes with specific DNA sequences in the 5′ flanking regions of target genes such as those encoding the mouse mammary tumor virus (MMTV) (Payvar et al., 1983; Scheidereit et al., 1983), chicken lysozyme (Renkawitz et al., 1984), and human metallothionein-IIa (hMT-IIa) (Karin et al., 1984) appears to be important for transcriptional activation of their respective promoters. However, the detailed mechanisms by which glucocorticoids regulate specific gene expression remain hazy, largely owing to our poor understanding of receptor structure and receptor interactions with the transcriptional apparatus. Moreover, many genes appear to be regulated in a complex fashion in which regulatory mechanisms distinct from direct transcriptional activation by the steroid-receptor complex may be involved. Lastly, the basis for the tissue-specific expression of a constellation of steroid-responsive genes observed within a particular cell type remains to be elucidated.

In this chapter we investigate the structure and function of glucocorticoid receptors by analysis of mutations that alter receptor integrity and characterize two genes that are regulated in distinct fashions by glucocorticoids.

Characterization of Glucocorticoid Receptor Mutants

We have previously isolated glucocorticoid unresponsive clones of the MMTV-infected cell line M1.19 using a fluorescence-activated cell sorter (Grove et al., 1980). Such cells (e.g., MSN 5.3) contain less than 10% of the glucocorticoid receptor present in wild-type cells as determined by binding of [^{3}H] dexamethasone. To test the possibility that MSN 5.3 cells produce a receptor protein that is defective in hormone binding we have assayed extracts of these cells with a monoclonal antibody that recognizes both mouse and rat receptors (Gametchu and Harrison, 1984). Figure 1 shows an analysis of cytosolic extracts of wild-type (MSC 1) and variant (MSN 5.3) cells; proteins were separated by sodium dodecyl sulfate (SDS) polyacrylamide gel electrophoresis, transferred to nitrocellulose paper, and receptor was identified by incubation of the immobilized proteins with monoclonal antireceptor followed by ^{125}I-protein A. The results parallel hormone-binding assays in that the mutant cells harbor vastly reduced levels of immunoreactive glucocorticoid receptor migrating at approximately 94 kilodaltons. Therefore the defect in these cells appears to be one of production or stability of receptor protein. We also note that the residual receptor appears indistinguishable, at least on the basis of size, from that in wild-type cells.

Figure 2 shows an immunological analysis of extracts from wild-type and r$^-$ (receptor minus) cells derived from the W7.2 lymphoma cell line by sequential ethyl-methanesulfonate mutagenesis and growth in 1 μm dexamethasone. Danielsen and Stallcup (1984) have reported that the r$^-$ cells contain from one fourth (ADR38) to one half (ADR6) the number of

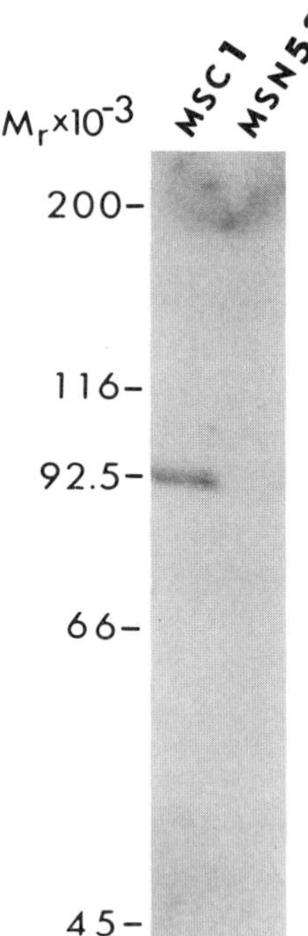

Fig. 1. Immunoblot of HTC cell cytosols electrophoresed on a $NaDodSO_4$ gel. Cytosols were prepared and analyzed as described elsewhere (Northrop et al., 1985). ^{125}I-protein A was used to detect the antireceptor monoclonal antibody. The glucocorticoid-unresponsive cell line, MSN5.3, contains less than 10% of the glucocorticoid receptor present in the responsive cell line, MSC1. Molecular weight markers are myosin (200,000), β-galactosidase (116, 250), phosphorylase B (92,500), bovine serum albumin (66,200), and ovalbumin (45,000).

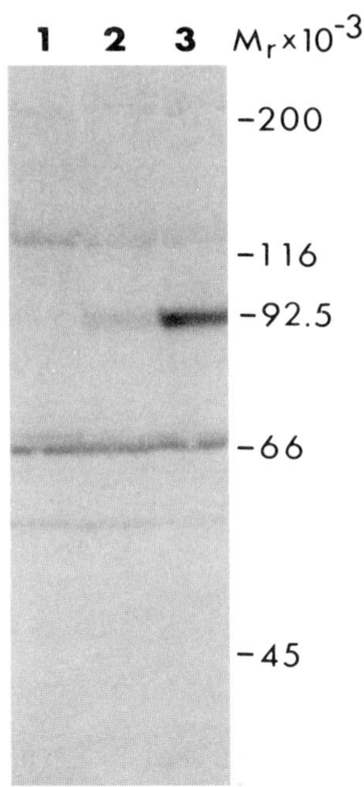

Fig. 2. Horseradish peroxidase-stained immunoblot of WEHI7 cell cytosols. Detection of the antireceptor monoclonal antibody was accomplished using a biotinylated horse antimouse IgG and subsequent incubation with avidin-biotin-peroxidase complex and development in diaminobenzidine solution. Bands detected at higher and lower molecular weights are due to nonspecific binding of the second antibody as they are present even when the monoclonal antibody is omitted (data not shown). Lane 1 (ADR6.M189D), Lane 2 (ADR38), Lane 3 (W7.2). Marker proteins are those used in Fig. 1.

receptors as the sensitive parental cells. Clone ADR6, after infection with MMTV and reselection in dexamethasone, yields clones (e.g., ADR6.M189D) that contain no detectable hormone-binding activity (Danielsen, 1984, unpublished). Our results show a very close parallel between the hormone-binding activity of such cells and the presence of immunologically reactive receptor protein. Clone ADR38 contains reduced levels of a 94 kilodaltons receptor protein as compared with the parental W7.2 cells, and no immunologically reactive receptor can be detected in the ADR6.M189D cells (Fig. 2). Thus, as with the rat hepatoma variants which were selected by completely different protocols, defects in hormone-binding activity of W7.2 mutants appear to be related to decreased production or stability of receptor rather than production of receptor protein with a defect in hormone-binding activity.

Analysis of receptor mRNA levels and gene structure in the HTC and W7 receptor defective cells was undertaken using a cDNA probe corresponding to the rat glucocorticoid receptor (Miesfeld et al., 1984). The results shown in Fig. 3 clearly demonstrate that the loss of receptor protein in these cases is due to a reduction in the steady-state level of receptor mRNA. Somewhat surprisingly, however, the MSN 5.3 and MSN 6.10 cells have undergone a parallel loss of receptor DNA sequences. Thus the selection of HTC receptor-defective "mutants" appears to be due to an instability of the receptor gene and/or chromosome whereas in W7 cells the lesion responsible for loss of receptor mRNA appears to be a subtle alteration in gene expression.

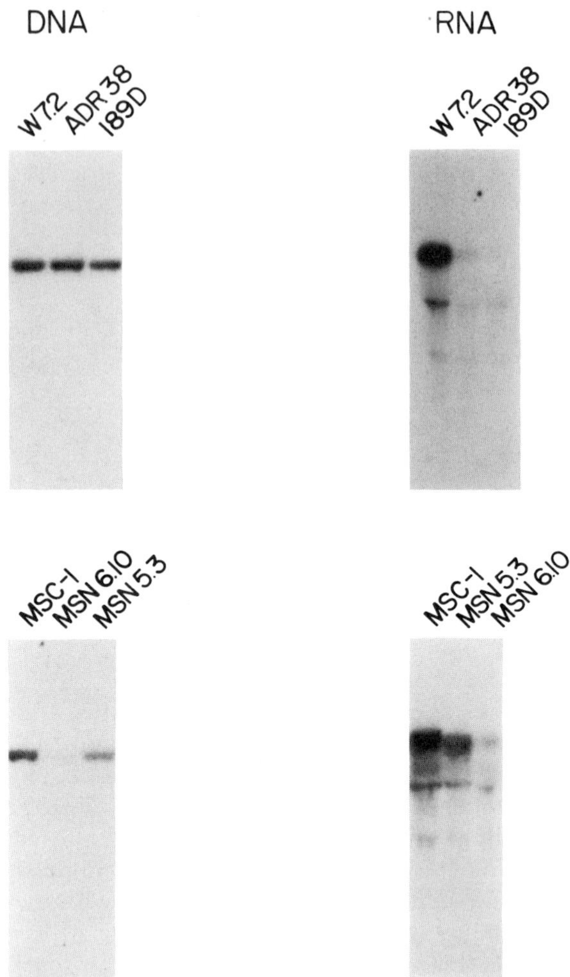

Fig. 3. Analysis of glucocorticoid receptor mRNA and DNA. **Right:** Total cell RNA was isolated from the indicated cell lines using the protocol of Chirgwin et al. (1979). Approximately 15 μg of each were separated in a 1% agarose gel, transferred to nitrocellulose and hybridized with a nick-translated receptor cDNA (Miesfeld et al., 1984). The major species of RNA seen is about 7kb in length. **Left:** Total cellular DNA was digested with *Eco*RI and analyzed by the procedure of Southern (1975). Filters were probed with a nick-translated cDNA corresponding to the 3′ untranslated region of the receptor gene. Similar quantitative results (although different sized bands) were obtained when DNAs were digested with Hind III. Analysis with probes from the coding region of the receptor cDNA also yielded quantitatively similar results as those shown here (Northrop and Danielsen, 1986, unpublished).

Tomkins and his colleagues (Sibley and Tomkins, 1974; Gehring and Tomkins, 1974) were the first to utilize the lympholytic action of glucocorticoids to select for mutants that exhibit defects in glucocorticoid responsiveness. Their experiments, which focused on the mouse T-cell line, S49, revealed that the vast majority of glucocorticoid-resistant S49 cells are of the r^- phenotype in that they lack wild-type levels of hormone binding activity. Characterization of such cells with the antireceptor monoclonal, however, reveals that they contain significant amounts of immunologically detectable receptor protein (Fig. 4, lane 5), When compared with its wild-type parent, S49.A2 (Fig. 4, lane 2), the r^- clone S49.1A.7R contains approximately one third the amount of immunologically cross-reactive receptor, yet little more than 10% the level of hormone-binding activity. In addition, two different clones of wild-type S49 cells contain approximately twice as much immunoreactive receptor per unit of hormone-binding activity as compared with W7.2 cells. These results strongly suggest that all clones of S49 cells express a mutant form of the glucocorticoid receptor which is unable to bind hormone; in r^- cells it appears that the second functional allele has been inactivated.

Two particularly intriguing mutations found in S49 cells are the nt^- (nuclear transfer minus) and nt^i (nuclear transfer increase) forms of receptor (for review see Yamamoto et al., 1976). The nt^- receptor binds

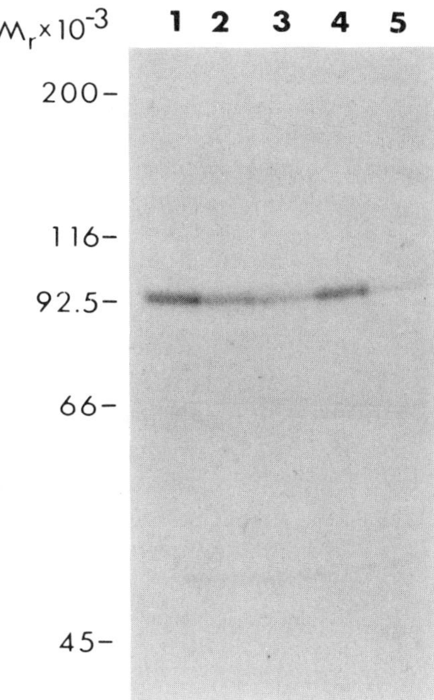

Fig. 4. Horseradish peroxidase-stained immunoblot of S49 cell cytosols. Peroxidase-coupled antimouse IgG and development in diaminobenzidine solution was used to detect the antireceptor monoclonal antibody. Lane 1 (S49.JN, wild type), Lane 2 (S49.A2, wild type), Lane 3 (nt^i), Lane 4 (nt^-), Lane 5 (r^-). Marker proteins are those used in Fig. 1.

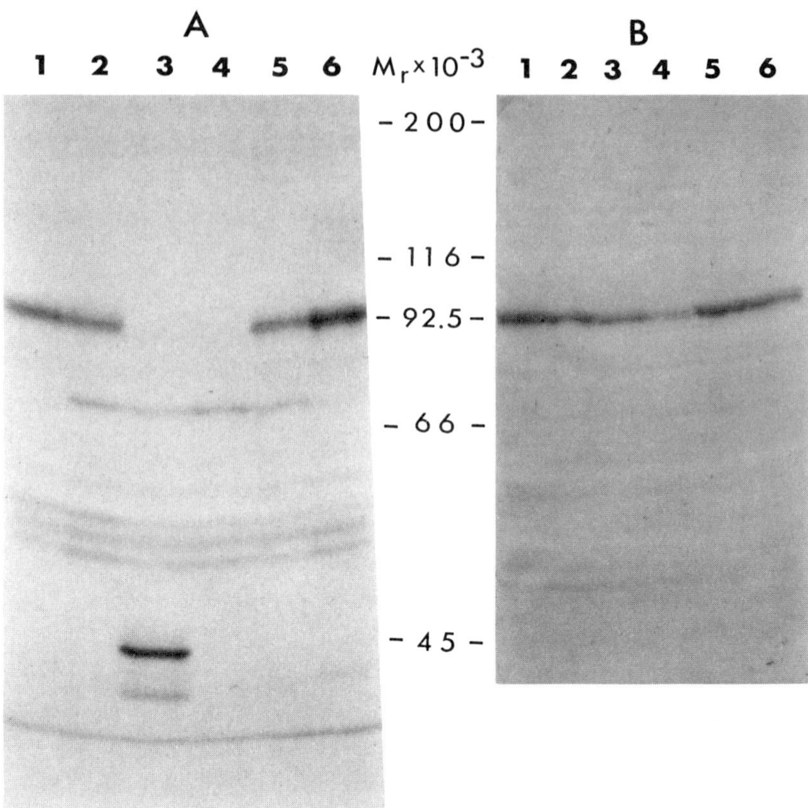

Fig. 5. NaDodSO_4-gel electrophoresis and immunoblot of affinity-labeled S49 and W7.2 cell cytosols. **Panel A:** Cytosols were affinity labeled with [^{3}H] dexamethasone mesylate without addition of unlabeled hormone. **Panel B:** Immunoblotting was done as described in Fig. 3. Lanes for both Panel A and B are as follows: Lane 1 (S49.JN, wild type), Lane 2 (S49.A2, wild type), Lane 3 (nt^i), Lane 4 (r^-), Lane 5 (nt^-), Lane 6 (W7.2). Marker proteins are those used in Fig. 1.

hormone more or less normally but localizes poorly within nuclei and has reduced DNA-binding activity in vitro. The nt^i receptor binds hormone normally, has an increased affinity for DNA, and is smaller than wild-type receptor. Whereas wild-type receptor has a molecular weight of about 94 Kd, the nt^i receptor's is approximately 40 Kd.

A more detailed analysis of the nt^i cells reveals that the only immunologically detectable protein migrates at approximately 94 Kd (Fig. 4, lane 3); however, the level is approximately half that seen in wild-type cells. This result is consistent with the presence of a preexisting (non-hormone binding) mutant receptor of 94 Kd in all S49 cells. We have not detected the presence of a smaller form of the glucocorticoid receptor immunologically, although a 40 Kd form is easily detectable by hormone affinity labeling (see below).

Simons and Thompson (1981) have used the α-keto mesylate derivative of dexamethasone to affinity label glucocorticoid receptors from rat hepatoma cells. We have used this reagent to both quantify and estimate the size of the glucocorticoid receptor in W7.2, S49, and various mutant cell lines. Figure 5A shows an autoradiogram of [^{3}H]dexamethasone mesylate labeled cytosol proteins analyzed by SDS-gel electrophoresis. A prominent band appears at approximately 94 kDa in cytosols from wild type and nt$^-$ cells (lanes 1,2,5, and 6), which is missing in the r$^-$ S49 cells (lane 4).

In the case of nt^i cells, dexamethasone mesylate predominantly labels a 40–45 kDa species of receptor (Fig. 5, lane 3). No labeling of a 90–95 kDa protein can be detected. Incubation of the extracts with dexamethasone mesylate in the presence of a large excess of unlabeled dexamethasone abrogates the labeling of the 94 kDa receptor in wild-type cells and of the 40–45 kDa form in the nt^i cells (Fig. 5A). Additional bands that are nonspecifically labeled by the dexamethasone mesylate (e.g., the band at about 70 kDa) cannot be competed by dexamethasone.

Figure 5B shows an immunoblot of the dexamethasone mesylate labeled cytosols shown in Fig. 5A. Thus the labeling procedure itself does not affect the ability of the monoclonal antibody to recognize the receptor. By comparison of these analyses it is clear that both r$^-$ and nt^i cells contain significant amounts of immunoreactive receptor of approximately 94 kDa that is incapable of binding hormone.

The analyses presented above indicate that no hormone-binding receptor of 94 kDa can be found in nt^i cells. Since the previous results were obtained using cytosols prepared from each of the cell lines, it seemed conceivable that we might be missing a precursor present in nuclei or that in the process of preparing the extracts we might be degrading a larger precursor. We have tried to address these issues by labeling the receptors in nt^i (and wild type) cells in vivo. Figure 6B shows an SDS gel of the proteins labeled in intact cells. Although there is a significant increase in nonspecific labeling of proteins, it is clear that a prominent band at about 94 kDa is seen from wild-type cells whereas a 40–45 kDa band is observed from nt^i cells. These results indicate that the only stable hormone-binding form of glucocorticoid receptor is nt^i cells is indeed the truncated form. Moreover, we have not been able to detect, by immunoblotting, any other receptor fragments in extracts of nt^i cells (Fig. 4 and 5). Thus it seems likely that the nt^i receptor is produced as a truncated protein rather than arising from a proteolytically sensitive receptor of wild-type size. Consistent with this interpretation, Miesfeld et al. (1984) and Northrop, Danielsen, and Ringold (1985, unpublished) detect a novel receptor mRNA lacking 5′ coding sequences (i.e., corresponding to the NH_2 terminal portion of the receptor) in S49 nt^i cells.

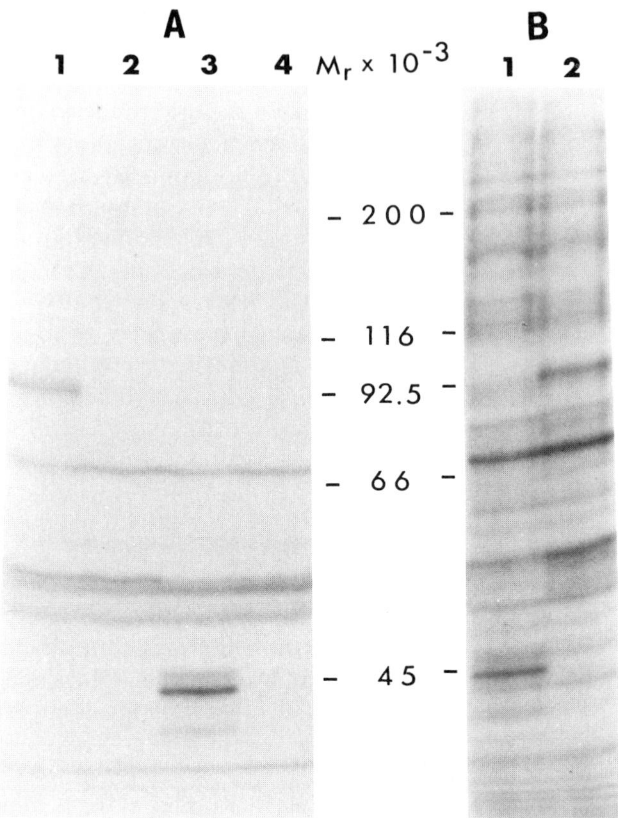

Fig. 6. $NaDodSO_4$-gel electrophesis of affinity-labeled S49 cell cytosols and whole cells. Affinity labeling was done as described (Northrop et al., 1985). **Panel A:** Cytosols were labeled with [^3H] dexamethasone mesylate plus or minus a 200-fold excess of unlabeled dexamethasone. Lanes 1 and 2 (S49.A2, wild type), Lanes 3 and 4 (nt^i). Excess unlabeled hormone was used in lanes 2 and 4. **Panel B:** Whole cells were labeled in vivo in the presence of 2×10^{-7} *M* [^3H] dexamethasone mesylate without addition of unlabeled hormone. Lane 1 (nt^i), Lane 2 (S49.A2, wild type). Marker proteins are those used in Fig. 1.

Regulation of α_1 Acid Glycoprotein Gene Expression

The acute phase reactant, α_1-acid glycoprotein (AGP), also called orosomucoid, is one of several plasma proteins synthesized by the liver in response to various stressful stimuli. Physical trauma such as surgery or wounding, bacterial infection, or nonspecific inflammatory stimuli such as subcutaneous injection of turpentine elicit the so-called acute-phase response (for review see Koj, 1974). In addition to AGP, whose physiological function remains obscure, a variety of protease inhibitors (e.g., α_1-anti-trypsin and ceruloplasmin) and iron scavengers (e.g., haptoglobin and hemopexin) are also induced. The mechanisms by which acute-phase reactants are induced are poorly understood. Macrophage factors, in particular interleukin I (Sipe et al., 1979), and glucocorticoids (Grieninger et al., 1978) have both been implicated in induction of some acute-phase reactant proteins. We (Vannice et al., 1983), Baumann et al. (1983), and Feinberg et al. (1983) have recently shown that AGP and its mRNA are induced several hundredfold by glucocorticoids in the rat hepatoma cell line (HTC).

The kinetics of AGP mRNA induction by dexamethasone in the MMTV-infected HTC cell line JZ.1 is shown in Fig. 7. A single RNA species of about 850 bases in length accumulates several hundredfold with a half-

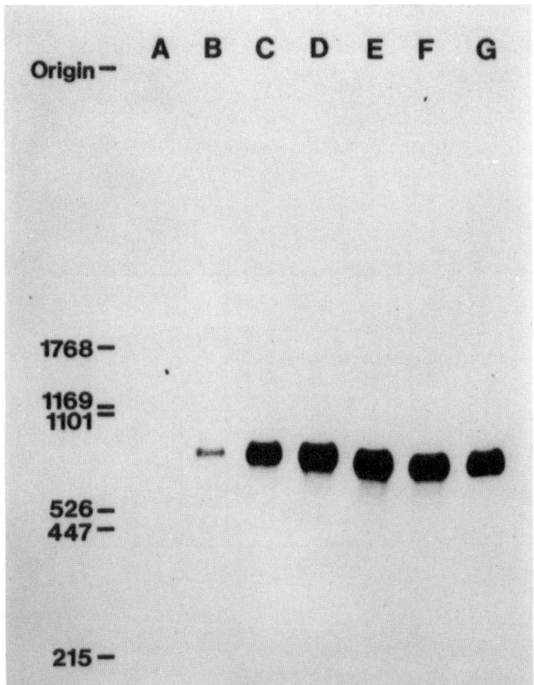

Fig. 7. Northern blot of the time course of induction of α_1-AGP mRNA by dexamethasone in JZ.1 cells. Cells were treated with 1 μM dexamethasone for various times, and the RNA was isolated by the method of Chirgwin et al. (1979). Each lane contains 15 μg of total RNA from cells treated with dexamethasone for 0, 6, 12, 24, 48, 72, and 96 h (lanes A–G, respectively). Molecular weight markers are [^{32}P]-labeled Hind III restriction enzyme-digested SV40 DNA.

time of 6–12 h. Unlike the classical induction of MMTV RNA, which is due to a direct interaction of the glucocorticoid-receptor complex with regulatory sequences in the MMTV LTR (for review see Ringold, 1985), the induction of AGP RNA is dramatically inhibited by treatment of cells with cycloheximide (data not shown). This requirement for ongoing protein synthesis can be interpreted in several ways: 1) the induction of AGP mRNA is dependent on the prior induction of another protein whose gene is directly regulated by glucocorticoids; 2) a labile preexisting protein is required for the induction of AGP mRNA, working either in concert with or independent of the receptor; or 3) inhibiting translation of AGP mRNA destabilizes it, resulting in lower steady-state levels. Additional experiments will be required to determine the basis for the effect(s) of protein synthesis inhibitors on accumulation of AGP mRNA.

The increase in abundance of a specific mRNA species in a cell may be due to an increase in its rate of synthesis, a decrease in its rate of degradation, or a combination of these processes. To determine if the rate of synthesis of AGP mRNA changes in response to glucocorticoids, we performed in vitro transcription assays in isolated nuclei from treated and untreated cells (Vannice et al., 1984). Figure 8 shows the results of such

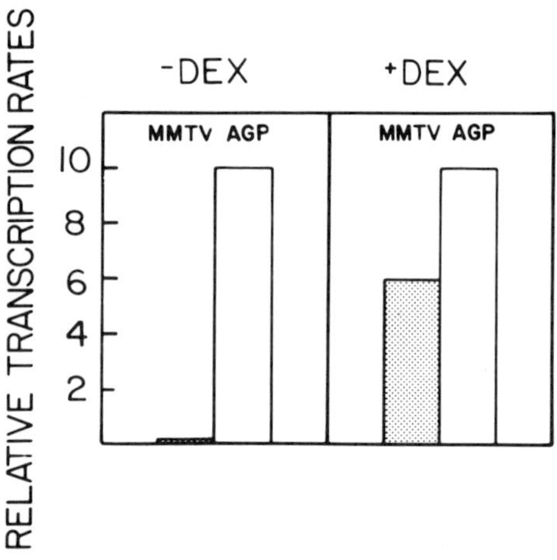

Fig. 8. Nuclear transcription assays of AGP and MMTV genes. JZ.1 cells were grown in suspension culture and treated without or with 1 μM dexamethasone for 1 h (or 4 h). Nuclei were prepared and in vitro transcription assays were performed as described (Vannice et al., 1984); α-^{32}P-labeled RNA (10^7 disintegration per minute, dpm) was incubated with filters containing 3 μg of denatured MMTV or AGP plasmids for 4 days, washed, and autoradiographed. The resulting hybridization signals were scanned by densitometry to determine the relative extents of transcription. Results at 1 h and 4 h of dexamethasone treatment were the same.

an assay. JZ.1 cells exhibit an increase in transcription rate of MMTV RNA in response to 1 μM dexamethasone while both actin (not shown) and AGP message synthesis remain nearly constant. Furthermore, MSC1 cells, which contain ten MMTV proviral copies (as opposed to the single copy in JZ.1 cells), show a proportional increase in MMTV mRNA synthesis over JZ.1 cells in both the presence and absence of dexamethasone while AGP transcription rate is constant and equal to that seen in JZ.1 cells (not shown). A time course of the effect of dexamethasone on the transcription rate of these two genes was performed. The synthetic rate of MMTV mRNA is maximal between 30–60 min after addition of hormone; however, AGP mRNA shows little, if any, increase in synthesis for times up to 4 h. Induction of mature AGP mRNA has occurred in the hormone-treated cells, however, since cytoplasmic RNA isolated from these same cells at the time of preparing the nuclei exhibits a dramatic increase in AGP mRNA abundance (Vannice et al., 1984).

The results of these experiments indicate that RNA polymerases are transcribing the AGP gene even in the absence of glucocorticoids. It is puzzling, therefore, that steady-state accumulation of AGP mRNA is detected only in hormone-treated cells. A plausible explanation is that glucocorticoids are inducing an RNA processing/stabilizing factor that specifically prevents the degradation of the constitutively synthesized transcripts. Alternatively, the bulk of the AGP RNA being synthesized in the nuclear transcription assays is not reflecting the production of bona fide AGP mRNA precursors. Rather, fortuitous transcription read-through from an upstream gene could yield this unusual result.

We have tested, using alternative methods, the possibility that transcriptional activation of the AGP gene may indeed be of significance in the glucocorticoid effect. Introduction of a chimeric gene containing the AGP 5′ flanking region from −763 to +20 fused to the chloramphenicol acetyl transferase (CAT) gene into JZ.1 cells was accomplished by cotransfection with an SV40 plasmid encoding a neomycin resistance marker. Measurement of CAT activity in clones of transfected cells (Fig. 9) reveals that the linked marker is glucocorticoid inducible (in most cases), thereby indicating that at least a part of the hormonally dependent expression of AGP is conferred by the 5′ flanking region; similar results have been obtained with a molecule lacking the sequences between +1 and +20 (Klein, 1986, unpublished). Although we have not yet resolved the paradox between the results of the nuclear transcription assays and the transfection assays, we are led to believe that the constitutive transcription observed in isolated nuclei does not reflect the production of authentic AGP RNA. Further analysis of the origin of this RNA and of the sequences involved in transcriptional activation of the AGP promoter is indicated.

In summary, the massive induction of AGP mRNA by glucocorticoids in HTC cells appears to be the consequence of complex events, some of

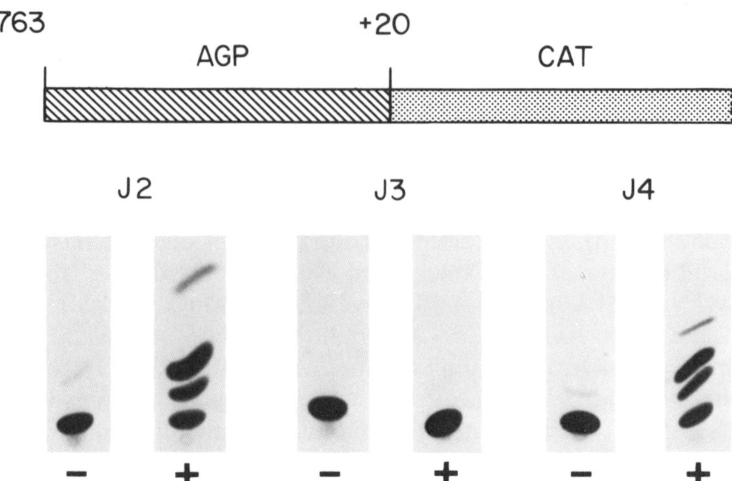

Fig. 9. Induction of CAT activity in JZ.1 cells transfected with an AGP promoter-CAT fusion plasmid. A fragment of the AGP gene from −763 to +20 (relative to the start of transcription) (Reinke et al., 1985), was inserted into pSVO-CAT (Gorman et al., 1982). JZ.1 cells were cotransfected with the resultant plasmid, pAGP-CAT and pSV2neo (Southern and Berg, 1982); G418 resistant-colonies were isolated and assayed for expression of CAT activity in the presence and absence of dexamethasone (1 μ*M* for 24 h). CAT assays were performed as described (Gorman et al., 1982). The results from three clones are shown here; out of a total of nine clones analyzed, seven exhibited glucocorticoid inducible expression of CAT activity.

which may involve secondary actions of the hormone on RNA processing and/or stability. Transfection analyses point to effects of the hormone at the transcriptional level; however, the sequences involved and the possible role for direct interactions with the glucocorticoid-receptor complex remain to be explored. Nevertheless, the prospect of unraveling the mysteries of AGP gene expression stands as an enticing challenge for the enterprising and strong-hearted devotee of glucocorticoid hormone action.

Glucocorticoid Control of an Adipocyte-Specific P450 Gene

The mechanisms that allow the tissue-specific expression of genes have only recently become amenable to analysis at the molecular level. Similarly, the biochemical basis for restricting the expression of a given set of genes within a particular cell type can at present only dimly be imagined. These issues are of prime concern when considering why it is that a particular gene can be expressed and respond to a steroid hormone in one tissue but not in another. Here we describe a gene that is expressed in a differentiation-dependent manner and that exhibits glucocorticoid inducibility in differentiated adipocytes but not in pre-adipocytes. Detailed characterization of *cis*-acting regulatory elements and structural alterations in such a gene may yield significant insights into the questions raised above.

We have recently isolated cDNA clones corresponding to mRNAs that are induced during adipogenesis of the tissue culture line, TA1 (Chapman et al., 1984). Among these we have identified one (clone 10) that fortuitously is expressed at higher levels in glucocorticoid-treated TA1 adipocytes. Sequence analysis of this cDNA clone reveals that it is a member of the cytochrome P450 gene family (Knight, 1985, unpublished data). Figure 10 shows that the RNA for clone 10 is developmentally regulated in the absence of hormones and that exposure of TA1 cells (during differentiation to adipocytes) results in a 10- to 20-fold induction of clone 10 RNA over the nonhormone-treated levels (Fig. 11). As previously described (Chapman et al., 1984), one also sees that the developmental activation of clone 10 RNA, as well as other adipose-specific gene products such as clone 1 RNA, occurs precociously in dexamethasone-treated cells. This general effect of glucocorticoids on triggering the differentiation of TA1 adipocytes, however, will not be discussed in any greater detail here.

We have analyzed the dexamethasone dose requirement for induction of clone 10 RNA in differentiated TA1 adipocytes. The half maximal response (data not shown) occurs with approximately 10 n*M* dexamethasone, a value characteristic of glucocorticoid receptor mediated responses. In addition, clone 10 RNA induction by dexamethasone is significantly reduced by addition of excess progesterone, a competitive inhibitor of glucocorticoid binding to its receptor (data not shown). Thus the induction

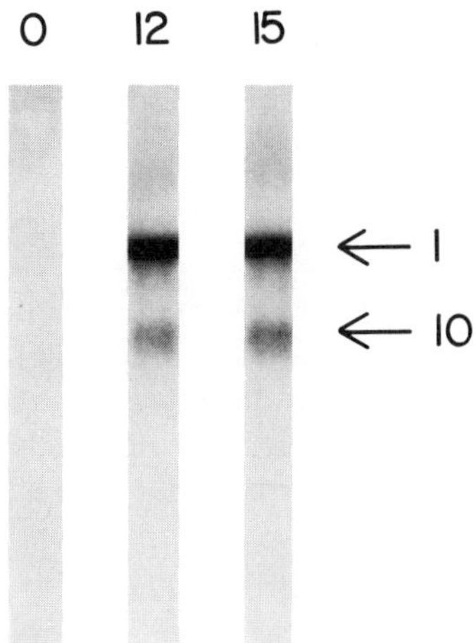

Fig. 10. Developmental activation of clone 1 and clone 10 RNA expression in TA1 adipocytes. RNA was isolated from nonhormone-treated TA1 cells at confluence (day 0) and at 12 and 15 days postconfluence by which time adipogenic conversion was complete. Ten μg of total RNA were electrophoresed in an agarose formaldehyde gel, transferred to nitrocellulose, and hybridized with ^{32}P nick-translated clone 1 and clone 10 DNAs (see Chapman et al., 1984). Clone 1 RNA is 4.3 kb and clone 10 RNA is 2.3 kb. Exposure of the film was for two weeks.

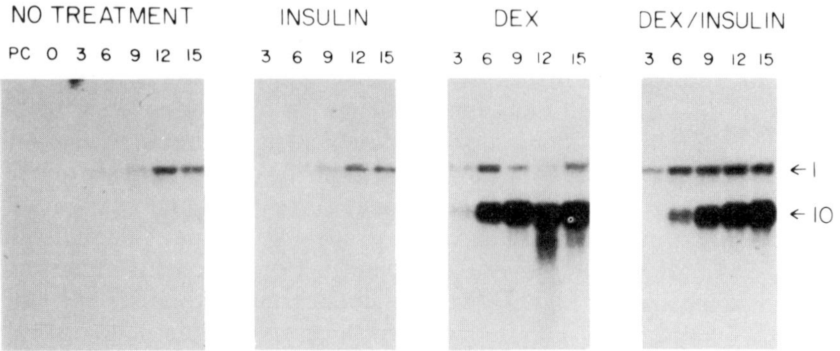

Fig. 11. Time course of clone 1 and clone 10 RNA accumulation in control and hormone-treated cells. TA1 preadipocytes were treated at confluence with either 1 μ*M* dexamethasone, 5μg/ml insulin, or a combination of both. RNA was isolated from cells one day prior to confluence (PC), at confluence (O), and every three days postconfluence with continual hormone treatment (3–15). Ten μg of total RNA were electrophoresed in an agarose formaldehyde gel and transferred to a nitrocellulose filter. This blot was then hybridized to nick-translated clone 1 and clone 10 DNA. Exposure was for three days.

of this RNA appears to be mediated through the well characterized glucocorticoid receptor.

To test whether clone 10 RNA is glucocorticoid inducible prior to its differentiation-dependent activation, 10T1/2 cells and TA1 preadipocytes (preconfluent) were treated for three days with 1 μ*M* dexamethasone. Ten μg of poly(A)-containing RNA from such cells (corresponding to approximately 300 μg of total RNA) were fractionated on an agarose formaldehyde gel and transferred to nitrocellulose paper. For comparison, 10 μg of total RNA from TA1 adipocytes that were treated for three days with dexamethasone were also analyzed (Fig. 12). There is no detectable clone 10 mRNA in treated or control 10T1/2 cells despite the more than 30-fold excess of poly(A)-containing RNA when compared with the glucocorticoid-treated mature adipocytes. Similarly, confluent 10T1/2 cells treated with the identical hormone regimens described in Fig. 2 for TA1 cells show no clone 10 RNA expression (1985, unpublished results). Analysis of MMTV expression in 10T1/2 cells indicates, however, that these cells contain a full complement of functional glucocorticoid receptors.

TA1 preadipocytes do show a small amount of clone 10 mRNA in the dexamethasone-treated cells. Calculations using results from densitometry scanning, which take into account the relative excess of poly(A) containing RNA in these samples, show that the level of clone 10 RNA expression in these cells is approximately 9,000-fold lower than that seen in the mature TA1 adipocytes similarly treated with dexamethasone. Probing of this filter

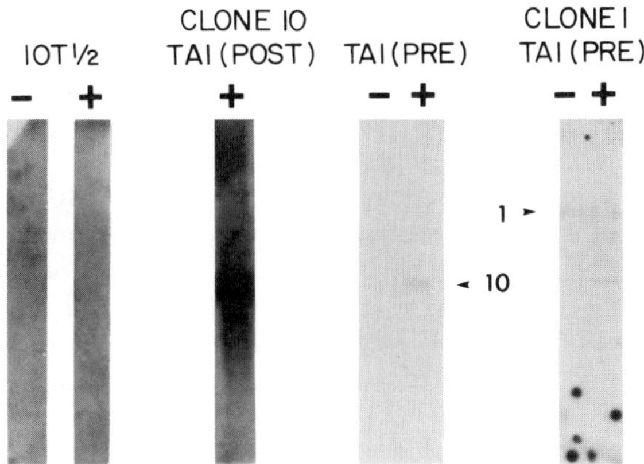

Fig. 12. Analysis of clone 10 RNA expression and dexamethasone inducibility in 10T1/2 cells and TA1 preadipocytes. Preconfluent 10T1/2 cells and TA1 preadipocytes (PRE) were grown in the presence (+) or absence (−) of 1 μM dexamethasone. Ten μg of poly(A)-containing RNA was isolated from these subconfluent cells and electrophoresed in an agarose formaldehyde gel as described (Chapman et al., 1984). Ten μg of total RNA isolated from mature TA1 adipocytes which had been treated for three days with dexamethasone (POST) was included for comparison. RNA was transferred to a nitrocellulose filter and probed with nick-translated clone 10 DNA. After exposure to film, the filter was washed free of probe and hybridized with nick-translated clone 1 DNA.

with nick-translated clone 1 DNA reveals a similar low level of clone 1 RNA expression in both treated and control preadipocytes, which could reflect either a very low level of expression in all preadipocytes or a small background of differentiating adipocytes in the preadipocyte population. We feel the latter is more likely because small areas of confluence can often be detected in subconfluent cultures, and TA1 differentiation is not synchronous. The low levels of clone 10 RNA thus more likely reflect dexamethasone-mediated induction in a small number of differentiated adipocytes rather than expression in preadipocytes.

Acquisition of Glucocorticoid Responsiveness by Clone 10 RNA During Adipocyte Differentiation

To summarize, during TA1 adipocyte differentiation, clone 10 mRNA is induced in the absence of hormone treatment, reaching a plateau level within 12 days after confluence. Its expression during differentiation in the presence of dexamethasone is thus affected by at least two factors: the hormonal induction of the general differentiation mechanism and the specific hormonal induction of clone 10 RNA. The relative importance of each factor is difficult to determine given the likelihood that clone 10 RNA is not inducible by glucocorticoids before the adipogenic conversion. Thus, differentiation is accompanied by some change that confers hormone responsiveness to the clone 10 gene. This change is not mediated by the induction of previously absent receptors, as both 10T1/2 and TA1 preadipocytes have functional glucocorticoid receptors. Rather, differentiation may lead to an alteration of chromatin structure or generation of a transcriptional factor(s) that allows clone 10 induction and expression.

Recent studies on MMTV have revealed a correlation between gene activity and hormonal inducibility which is reflected in an altered chromatin DNAase sensitivity (Feinstein et al., 1982; Zaret and Yamamoto, 1984). Studies of the chicken vitellogenin gene have revealed DNAase hypersensitive sites that exist prior to estrogen-mediated expression and that correlate in tissue distribution with VTGII hormone inducibility (Burch and Weintraub, 1983). It has not yet been possible, though, to observe and study the generation of such sites that mark future hormone inducibility during differentiation. TA1 cells may offer an advantage in characterizing such changes since the precursor cells in which clone 10 RNA is not inducible are readily available. Analysis of clone 10 chromatin structure is underway in an attempt to find changes that correlate with transcriptional activity, potential hormone inducibility, and hormone induction itself. If the induction of clone 10 transcription is indeed a primary glucocorticoid response, it will be particularly interesting to identify DNA-binding sites for the glucocorticoid-receptor complex and analyze the factors that affect their chromatin structure during differentiation.

Acknowledgments. We thank our many colleagues for assistance in various aspects of the work described here. In particular we are indebted to Drs. B. Gametchu and R. Harrison for the antireceptor monoclonal antibody, Drs. R. Reinke and P. Feigelson for the clone of the AGP gene, and Drs. R. Miesfeld and K. Yamamoto for the glucocorticoid receptor cDNA. This work was supported by grants from the National Institutes of Health and the March of Dimes. M. Danielsen was a scholar of the Leukemia Society and G.M. Ringold was supported by an Established Investigator Award from the American Heart Association.

References

Baumann H, Firestone GL, Burgess TL, Gross KW, Yamamoto KR, Held WA (1983) J Biol Chem 258:563–570

Burch JBE, Weintraub H (1983) Cell 33:65–76

Chapman AB, Knight DM, Dieckmann BS, Ringold GM (1984) 259:15548–15555

Chirgwin JM, Przybyla AE, MacDonald RJ, Rutter WJ (1979) Biochemistry 18: 5294–5299

Danielsen M, Stallcup MR (1984) Mol Cell Biol 4:449–453
Feinberg RF, Sun L-HK, Ordahl CP, Frankel FR (1983) Proc Natl Acad Sci USA 80:5042–5046
Feinstein SC, Ross SR, Yamamoto KR (1982) J Mol Biol 156:549–565
Gametchu B, Harrison RW (1984) Endocrinology 114:274–279
Gehring U, Tomkins GM (1974) Cell 3:301–306
Gorman CM, Moffat LM, Howard BH (1982) Mol Cell Biol 2:1044–1051
Grieninger G, Hertzberg KM, Pindyck J (1978) Proc Natl Acad Sci USA 75:5506–5510
Grove JR, Dieckmann BS, Schroer TA, Ringold GM (1980) Cell 21:47–56
Karin M, Haslinger A, Holtgreve H, Cathala G, Slater E, Baxter JD (1984) Cell 36:371–379
Koj A (1974) In: Allison AC (ed) Structure and Function of Plasma Proteins, 1, Plenum Press, London, pp 73–125
Miesfeld R, Okret S, Wikstrom A-C, Wrange O, Gustafsson J-A, Yamamoto KR (1984) Nature 312:779–781
Northrop JP, Gametchu B, Harrison RW, Ringold GM. (1985) J Biol Chem 260:6398–6403
Payvar F, DeFranco D, Firestone GL, Edgar B, Wrange O, Okret S, Gustafsson J-A, Yamamoto KR (1983) Cell 35:381–392
Reinke R, Feigelson P (1985) J Biol Chem 260:4397–4403
Renkawitz R, Schutz G, von der Ahe D, Beato M (1984) Cell 37:503–510
Ringold GM (1985) Annu Rev Pharmacol Toxicol 25:529–566
Scheidereit C, Geisse S, Westphal HM, Beato M (1983) Nature 304:749–752
Sibley CH, Tomkins GM (1974) Cell 2:213–220
Simons SS Jr, Thompson EB (1981) Proc Natl Acad Sci USA 78:3541–3545
Sipe JD, Vogel SN, Ryan JL, McAdam KPWJ, Rosenstreich DL (1979) J Exp Med 150:597–606
Southern EM (1975) J Mol Biol 98:502–517
Southern PJ, Berg P (1982) J Molec Appl Genet 1:327–341
Vannice JL, Ringold GM, McLean JW, Taylor JM (1983) DNA 2:205–212
Vannice JL, Taylor JM, Ringold GM (1984) Proc Natl Acad Sci USA 81:4241–4245
Yamamoto KR, Alberts BM (1976) Annu Rev Biochem 45:721–746
Yamamoto KR, Gehring U, Stampfer MR, Sibley CH (1976) Recent Prog Horm Res 32:3–32
Zaret KS, Yamamoto KR (1984) Cell 38:29–38

Discussion of the Paper Presented by G. Ringold

O'MALLEY: Concerning the AGP paradox, let's assume that your initial data are correct. That is, there is a small inductive effect of glucocorticoids and a major posttranscriptional effect leading to the large accumulation of message. Why not postulate that in your CAT construct you simply either have not included an enhancer that is naturally there or it is not active in the cell transcription. That would explain all of the data because the enhancer then would probably overide much of the glucocorticoid effect.

RINGOLD: Yes, that is a possibility. We would still, however, be faced with the problem of explaining how one obtains major posttranscriptional regulation of the AGP RNA. How is it that in the natural situation there appears to be high levels of transcription, yet no RNA accumulates. If so, one imagines that an enhancer drives high levels of transcription, but the RNA, which is made then, has to be degraded instantly (or extremely rapidly).
O'MALLEY: But, that would be the interesting part of the system, correct? If not, it would be more like a standard inducible gene.
RINGOLD: I think we now have on hand the reagents to start answering that question. First, we have additional constructions that remove the first 20 nucleotides; as such, we can't have any of the structural part of the AGP RNA remaining. The second thing we have is the AGP-CAT construct in stable transformants. We now need to do run-off assays in the standard way and compare the level of transcription from this gene versus the endogenous gene. It turns out we have to subclone all of these fragments and CAT fragment in *M* 13 because if one uses a PBR-based plasmid probe, we see very high levels of expression of PBR sequences in these transformed cells.
O'MALLEY: I realize the results are still preliminary, but it is an astounding deduction, that there might be 10 copies of this gene. You postulate then only one was active? Is that correct?
RINGOLD: In the HTC cells I would argue that there are multiple, functional copies of the gene based on the fact that you see 10 times more receptors in the wild type than you do in the mutant, and as you progressively lose copies of receptor DNA, you see an equivalent decrement in the amount of functional hormone-binding activity and immunoreactive material. So, in the HTC cells there appear to be multiple functional copies of the glucocorticoid receptor gene.
O'MALLEY: How about in S49 cells?
RINGOLD: We don't know yet; nor do we know the copy number in normal tissue. The S49 and W7 signal that we get on the Southern blots, however, looks very similar to that which we see from HTC cells, suggesting there may be multiple copies of the gene.
O'MALLEY: But, one can get a one-step mutation leading to receptor defectiveness in some of these cells. Correct?
RINGOLD: Yes.
O'MALLEY: So in those cells you have to postulate that all copies were not active, and if they're not all active, it is unlikely that restriction sites would be highly conserved in that number of copy of genes. Wouldn't you expect more heterogeneity in your restriction patterns?
RINGOLD: Probably. At this point all we can say with any level of certainty is that there are multiple copies of the receptor gene in the HTC cells.
O'MALLEY: Is there anything that you could imagine that could have caused amplification?
RINGOLD: Well, HTC cells are unusual. They are subtetraploid in general but individual genes may be differentially amplified. We haven't performed a karyotypic analysis, or looked for evidence of amplified loci.
O'MALLEY: Were these cells cloned as high responders?
RINGOLD: This was an MMTV-infected clone we selected randomly.
BARDIN: Did you want to comment on that Dr. Milgrom:
MILGROM: Yes, I wanted to simply follow up the question of the number of the

gene copies. Did you completely eliminate the possibility of having a heterogeneous population of cells, some of which contain the normal gene and most of them which have lost it? If you subclone the cells, will you get constantly the same pattern?
RINGOLD: Generally yes.
MILGROM: You've done that and compared among subclones?
RINGOLD: We cloned the isolates to start with and have recently subcloned them.
MILGROM: These give the same signal? So it can't be you started with a heterogenous cell population?
RINGOLD: It appears not to be heterogeneous.
MILGROM: If you compare the hybridization signal with a signal for a gene that has a single copy using a similar size probe, would you get a 10-fold difference?
RINGOLD: That is being done right now; we don't know the answer to that experiment yet.
THOMPSON: Just an additional comment with respect to the gene copy issue. It must be complicated by the fact that the 5 kb and the 7 kb messenger RNAs are expressed more or less at the same levels in the S49 nt^{i} cells as well. If they have 10 genes, they must have 5 wild-type and 5 nt^{i}.
RINGOLD: I agree that raises a major problem; however, at this point I only suggested it as a possibility. We have no data that would suggest that the gene is polyploid in any cell other than the HTC cell.
MUELLER: I was very interested in your cycloheximide effect, Gordon, did you also test puromycin in those experiments?
RINGOLD: Yes. Cycloheximide and puromycin gave the same result.
MUELLER: I wonder if there is a possibility that in this case you're running into an alteration in the translational process. In other words, it may be that the message is not able to be retained in a functional state unless translation is actually occurring. That may account for some of the posttranscriptional modification of the message. I think this comes back to some things we saw many years ago in terms of histone message and its survival. This might be an interesting proposition and one that might redirect people's interest toward determining whether glucocorticoid receptors can work at alternate sites in the cell.
RINGOLD: That's an intriguing possibility. We have to look at it.
MUELLER: I think an interesting experiment would be to analyze the fate of the message that is in the cytoplasm.
RINGOLD: One could look at the half-life of the message in the presence or absence of cycloheximide.
MUELLER: Yes, and also look at its size characteristics.
GUSTAFSSON: I was somewhat surprised by your Western immuno blots of the nt^{i} receptor. The Harrison antibody did not pick up the 40 kilodalton nt^{i}, receptor, whereas Harrison had a paper in which he showed that his antibody specifically binds to the DNA-binding receptor fragment. Do you have a comment?
RINGOLD: Yes. We haven't actually done Harrison's experiment. It is quite clear, however, that two of his monoclonals (although they may be seeing the same epitope) do not react with the nt^{i} receptor. I think our results are unambiguous. We have not, however, looked at the chymotryptic fragment or smaller DNA-binding fragments to see whether the antibodies recognize them. I don't want to try to explain his data; however, my feeling is that those antibodies do not see the DNA-binding site. One could of course make the argument that the structure

of the nt[i] receptors is more complicated then we imagined and that it is really not identical to the chymotryptic fragment.
GUSTAFSSON: So your conclusion would be that the Harrison antibody also would recognize the domain C, the modulating region?
RINGOLD: Yes. That's the way I would interpret our data.

Discussants: W. BARDIN, J.A. GUSTAFSSON, E. MILGROM, G. MUELLER, B. O'MALLEY, G. RINGOLD, and A. THOMPSON

Chapter 13

Activation and Regulation of the Vitellogenin Gene Family

J.R. TATA, W.C. NG, A.J. PERLMAN, AND A.P. WOLFFE

Introduction

Steroid hormone-regulated systems have played an important role in furthering our understanding of how the interaction between the hormone-receptor complex and the genome might initiate or modulate the physiological changes in the hormone's target cells. With the advent of recombinant DNA technology, major emphasis has been placed on DNA sequences that are important for regulation by the hormone-receptor complex, best illustrated by the recent work on regulation by estrogen, androgen, and glucocorticoids of genes encoding egg proteins, mammary tumor virus, and α_{2u}-globulin (for several recent reviews see Eriksson and Gustafsson, 1983; Chambon et al., 1984; Renkawitz et al., 1984; Payvar et al., 1983). Attention is now being focused on the implications of the structure of the nucleus in modulating specific gene activity, such as the association of steroid hormone receptors and hormone-regulated genes with the target cell's nuclear matrix or scaffold (Barrack and Coffey, 1983; O'Malley et al., 1985). This chapter is largely based on recent work from the authors' laboratory on the regulation of *Xenopus* vitellogenin genes by estrogen. Much of the work involves the manipulation of this gene family in primary cell cultures in which the full physiological process can be reversibly reproduced. It also describes how heat shock and cellular stress produced in setting up cell cultures can modify the action of estrogen. Finally, some recent results obtained in our laboratory on specific switching on in vitro of the silent vitellogenin genes in nuclei isolated from male *Xenopus* hepatocytes by soluble extracts are also described. In addition, it is intended to point out those areas of possible fruitful investigation in the future that are likely to further our understanding of the different elements that determine hormonal regulation of gene expression.

Advantages of the Vitellogenin System

In all oviparous vertebrates egg yolk protein synthesis is obligatorily and exclusively under the control of estrogen (Tata and Smith, 1979). A single administration of the hormone causes a massive accumulation of vitellogenin mRNA not only in female but also in male liver. The possibility of inducing vitellogenin genes in male hepatocytes, i.e., in cells in which they would be permanently silent, offers a unique opportunity of studying both the de novo activation and maintenance of transcription of a specific gene by a steroid hormone. As with other steroid hormones, estrogen has to be continuously present to maintain the induced gene activity, and its action is rapidly reversed upon withdrawal; this reversibility allows one to study both the induction and de-induction processes. For these and other reasons, there has been a growing interest in several laboratories to analyze the expression of vitellogenin genes in chicken and *Xenopus* liver (Tata and Smith, 1979; Wyatt, 1980; Wahli et al., 1981; Tata, 1982; Ryffel and Wahli, 1983; Shapiro and Brock, 1985). In our laboratory we chose the amphibian species as a model because of the following advantages: 1) the dormant genes can be reversibly activated by estrogen in male hepatocytes; 2) both qualitatively and quantitatively the full physiological process can be reproduced in primary cell cultures; 3) there is the absence of any DNA synthesis or cell proliferation that could complicate the interpretation of the early phase of hormone-induced changes in the conformation and transcription of genes.

The Xenopus Vitellogenin Gene Family

As shown in Fig. 1, there are four vitellogenin genes that are actively expressed in the presence of estrogen in *Xenopus laevis* liver (Wahli et al., 1979). These have been termed A1, A2, B1, and B2, the two genes of A and B groups sharing a 95% coding sequence homology, with 80% homology between the two groups. All four genes comprise 34 exons but there is a large variation in intron size between the genes. The A1 and B1 genes are linked with a 15-kb DNA segment, but it is not known if the other two genes are (Wahli et al., 1982). Wahli's group has also sequenced the 5′ upstream flanking DNA for each of the four genes (Germond et al., 1984; Walker et al., 1984). These have revealed similarities and differences within this *Xenopus* gene family, and have emphasized the important regions with which the hormone-receptor complex is likely to interact.

An interesting feature concerning the organization and evolution of the *Xenopus* vitellogenin gene family has come to light in a recent study from our laboratory in which we compared these genes and their products in a closely *(X. borealis)* and a distantly *(X. tropicalis)* related species to *X. laevis* (Baker et al., 1985). The more ancient *X. tropicalis* has half the DNA content and number of chromosomes as the other two species of *Xenopus,* and it was previously shown that it has two instead of four expressed vitellogenin genes in *X. laevis* (Jaggi et al., 1982). In our studies (Fig. 2), restriction endonuclease digestion patterns revealed that although the coding sequences of the individual genes were highly conserved both within and between different species (Wahli et al., 1979, 1981; Walker et

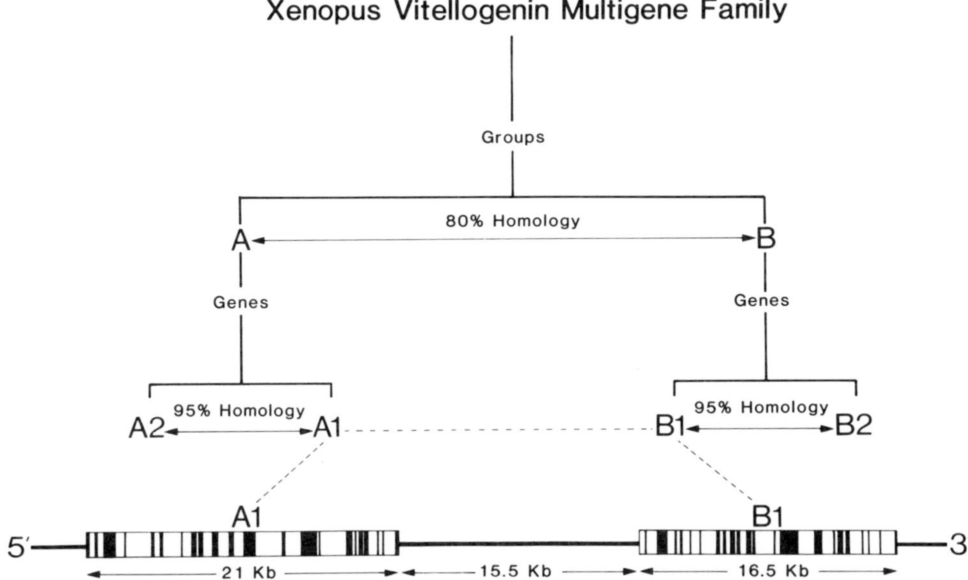

Fig. 1. The four actively expressed vitellogenin genes in *Xenopus laevis* and their classification as two members each of the A and B groups. The figures as % denote the coding sequence homology between individual genes and the two groups (Wahli et al., 1979).

al., 1983), there were significant differences in the organization of the individual genes. Thus, there is significant rearrangement within the *Xenopus* multigene family during evolution.

An important question that arises concerning the vitellogenin multigene family is to know whether or not the individual gene members are equally and coordinately expressed and how they would respond at different times of hormonal stimulation. Clearly, experiments in whole animals are unlikely to fully resolve these questions, which is why our laboratory has devoted much effort in developing a primary *Xenopus* hepatocyte culture system that mimicks in every respect the de novo activation as well as the secondary modulation of vitellogenin gene expression seen in vivo (Tata et al., 1985).

Expression of Vitellogenin Genes in Primary *Xenopus* Hepatocyte Cultures

Advantages

Among the many advantages of cell culture over whole animals in studying hormonal regulation of gene expression are: 1) ability to accurately control hormone concentration; 2) better analysis of the early events associated

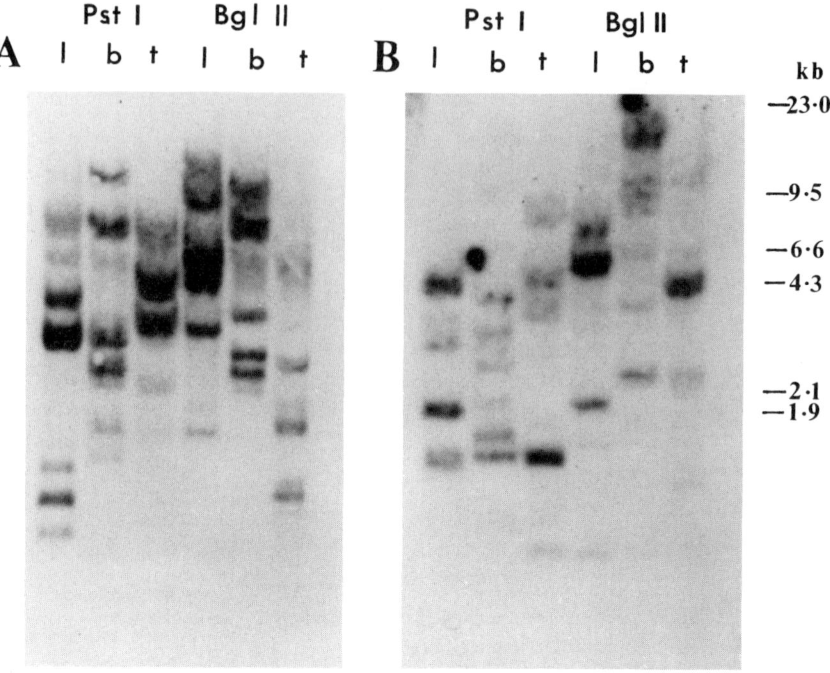

Fig. 2. Southern blot analysis of restriction digests of DNA from *X. laevis* (l), *X. borealis* (b) and *X. tropicalis* (t) probed with cloned ^{32}P-labeled cDNA corresponding to (**A**) *X. laevis* vitellogenin gene A1 and (**B**) *X. laevis* vitellogenin gene B1. DNA digested with restriction endonucleases *Pst* I or *Bgl* II was resolved by electrophoresis on 0.8% agarose gel and transferred to nitrocellulose filters (Baker et al., 1985).

with hormone-receptor interaction with the regulatory elements; 3) rapid reversibility and de-induction of gene expression upon hormone removal; 4) analysis of single-cell types in heterogeneous tissues. Although a large amount of work has been carried out on hormone action in cultured cells, most of these studies have involved transformed cells or tumor-derived cell lines (Sato et al., 1982). Although interesting in themselves, the findings from such studies do not necessarily reflect the hormonal control mechanisms operating under physiological conditions or during development.

Despite much effort, generally the reproduction of the full physiological hormonal activity in primary cell cultures has proved to be deficient in many respects (Wolffe and Tata, 1984). Over the past few years our laboratory has succeeded in developing a primary *Xenopus* hepatocyte culture system in which the regulation by estrogen of vitellogenin gene expression is fully reproduced both qualitatively and quantitatively (Searle and Tata, 1981; Wolffe and Tata, 1983; Ng et al., 1984b). Among the many factors contributing to this success were the important considerations of cell density and the rapid metabolism of the hormone by primary cell cultures (Tenniswood et al., 1983). Most important of all was the realization of the initial refractory period during which the freshly prepared cells go into "culture shock" whereby they fail to respond to the hormone but that this refractoriness is reversible (Wolffe et al., 1984a).

Culture Shock and Optimization of Hormonal Response In Vitro

Figure 3 depicts the refractory period preceding the accumulation of vitellogenin mRNA as a function of time in primary culture of male *Xenopus* hepatocytes upon de novo activation of the genes. What is the cause of this refractoriness and the subsequent recovery? Although other investigators have also observed the lack of response of fresh primary cell cultures to a variety of hormones and other agents (Wolffe and Tata, 1984), no possible explanation for the phenomenon was available until now.

We discovered that the stress of isolation of cells from various tissues results in a large accumulation in the cells of stress or heat shock proteins (hsps), particularly hsp 70 (Wolffe et al., 1984a). The synthesis of these

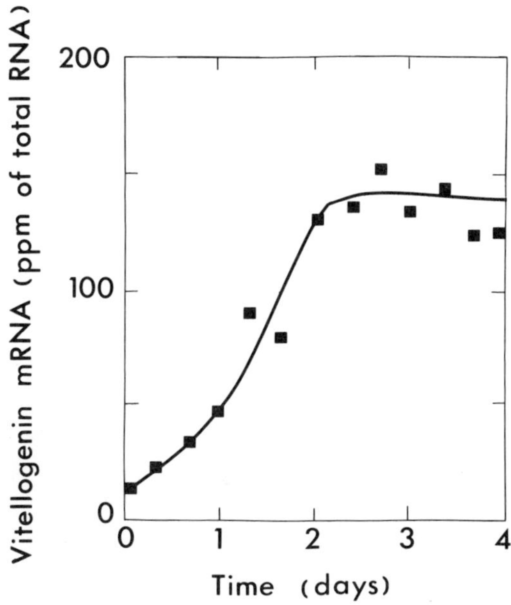

Fig. 3. Refractoriness and acquisition of full responsiveness of male *Xenopus* hepatocytes to estrogen during the first 4 days in primary culture at 26°C. Vitellogenin mRNA was measured by disc hybridization to cloned cDNA 8 h after the addition of $10^{-6}M$ estradiol-17β to hepatocytes at different times after setting up the cultures, as indicated (Wolffe et al., 1984a).

stress proteins declines markedly with time in culture and is paralleled by the reacquisition of response to external stimuli. Table 1 shows that the increasing ability of estrogen to activate vitellogenin genes, as also seen in Fig. 3, is a function of hsp level in the cells. It is now well known that the buildup and decline of hsp's is accompanied by a reversible restructuring of the cytoskeleton and modified cell-cell interactions (Schlesinger et al., 1982). It is therefore significant that loss and recovery of response to estrogen are accompanied by changes in cell shape and intercellular contacts (Table 1). This phenomenon of culture shock is not restricted to the vitellogenic response of isolated *Xenopus* hepatocytes but is likely to be generally applicable to to all inductive processes in primary cell culture (Wolffe and Tata, 1984). It will be shown later how the experimental induction of hsp's can itself be exploited as a tool to analyze the relationship between activation of vitellogenin gene transcription and accumulation of estrogen receptor in nuclei of cultured hepatocytes.

Once it was realized that freshly prepared *Xenopus* hepatocytes, isolated by collagenase treatment of the tissue, gave poor and irreproducible responses to estrogen, we developed a routine procedure of allowing the cells to recover from the stress during the first three days in culture. If the hormone was added at the end of this period, when the synthesis of hsp had declined to its lowest level, accumulation of vitellogenin mRNA occurred at very high rates over long periods of culture. Figure 4 shows that the accumulation of mRNA in culture followed a major physiological characteristic of estrogen-induced vitellogenin gene transcription, namely that the secondary response occurs more rapidly and is of a higher magnitude than the primary response (Tata and Smith, 1979). When the results of the Northern blot shown in Fig. 4 were expressed as molecules of mRNA

Table 1. Association Between hsp Synthesis, Vitellogenic Response to Estrogen, and Morphology of *Xenopus* Hepatocytes as a Function of Time and Temperature of Culture[a]

Culture conditions	^{35}S-hsp (% total)	Response to E_2 (ppm Vg mRNA)	Cell morphology (roundedness/ aggregation)
Day 0/26°C	9.1	20	+ + +/−
Day 1/26°C	4.0	50	+ +/+
Day 3/26°C	2.5	140	−/+ + +

[a]Male *Xenopus* hepatocytes, freshly prepared (0 days in culture) or cultured for 1 or 3 days at 26°C, were labeled with [^{35}S]methionine for the last 12 h in order to measure the extent of heat shock protein (hsp) synthesis. The response to estrogen (E_2) was quantitated as the amount of vitellogenin (Vg) mRNA accumulating during an 8-h period following the addition of $10^{-6}M$ estradiol at 0, 1, or 3 days in culture. Cell morphology was semi-quantitatively expressed as the proportion of cells seen by scanning electron microscopy on a relative scale of virtually all cells (+ + +) being rounded and disaggregated or being flattened and aggregated (−).
Source: Wolffe et al., 1984a.

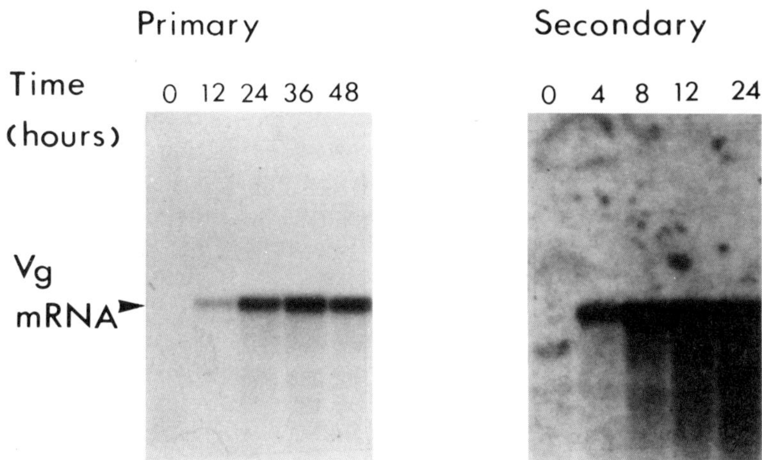

Fig. 4. Northern blot analysis of RNA from primary cultures of male *Xenopus* hepatocytes during primary and secondary response to estrogen in vitro. $10^{-6}M$ Estradiol-17β was added to naive male hepatocytes (primary response) or to male hepatocytes that had been previously exposed to estrogen but had ceased to synthesize vitellogenin mRNA (secondary response). RNA was extracted at different times indicated after the first addition of the hormone, resolved by electrophorsis, and the vitellogenin mRNA identified by hybridization to all four ^{32}P-labeled *Xenopus* vitellogenin cDNA, followed by autoradiography.

accumulating per cell, these turned out to approach those reported in whole animals (Baker and Shapiro, 1978). This optimization of the primary cell culture system has allowed us to answer questions concerning the dynamics of estrogen receptor content, the transcription of individual vitellogenin genes, and the stability of their mRNAs, hitherto not possible in vivo.

Differential Activation of *Xenopus* Vitellogenin Genes

In view of the simplicity of activating de novo the vitellogenin genes in male hepatocytes and the fidelity of their reversible induction with estrogen in tissue culture, it was thought particularly useful to monitor separately the conformational and transcriptional status of each group and member of this multigene family.

Transcription and Conformation of A and B Groups of Vitellogenin Genes

Because of the 20% coding sequence divergence between each pair of the four vitellogenin genes, our earlier studies were restricted to the relatively easier discrimination between the transcripts specified by the A and B groups of vitellogenin genes in *Xenopus laevis* (Wolffe and Tata, 1983; Williams and Tata, 1983). In one study in liver nuclei from male animals treated for the first time with estrogen, the B group genes were more sensitive at early stages to hormonal activation than were the A group genes (Williams and Tata, 1983). This was true for both general sensitivity of the genes to DNase I digestion of nuclei as well as in run-off transcription assays. These differences disappeared during secondary hormonal stimulation of males and were also not seen in females.

While run-off transcription rates declined very rapidly upon hormone withdrawal, the elevated DNase I sensitivity of vitellogenin genes was maintained for several weeks. It was only after 3–4 months that the active conformation or higher DNase I sensitivity reverted to the basal sensitivity found in naive male tissue (Fig. 5). The delayed return to the inactive gene conformation may be related to the slow turnover of liver parenchymal cells, while the rapid decay of transcriptional activation may parallel the rapid metabolism and loss of estradiol (Tenniswood et al., 1983).

The addition of estrogen directly to cultured hepatocytes also resulted in a more rapid accumulation of B group mRNA than A group transcripts at the early stages of hormonal induction. These culture experiments allowed us to establish the very rapid arrest of transcription upon hormone withdrawal and that the accumulated vitellogenin mRNA disappeared equally rapidly from the cells (Wolffe and Tata, 1983). The latter indicates that processing and maturation of mRNA must normally occur very rapidly

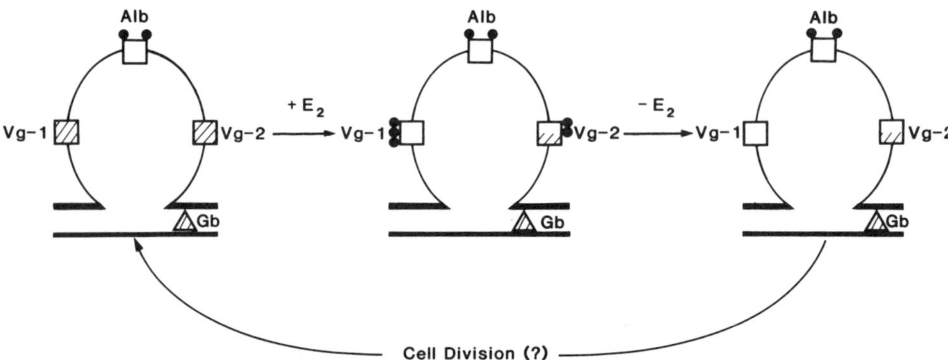

Fig. 5. Scheme depicting the enhanced overall DNase I sensitivity coupled to transcription of two *X. laevis* vitellogenin genes induced by estradiol-17β administration ($+E_2$) and the dissociation between gene conformation and transcription upon hormone withdrawal ($-E_2$). Genes Vg−1 and Vg−2 are two vitellogenin genes initially transcribed unequally in male *Xenopus* liver. The involvement of cell division in the DNase I sensitivity returning to control male levels is hypothetical. (Vg = vitellogenin; Alb = albumin; Gb = globin; ▨,◬ = inactive; □ = potentially active (DNase-sensitive); ⊓̈ = transcribed.) (Scheme based on datafrom Williams and Tata, 1983.)

without any significant intranuclear pool of unprocessed vitellogenin mRNA precursors. The changes in stability of vitellogenin mRNA coupled to those in its transcription are considered later.

Activation of Individual Vitellogenin Genes

By devising a disc assay based on very stringent "R-loop" conditions of hybridization and washing, it was possible to measure the transcripts of each individual vitellogenin gene, despite the 95% coding sequence homology between the two members of the A and B groups (Ng et al., 1984b). We were thus able to show that the rates of transcription and steady-state levels of each mRNA were not coordinately or equally regulated. In both adult male and female hepatocyte cultures the rate and extent of accumulation of vitellogenin mRNA increased in the order of B1 > A1 > A2 ≃ B2 upon the addition of estradiol (Fig. 6). Experimental manipulation of adult hepatocytes showed that this pattern of expression of the four *Xenopus* vitellogenin genes can be varied to some extent. Thus the differential expression of the A1 – B1 and A2 ≃ B2 pairs of genes was enhanced by varying the period of exposure to estradiol or the dose of the hormone (Ng et al., 1984b). It can be seen in Fig. 6 that at shorter time-intervals (up to 2 h) after addition of the higher dose of $10^{-6}M$ estradiol to cultures of naive male hepatocytes, the bulk of the mRNA was made up of A1 and B1 vitellogenin gene transcripts but the difference in concentrations of the four mRNAs was less noticeable at later times (compare Fig. 6A and B). If, however, the dose of the hormone was reduced to $10^{-8}M$, then virtually all the mRNA was made up of A1 and B1 gene transcripts, even at 12 h after hormone addition (compare Fig. 6A and C).

Since vitellogenin mRNA is very stable in the presence of estradiol (Brock and Shapiro, 1983; Wolffe et al., 1984b; Shapiro and Brock, 1985), the above pattern of initial increases in the steady-state levels of the individual mRNAs reflects the differential rate of transcription and processing of each mRNA. This was confirmed by direct measurement of the absolute rate of transcription of each gene, as shown in Table 2 (Ng et al., 1984b). It should be noted that the absolute rates of transcription shown in this table are for female hepatocytes to which estradiol was added 24 h before the measurements, thus representing a secondary induction with estrogen in contrast to the primary induction shown for male hepatocytes in Fig. 6. When similar measurements were made at earlier-times after addition of the hormone to naive male hepatocytes, the differences between the rates of transcription of the A1 – B1 and A2 – B2 pairs of genes were more noticeable than those shown in Table 2 (data not shown).

We also determined whether the unequal pattern of expression within the *Xenopus* vitellogenin multigene family is established early in devel-

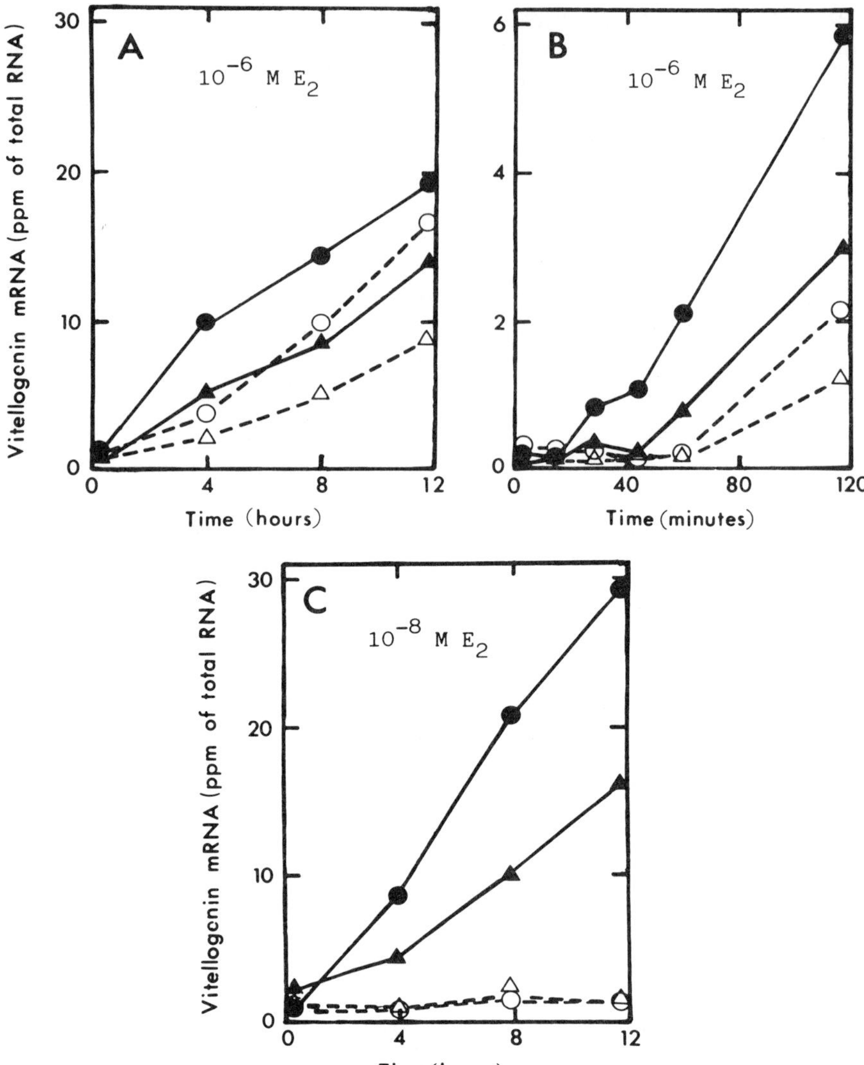

Fig. 6. Kinetics of accumulation of transcripts corresponding to the individual vitellogenin genes in male *Xenopus* hepatocytes cultures after primary induction with estradiol-17β. Hepatocytes were prepared from eight male *Xenopus* livers, cultured for 3 days before the addition of estradiol and RNA extracted from the cells (Wolffe and Tata, 1983) at the times indicated after hormone addition. Vitellogenin mRNA corresponding to genes A1 (▲), A2 (△), B1 (●), and B2 (○) was quantitated by filter disc hybridization to ^{32}P-nick translated *Hind*III excised cDNA. **A**: $10^{-6}M$ estradiol was added once at time zero; **B**: acccumulation of vitellogenin mRNAs during very early time intervals following induction by a single addition of 10^{-6} estradiol at time zero; **C**: $10^{-8}M$ estradiol was replenished in the culture medium every hour over 12 h. Note the different scales of abscissas and ordinates in A, B, and C (Ng et al., 1984b).

Table 2. Absolute Rate of Transcription of Individual Vitellogenin Genes Upon Estrogen Activation of Cultured Hepatocytes From Adult Female *Xenopus*[a]

Gene	Absolute rate of transcription (molecules $cell^{-1}h^{-1}$)
A1	265
A2	182
B1	670
B2	117

[a]RNA was labeled with 1-h pulse of [^{3}H]uridine, 24 h after the addition of $10^{-6}M$ estradiol to the cultured cells. The labeled RNA was extracted, hybridized to nonradioactive *Hind*III excised cloned cDNA corresponding to each of the four *Xenopus* vitellogenin genes, and the rate of transcription calculated as described by Wolffe and Tata, 1983.
Source: Ng et al., 1984b.

opment and whether or not it is rigidly maintained throughout development and in adult tissue. It was found that once the *Xenopus* larval hepatocytes acquire competence to respond to estrogen at about Nieuwkoop-Faber stage 61, the pattern of expression of B1 > A1 > A2 ≃ B2 genes can be discerned at all developmental stages (Fig. 7). Thus the unequal pattern of expression is maintained throughout life, although the absolute rate of transcription of all four genes increases considerably during development. The most likely explanation for the differential activation and the flexibility in the relative sensitivity of the individual members of this gene family may be different promoter strengths or variable intensities of interaction between the estrogen receptor and the transcription regulatory elements. It is therefore important to consider the relationship between estrogen receptor and vitellogenin gene conformation and transcription, as well as the similarities and differences in the sequences 5′ upstream of the individual genes.

Estrogen Receptor and Vitellogenin Gene Activation

There is now substantial evidence that the level of steroid hormone receptor determines the kinetics of regulation of transcription of specific genes in the hormonal target cell (see Eriksson and Gustafsson, 1983, for several reviews). Much of it is, however, based on transformed or neoplastic cells or in untransformed cells in which the gene is already expressed at a low level in the absence of the hormone. It has also recently been possible, with the use of defined, cloned DNA sequences, to identify the 5′ upstream regions that are likely sites of interaction with steroid

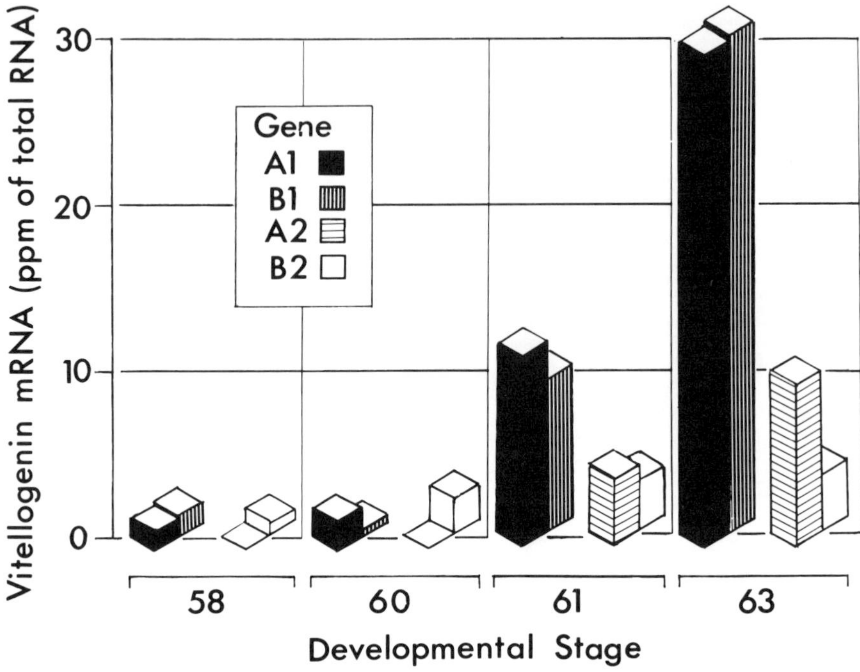

Fig. 7. Accumulation of the four vitellogenin mRNAs measured separately in the liver of *X. laevis* larvae at different stages of development in response to estradiol. Tadpoles and froglets at different developmental stages were immersed in water containing $10^{-6}M$ estradiol-17β for 3 days, after which the animals were staged, and the RNA extracted from 9–50 pooled livers. The vitellogenin mRNA levels were determined by hybridization to *Hind*III excised inserts of cloned cDNA specified by each of the vitellogenin genes, the values given being the average of two independent determinations (Ng et al., 1984b).

hormone receptor (Eriksson and Gustafsson, 1983; Chambon et al., 1984; Renkawitz et al., 1984; Jost et al., 1984; Payvar et al., 1983). These studies do not bear directly on the de novo activation of transcription nor do they establish a stoichiometric relationship between nuclear receptor and activation of gene expression under normal physiological conditions. The facts that estrogen can activate de novo the quiescent vitellogenin genes in male *Xenopus* hepatocytes in culture and that the physiological process can be faithfully and reversibly reproduced in primary cell cultures offer a unique opportunity to test directly the relationship between hormone receptor and gene transcription.

Adult male *Xenopus* liver has low levels of estrogen receptor, comprising only 200–500 molecules tightly bound to the nucleus per cell (Westley and Knowland, 1978; Hayward et al., 1980; Perlman et al., 1984). It has also been shown that treatment of naive male *Xenopus* with estrogen causes a five- to ten-fold increase in high-affinity liver nuclear receptor to reach levels found in female liver (Westley and Knowland, 1979; Hayward et al., 1980; Perlman et al., 1984). This elevated level of receptor in male hepatocytes persists for several weeks so that it may also explain the more rapid and extensive response to the hormone during secondary induction in addition to any long-lasting changes in the conformation of vitellogenin genes (Fig. 5). We have studied the upregulation by estradiol of its own receptor in primary cultures of male *Xenopus* hepatocytes and have measured the absolute rate of transcription of vitellogenin genes as a function of time (Perlman et al., 1984; Table 3, Fig. 8). Such an accurate analysis can only be possible in cell cultures, and the results depicted in Fig. 8 clearly show a stoichiometric relationship between nuclear receptor and activation of dormant genes. Stimulation with the hormone caused its own low receptor level in naive male cells to rise to those found in female cells. This was accompanied by similar enhancement of vitellogenin gene transcription to rates observed in female hepatocytes or in male cells upon secondary induction. Experiments with cycloheximide added at different times of culture showed that the small amount of receptor residing in male liver nuclei at the start of the experiment accounted for the activation of gene transcription in the first 4 h after which the increase in transcription required continuing protein synthesis for both processes (Fig. 9). They also demonstrated the reversibility of the relationship when the high receptor levels previously elevated by the hormone were depleted rapidly in the presence of the inhibitor. These experiments represent the first direct evidence for the rapid and reversible coupled induction of receptor and transcription by a steroid hormone in normal cells.

Considerable progress has been made recently with nuclease protection or "foot-printing" procedures for determining regions around steroid-regulated genes that interact with the relevant receptor. Thus, DNA sequences located between −100 and −700 bp upstream from the transcription initiation site have been implicated as the site of regulation by many steroid hormone receptors of a variety of genes, such as ovalbumin, lysozyme, uteroglobin, and MMTV (Eriksson and Gustafsson, 1983; Chambon et al., 1984; Renkawitz et al., 1984; Payvar et al., 1983; Dean et al., 1984). A consensus sequence located at −458 to −725 bp upstream in the 5′ flank

Table 3. Nuclear Estrogen (E_2) Receptor Levels and Absolute Vitellogenin Gene Transcription Rate in *Xenopus* Hepatocyte Primary Cultures[a]

Hepatocytes	Receptors/nucleus	Absolute transcription rate (mol/h/cell)
Naive male	100	180
Female	700	1800
E_2-stimulated male	700	1500

[a]Data adapted from Perlman et al., 1984.

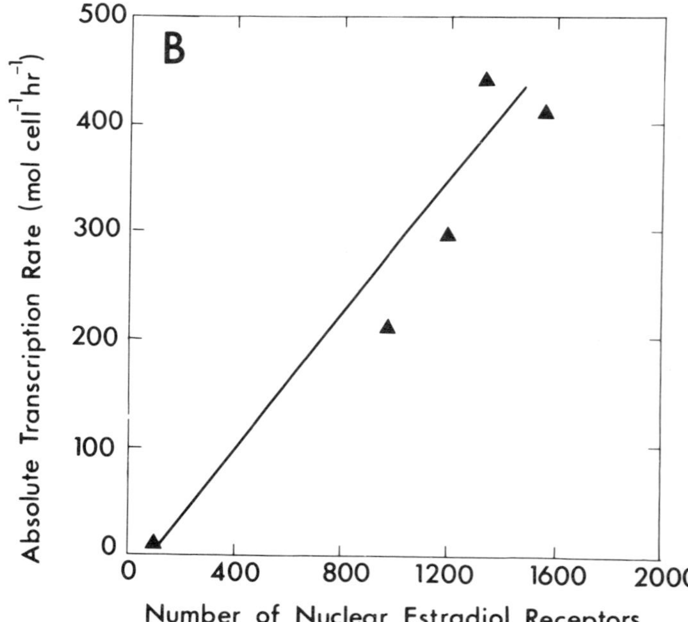

Fig. 8. Correlation of nuclear estrogen receptor levels with absolute rates of vitellogenin gene transcription. **A**: $10^{-6}M$ estradiol was added to naive male *Xenopus* hepatocytes every 4 h for varying times of primary hormonal stimulation. At the end of this period, the hepatocyte cultures were washed with fresh culture medium without estradiol and maintained for a further 36 h in the absence of estradiol. After this hormone withdrawal peroid the number of nuclear estrogen receptors in the hepatocytes (●) was determined. In parallel cultures, the hepatocytes were re-exposed to $10^{-6}M$ estradiol at the end of the withdrawal period and the absolute transcription rate of the vitellogenin genes determined during the second hour of this stimulation by pulse-labeling the cultures with [^{3}H]uridine (■). **B**: Relationship between nuclear estrogen receptor and absolute rate of transcription of vitellogenin genes in male *Xenopus* hepatocyte cultures during the second hour of secondary stimulation with estradiol following various periods of primary exposure to the hormone as in **A** (Perlman et al., 1984).

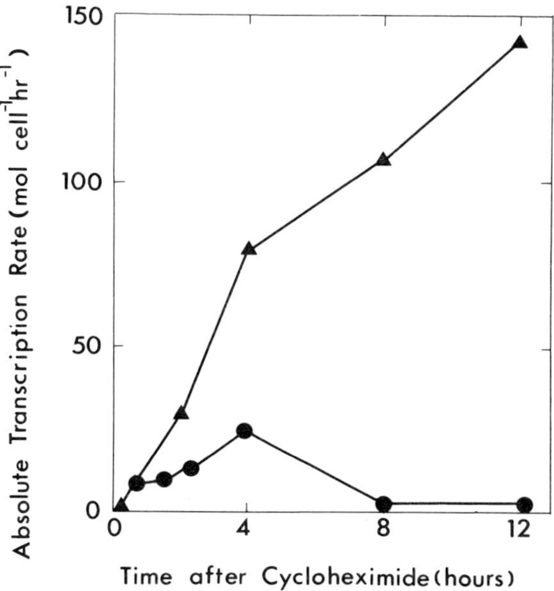

Fig. 9. Inhibition by cycloheximide of the increase in absolute transcription rate of vitellogenin genes during primary stimulation of male *Xenopus* hepatocyte cultures with estradiol. Male hepatocytes were stimulated with $10^{-6}M$ estradiol, replenished every 4 h in the presence (●) or absence (▲) of 10 μg/ml cycloheximide. At various times following estradiol stimulation the cells were pulse-labeled with [^{3}H]uridine and the absolute transcription rates determined (Perlman et al., 1984).

of the chicken vitellogenin gene II was found to be a site of interaction with estrogen receptor, as judged from DNase I protection assays (Jost et al., 1984). As shown in Fig. 10, similar core sequences have been detected in the 5′ flank of all four *Xenopus* vitellogenin genes (Walker et al., 1984). Also, the chicken apoVLDL gene, which is estrogen-regulated, but not induced de novo, in the liver has two similar sequences at about −300 bp upstream but not about −600 bp. Burch (1984) has also reported the presence of an SV40-like enhancer core sequence at a 5′ upstream nuclease hypersensitive site in the chicken vitellogenin II gene. Four 7–9 bp sequence elements were also found in this gene and apoVLDL gene, as well as in three estrogen-induced genes in the oviduct. In an ontogenic study in chick embryo liver, the apoVLDL and vitellogenin genes were not, however, simultaneously activated by estrogen (Elbrecht et al., 1984). What is of particular interest in the context of the transcription of the four *Xenopus* vitellogenin genes in our studies is the comparison of the number and location of these sequences. Whereas all four gene members have two such sequence blocks at about −310 to −375 bp, only the genes A1 and B1 have an additional element further upstream at about −663 and −554 bp, respectively. This finding raises the possibility that the greater

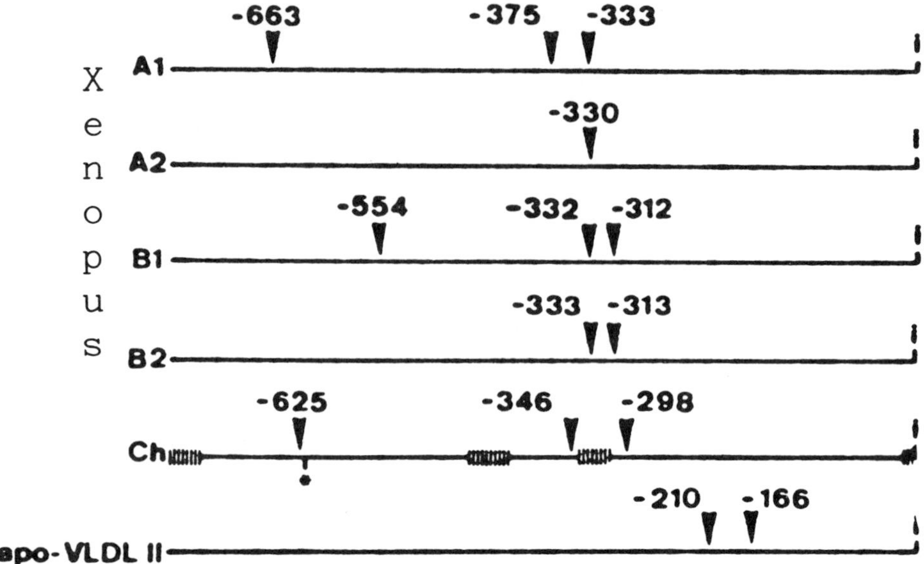

Fig. 10. Distribution of GGTCANNNTGACC element in DNA sequence 5′ upstream from transcription initiation site of *Xenopus* vitellogenin genes A1, A2, B1, and B2, as well as in chicken vitellogenin II and apoVLDL genes. (Data from Walker et al., 1984.)

sensitivity of this gene pair to activation by estrogen (see Figs. 6,7 and Table 2) may somehow result from a stronger interaction between the furthest upstream site and the estrogen receptor or that interaction with this site has an additive effect on that with the other two. It is, however, important to realize in drawing conclusions from DNA sequences alone that we know virtually nothing as yet about the influence of such factors as the higher order organization of genes in chromatin and the effect of possible distribution of different members of a gene family on different chromosomes on the regulation of gene activity.

The above discussion on estrogen receptor and transcription of vitellogenin genes is also relevant to structural considerations of the nucleus. For example, estrogen-sensitive genes when activated by the hormone have been shown to be preferentially enriched in the nuclear matrix fraction in contrast to the quiescent state of the genes, including vitellogenin genes (Barrack and Coffey, 1983; O'Malley et al., 1985; Robinson et al., 1983; Jost and Seldran, 1984). At the same time, estrogen receptor has also been found to be similarly enriched in the matrix fraction (Barrack and Coffey, 1983; Simmen et al., 1984). The exact significance of the localization of such a complex in the nuclear scaffold structure and how genes could be possibly translocated into and away from such structures by hormonal and other regulatory signals still remain to be elucidated.

The Exploitation of Heat Shock in Elucidating the Role of Estrogen Receptor in Vitellogenin Gene Expression

In our laboratory we recently exploited the thermal induction of hsp's in *Xenopus* hepatocyte cultures as a tool to dissect the early events leading to the activation by estrogen of vitellogenin genes (Wolffe et al., 1984b). Heat shock, which has been extensively studied in a variety of cells and species (Schlesinger et al., 1982), was deliberately applied to transiently arrest the accumulation of vitellogenin mRNAs. We have already seen how the competence of fresh primary cultures of *Xenopus* hepatocytes in their vitellogenic response to estrogen is affected by the accumulation of hsp-like proteins induced by cellular stress in setting up the cultures (Fig. 3, Table 1). Figure 11 shows the way in which the application of thermal shock to naive male *Xenopus* hepatocytes modifies the accumulation of vitellogenin mRNA induced by estrogen. Both the transcription and stability of the mRNA are determined by the temperature and duration of heat shock. The two effects can, however, be dissociated. At 31°C, or following a brief pulse of heat shock at higher temperatures, only transcription was affected, whereas above that temperature or for thermal shock exceeding 2 h, the vitellogenin mRNA already accumulated was rapidly degraded, even in the presence of estrogen. Direct measurement of absolute transcription rates confirmed specific transcriptional arrest by thermal stress (Fig. 12). The arrest of transcription, accompanied by an accelerated breakdown of vitellogenin mRNA, was directly correlated to the amount of newly synthesized hsp's in the cell cultures.

Other workers have shown that the rapid hormonal induction of specific proteins, such as ovalbumin, casein, and vitellogenin, results from a combination of changes in the rates of synthesis and stabilization of the mRNAs encoding these proteins (Palmiter, 1975; Matusik and Rosen, 1978; Brock and Shapiro, 1983). Indeed, in every instance the enhancement of mRNA stabilization is of a higher magnitude than its transcription. Figure 13A shows the marked stability of vitellogenin mRNA in cultured *Xenopus* hepatocytes in the continuous presence of estradiol. Upon withdrawal of hormone, the mRNA is destabilized and disappears with a $t_{1/2} \simeq 15$ h, in agreement with the findings of other workers (Brock and Shapiro, 1983; Shapiro and Brock, 1985). Interestingly, even the presence of the hormone fails to stabilize the vitellogenin mRNA at 34°, which is degraded with a $t_{1/2} \simeq 3$ h and which explains the results on steady-state levels of vitellogenin mRNA shown in Fig. 11. The stabilization of vitellogenin mRNA at the normal temperature of culture (26°) is highly specific for the induced message since it did not protect albumin mRNA similarly (Fig. 13B). In fact, in complete contrast to vitellogenin mRNA, the albumin mRNA was less stable in the presence of estrogen than in its absence (Wolffe et al., 1985), a finding that partly explains the well-known de-induction of albumin syn-

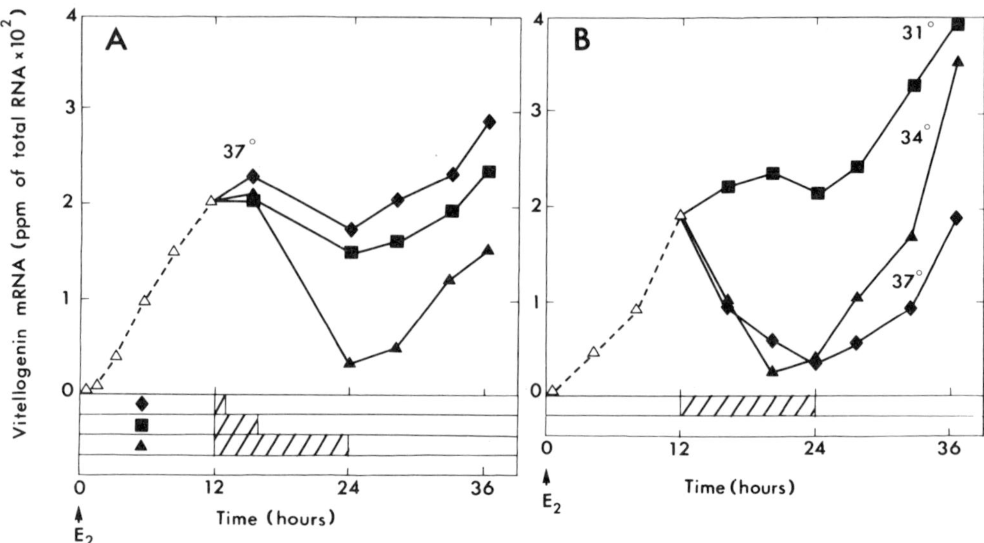

Fig. 11. Effect on estrogen-induced vitellogenin mRNA accumulation in hepatocyte cultures of **(A)** varying periods of heat shock at 37°C and **(B)** recovery from heat shock at different temperatures. Male *Xenopus* hepatocytes were treated with $10^{-6}M$ estradiol, replenished every 4 h throughout the experiment except where stated otherwise. Vitellogenin mRNA was quantitated at different times of incubation indicated. **A**: After 12 h of incubation with the hormone at the normal incubation temperature of 26°C, the cells were transferred to 37°C in the absence of the hormone. The heat shock peroid (□) was maintained with equal batches of cells for 1 (♦), 4 (■), or 12 (▲) h. Twenty-four hours after the beginning of the experiment the cells were returned to 26°C in the presence of the hormone. **B**: After 12 h at 26°C, equal batches of hepatocytes were incubated at 31°C (■), 34°C (▲), or 37°C (♦). Twelve hours later the cells were returned to 26°C in the presence of estradiol replenished every 4 h. Vitellogenin mRNA accumulation was measured at every 4-h interval (Wolffe et al., 1984b).

thesis accompanying the induction by estrogen of vitellogenin synthesis (Tata and Smith, 1979; Farmer et al., 1978; Kazamaier et al., 1985).

The marked inhibition of vitellogenin gene transcription by heat shock was also correlated with a striking effect on estrogen receptor level or activity in hepatocyte nuclei. Experiments described in Table 4 show how heat shock applied in the absence of the hormone causes a total loss of estrogen receptor in male cells in which the receptor had previously been upregulated by the hormone. The presence of the hormone during heat shock or exposure to it just prior to the application of the stress substantially protects the receptor. The exact cause or significance of this phenomenon of protection against thermal shock is not clear. Whatever these may be, heat shock can be a valuable tool in manipulating the level of steroid receptor and steroid-induced mRNA in studies designed to understand more fully the role of hormones in regulating gene expression.

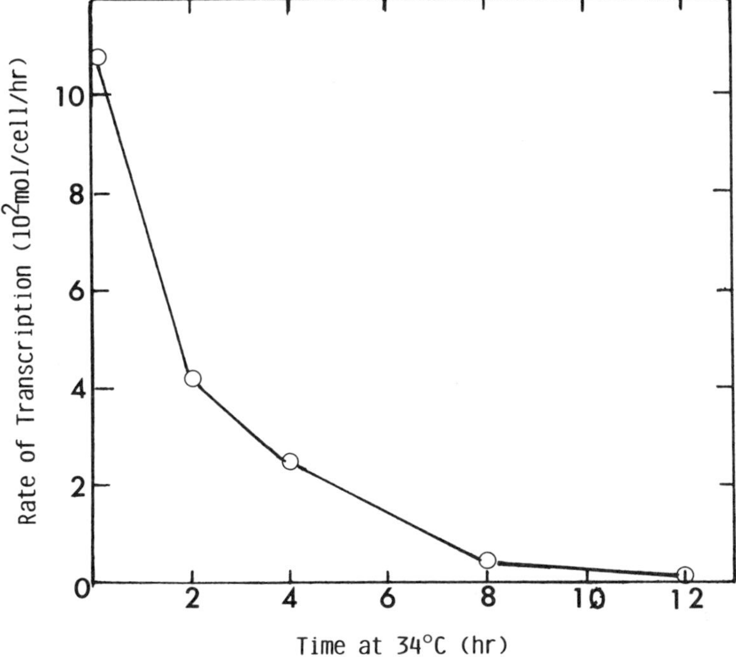

Fig. 12. Arrest of transcription of estrogen-induced vitellogenin genes by heat shock in primary cultures of male *Xenopus* hepatocytes. Twelve hours after the addition of $10^{-6}M$ estradiol at 26°C, cells were shifted to 34°C for different periods of time indicated and the absolute transcription rate determined as described by Wolffe et al., 1984b.

Switching on Silent Vitellogenin Genes in Isolated Nuclei

We have seen in Fig. 9 that the primary activation of vitellogenin gene transcription by estrogen added to male *Xenopus* hepatocyte cultures is highly sensitive to cycloheximide. One interpretation of this result would be that inhibition of protein synthesis blocks the buildup of estrogen receptor necessary to sustain an increasing accumulation of vitellogenin mRNA. It does not however rule out the possibility that an early effect of estrogen is to induce a protein, probably short-lived, that is essential for specifically activating or maintaining in an active conformation the induced genes. Such a process would be compatible with early ideas on "initial protein" rapidly induced by estrogen in mammalian uteri (Gorski et al., 1965; Baulieu, 1972).

In an attempt to determine more directly the possible presence of hormone-specific transcriptional factors, besides the enhanced accumulation of the receptor, we recently tested the effects of tissue extracts from con-

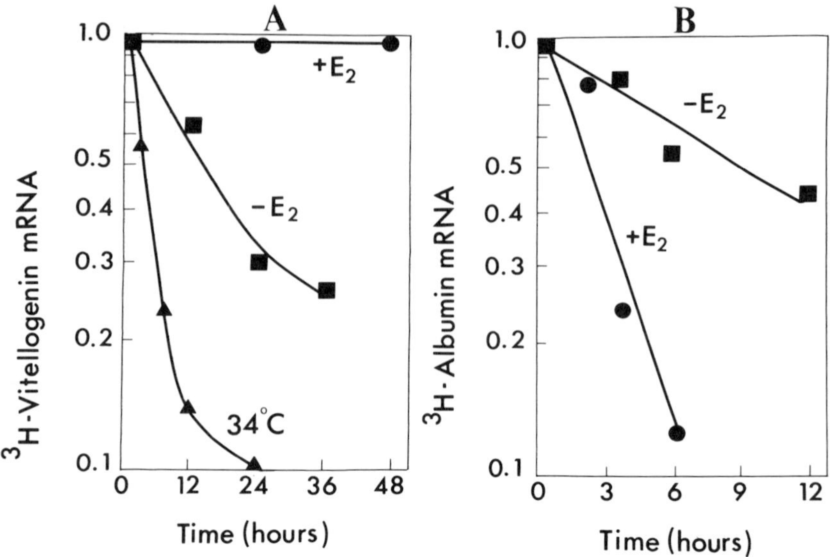

Fig. 13. **A**: Rate of disappearance of estrogen-induced vitellogenin mRNA in the presence and absence of estrogen and during heat shock. Male hepatocytes were incubated for 12 h at 26°C with a single addition of $10^{-6}M$ estradiol (E_2) in 2 ml of culture medium containing [^{3}H]uridine. The cultures were then transferred to fresh medium containing $10^{-6}M$ estradiol, replenished every 4 h, and incubated at either 26°C (●), or 34°C (▲). Another batch of hepatocytes (■) was transferred to fresh estradiol-free culture medium and incubated at 26°C. One hour later, total RNA was extracted at the times indicated and the amount of radioactive vitellogenin mRNA present per unit mass determined by hybridization to cloned vitellogenin cDNA. **B**: Destabilization of the 74 (kilo daltons) albumin mRNA in *Xenopus* hepatocytes exposed to estradiol. Male *Xenopus* hepatocyte cultures were incubated for 11 h without estrogen in 2-ml culture medium containing [^{3}H]uridine. The cultures were then transferred to fresh medium either containing 1 μM estradiol, replenished every 4 h (▲), or not (△). RNA was extracted at the times indicated and the amount of radioactive 74 kDa albumin mRNA per unit mass determined by hybridization to cloned 74 kDa albumin cDNA (Wolffe et al., 1985).

trol and estrogen-treated male and female *Xenopus* on in vitro transcription of several active and silent genes in isolated nuclei (Tata and Baker, 1985). The tissue extracts used were "soluble" (supernatents derived after centrifugation at 100,000 *g* for 24 h) and termed S-100. Figure 14 summarizes the general procedure adopted to assay for specific transcription-promoting activity in these extracts. Two features of the assay found to be critical for its success were 1) the amount of S-100 added to the nuclei, and 2) the requirement of a preincubation period of 45–90 min of nuclei and S-100 before the nucleotides were added to start the transcription reaction.

Figure 15 summarizes the results of a series of experiments in which it

Table 4. Effect of Heat Shock on Total and Nuclear Estrogen Receptor Levels in the Presence and Absence of Estradiol in Cultured Male *Xenopus* Hepatocytes[a]

Treatment	Receptors per cell	Receptors in nuclei
None	1200	420
Estradiol alone	1200	1200
Heat shock alone	0	0
Heat shock with E_2	1200	500
E_2 for 1 h, then heat shock with E_2	900	680
Heat shock alone, then recovery at 26°C for 4 h	200	120
Heat shock alone, then recovery at 26°C for 20 h	600	480

[a]After 3 days in estrogen-free medium, male hepatocyte cultures were incubated for 6 h with $10^{-6}M$ estradiol (E_2). The cells were transferred to hormone-free medium and allowed to incubate for 12 h to enable the cells to metabolize completely any remaining estradiol. The cells were then incubated in estrogen-free or $10^{-6}M$ estrogen-supplemented medium at normal temperature of 26°C or at the heat-shock temperature of 34°C, as indicated, for 12 h.
Source: Wolffe et al., 1984b.

was possible to demonstrate a specific activation de novo of transcription of the silent vitellogenin genes by nuclei isolated from control male *Xenopus* hepatocytes. Thus, preincubation of the homologous nuclei and S-100 from untreated male *Xenopus* revealed albumin as the only major newly synthesized mRNA without any transcription of vitellogenin sequences (Fig. 15a). However, preincubation and incubation of the same nuclei with S-100 from hormonally treated male *Xenopus* switched on the vitellogenin genes (Fig. 15b). That this effect was due to de novo initiation of transcription in vitro can be further seen from the marked inhibition of vitellogenin, but not albumin, gene transcription in the presence of heparin, a widely used inhibitor of initiation of transcription (Fig. 15c).

The above switching on in vitro is tissue-specific to some extent. The competent S-100 (from estrogen-stimulated male liver) failed to activate vitellogenin mRNA synthesis in erythrocyte (Fig. 15d,e) and oviduct (not shown) nuclei. In other experiments on tissue specificity, we also analyzed in vitro synthesized nuclear transcripts for an uncharacterized oviduct-specific estrogen-inducible mRNA termed "6G" (James et al., 1985). As shown in Table 5, the hormonally competent liver S-100 failed to activate the dormant vitellogenin genes in oviduct nuclei, although 6G continued to be transcribed. Conversely, an S-100 from adult *Xenopus* oviduct, in which 6G gene is expressed, also failed to induce the transcription of this gene in male liver nuclei. In a recent study, the transcription by embryonic chicken liver nuclei of a hybrid gene containing the chicken vitellogenin

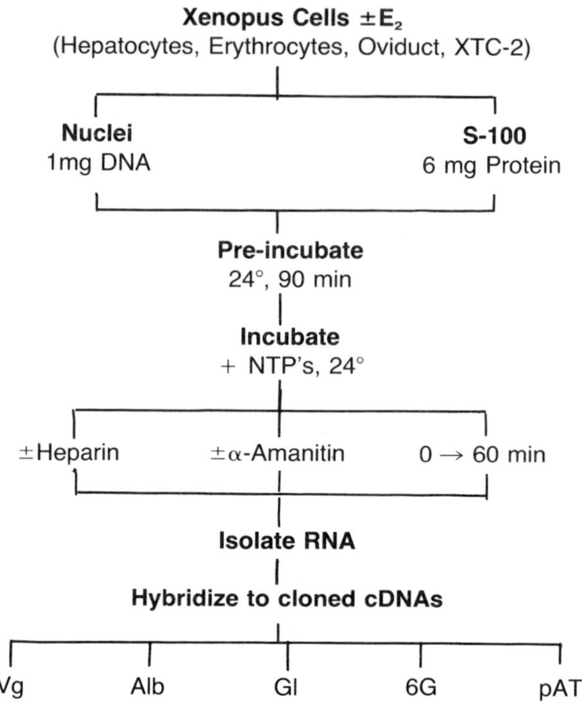

Fig. 14. Scheme summarizing the procedure for studying the hormone- and tissue-specific effects of S-100 extracts on transcription of specific genes in isolated nuclei. XTC-2 is a *Xenopus* embryo derived cell line. Vg = vitellogenin; Alb = albumin; Gl = globin; 6G = a cloned cDNA to an unidentified, estrogen-regulated gene expressed in the oviduct; pAT = plasmid DNA alone. (For details see Tata and Baker, 1985.)

gene was enhanced by the addition of nuclear estrogen receptor but the endogenous vitellogenin gene in the nuclei was not activated by the estrogen receptor preparation (Jost et al., 1985). Until purified *Xenopus* estrogen receptor is available, studies such as those above, or those based on microinjection of crude receptor-containing preparations into *Xenopus* oocytes (Tata et al., 1983; Knowland et al., 1984), do not rule out the possibility that both receptor and some other transcriptional factor(s) are required together to regulate hormone-specified genes. Since both *Xenopus* liver and oviduct are targets for estrogen, it is unlikely that the hormone receptor is the only active component in our extracts responsible for switching on dormant vitellogenin genes, but that some other factor(s), probably tissue specific, must also be involved.

These in vitro experiments on transcription raise several new questions and offer new experimental approaches to earlier questions. For example, it would be important to characterize the factor(s) conferring the expression

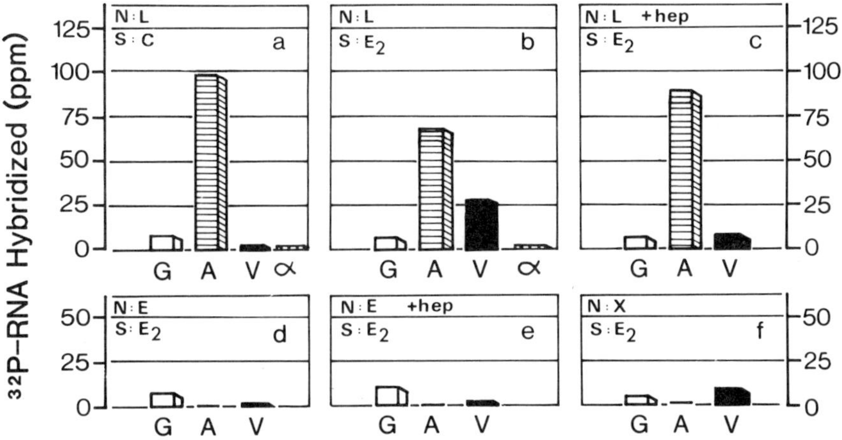

Fig. 15. In vitro synthesis of globin (□, G), albumin (□, A), and vitellogenin (■, V) mRNA by nuclei from control male *Xenopus* liver, XTC-2 cells, and erythrocytes following preincubation with S-100 extracts from control and estrogen-treated male *Xenopus* liver. Different combinations of nuclei (1.0–1.2 mg DNA) and S-100 (6 mg protein) were preincubated at 24°C for 90 min before incubation at the same temperature for 45 min upon the addition of the ^{32}P-labeled and unlabeled nucleotides. ^{32}P-RNA hybridizing to the different cloned cDNAs was then measured as described in the text. Nuclei: control male liver (**a, b, c**); erythrocytes (**d, e**); XTC-2 cells (**f**). S-100: control male liver (**a**); estrogen-treated male liver (**b, c, d, e, f**). Heparin was added to incubations (**c**) and (**e**) to reduce initiation of transciption (Tata and Baker, 1985).

Table 5. Tissue Specificity of Activation of Vitellogenin and 6G Genes as Seen by Coincubation of Nuclei and S-100 Fractions From *Xenopus* Oviduct and Male Liver[a]

		Rate of transcription (ppm)	
Nuclei	S-100	Vitellogenin	6G
Male liver	Male liver	1	0
Male liver	Oviduct	2	0
Oviduct	Male liver + E_2	0	75

[a]Nuclei and S-100 were preincubated for 90 min before the nucleotides were added and the transcription reaction carried out for a further 45 min in the absence of heparin. 6G denotes an estrogen inducible gene expressed in *Xenopus* oviduct but not in the liver. S-100 denotes the post-nuclear supernatent obtained after centrifugation at 100,000g for 24 h. (For details see Tata and Baker, 1985.)

status on vitellogenin genes. Our in vitro system provides a valuable new assay system for this characterization and for following the fractionation of the active components. Although the S-100 preparation is a soluble extract, we do not know if the regulatory factor(s) is of nuclear or cytoplasmic origin. In fact, estrogen receptor itself, which is located in the nucleus, can be easily extracted into a S-100 fraction (King and Greene, 1984). It is also relevant to note that nuclear factors have been extracted which confer transcriptional specificity and DNase I hypersensitivity on 5′ upstream regions of the adult chicken β-globin gene (Emerson et al., 1985). An important question to resolve is to know if estrogen induces the factor(s) present in the competent S-100 extract and its stability upon withdrawal of the hormone. Our in vitro transcription system, in combination with primary cell cultures, is ideally suited for further analysis of the tissue specificity of steroid hormonal regulation of gene expression.

Concluding Remarks

The activation de novo of normally permanently silent vitellogenin genes in male *Xenopus,* followed by their sustained transcription, inactivation and re-induction, by manipulation with estrogen, has provided a unique experimental system toward furthering our understanding of steroid hormonal regulation of gene expression. In particular, the qualitative and quantitative reproduction of the physiological process in primary cell culture has further enhanced its value. The latter was made possible by our demonstration of the important role played by stress proteins during the setting up of primary cultures, while the artificial induction of hsp's itself can be a potentially valuable tool in dissecting the early events associated with the dynamics of estrogen receptor and regulation of transcription and stability of specific genes and their products.

In this article we have considered steroid receptors, gene structure and transcription without any reference to the organization of the cell nucleus. Yet, it is clear that the topology and structural complexity of the nucleus play a major role in determining gene structure and function. For example, there is growing evidence now that expressed genes may form looped out structures attached at both the 5′ and 3′ ends to a rigid nuclear scaffold, matrix, or cage (Barrack and Coffey, 1983; Hancock, 1982; Mirkovitch et al., 1984). The major regulatory elements controlling gene transcription are located in the first 200–300 bp of 5′ upstream flanking sequence of most eukaryotic and viral genes (see O'Malley, 1982, for review). It is therefore significant that a recent study found that the attachment of histone and hsp genes to the nuclear scaffold would be located further upstream from such elements (Mirkovitch et al., 1984). It may also be relevant that many steroid hormone receptor binding sites are located in the −300 to −700 bp region (Eriksson and Gustafsson, 1983; Chambon et al., 1984; Renkawitz et al., 1984; Payvar et al., 1983; O'Malley et al., 1983; Walker et al., 1984; Dean et al., 1984; Jost et al., 1984), i.e., between the point of attachment to the matrix and the transcription regulatory elements. The generality of such an organization at the 5′ attachment site can be tested with a number of cloned steroid hormone regulated genes that are now available. On the other hand, we know virtually nothing about the possible 3′ attachment site or the regulatory elements downstream from the coding part of the gene. In the light of recent findings on enhancer sequences (Banerji et al., 1983; Yaniv, 1984), it is also worth considering the possibility that sequences implicated in development or tissue-specific regulators may lie elsewhere than in the canonical 5′ flank of genes. Meanwhile, it is tempting to speculate that whereas the 5′ end of gene in an active, matrix-associated loop may contain the already well-defined transcription regulatory signals, those at the 3′ end of the loop may contain some elements determining developmental or tissue-specific signals (Ng et al., 1984a).

The establishment of an in vitro transcription system, described earlier in this chapter, is particularly important for an eventual explanation of the molecular basis for the tissue specificity of steroid hormone action. This problem is illustrated in Fig. 16 for *Xenopus* liver and oviduct. Thus, the same hormonal signal impinging on two adult tissues induces two different sets of genes. As we have seen in this chapter, estrogen activates

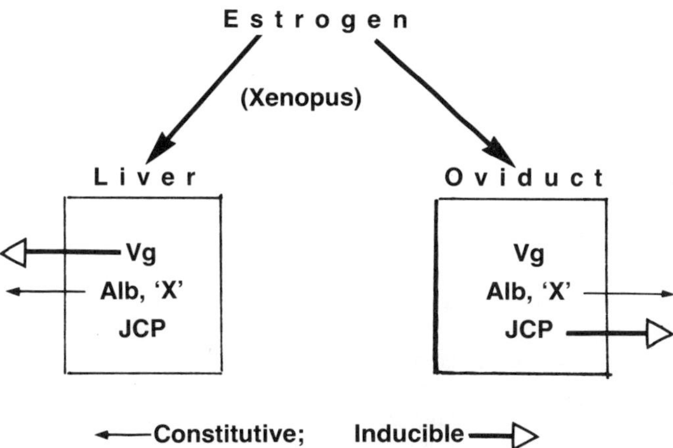

Fig. 16. Scheme depicting the tissue specificity of regulation of gene expression by estrogen in *Xenopus* liver and oviduct. Heavy *arrows* relate to genes induced by estrogen while light *arrows* indicate genes that are constitutively expressed in each tissue. Genes without an *arrow* are silent. Vg = vitellogenin; JCP = jelly coat protein; Alb = albumin; X = unknown gene constitutively expressed in the oviduct.

the vitellogenin genes in *Xenopus* hepatocytes but not in the oviduct, whereas in the latter tissue it activates genes coding for as yet poorly defined jelly coat proteins (James et al., 1985). In the chicken the egg white proteins would be somewhat analogous to amphibian jelly coat proteins, so that it is interesting to note the similarities and differences in the conformation of vitellogenin and ovalbumin genes in chicken liver and oviduct nuclei (Burch and Weintraub, 1983). It is generally accepted that steroid receptors are identical in different tissues, hence the differential regulation of gene expression must result from the action of some tissue-specific transcription factor(s) or structural chromatin component(s). Both sets of factors can be conveniently studied by the in vitro "reconstitution" system described here. An eventual explanation of tissue specificity would not only further our understanding of steroid hormone action but also of the acquisition of phenotypic distinctiveness during cellular differentiation.

Acknowledgments. We wish to thank Ena Heather for the preparation of this manuscript.

References

Baker BS, Steven J, Tata JR (1985) Comp Biochem Physiol 821:497
Baker HJ, Shapiro DJ (1978) J Biol Chem 253:4521
Banerji J, Olson L, Schaffner W (1983) Cell 33:729
Barrack ER, Coffey DS (1983) In: Eriksson H, Gustafsson J-A (eds) Steroid Hormone Receptors: Structure and Function. Elsevier, Amsterdam. p 221
Baulieu E-E (1972) Proc IV Internat Congress Endocrinol. p.30
Brock ML, Shapiro DJ (1983) Cell 34:207
Burch JBE (1984) Nucl Acids Res 12:1117
Burch JBE, Weintraub H (1983) Cell 33:65
Chambon P, Dierich A, Gaub M-P, Jakowlev S, Jongstra J, Krust A, LePennec J-P, Oudet P, Rendelhuber T (1984) Recent Prog Horm Res 40:1
Dean DC, Gope R, Knoll BJ, Riser ME, O'Malley BW (1984) J Biol Chem 259:9967
Elbrecht A, Lazier CB, Protter AA, Williams DL (1984) Science 225:639
Emerson BM, Lewis CD, Felsenfeld G (1985) Cell 41:21
Eriksson H, Gustafsson J-A (1983) Steroid Hormone Receptors: Structure and Function. Elsevier, Amsterdam
Farmer SR, Henshaw EC, Berridge MV, Tata JR (1978) Nature (London) 273:401
Germond J-E, Walker P, ten Heggeler B, Brown-Luedi M, de Bony E, Wahli W (1984) Nucl Acids Res 12:8595
Gorski J, Noteboom WD, Nicolette JA (1965) J Cell Comp Physiol 66:91
Hancock R (1982) Biol Cell 46:105
Hayward MA, Mitchell TA, Shapiro DJ (1980) J Biol Chem 255:11308
Jaggi RB, Wyler T, Ryffel GU (1982) Nucl Acids Res 10:1515
James TC, Maack CA, Bond UM, Champion J, Tata JR (1985) Comp Biochem Physiol 80B:89
Jost J-P, Seldran M (1984) EMBO J 3:2005

Jost J-P, Seldran M, Geiser M (1984) Proc Natl Acad Sci USA 81:429
Jost J-P, Geiser M, Seldran M (1985) Proc Natl Acad Sci USA 82:988
Kazamaier M, Bruning E, Ryffel GU (1985) EMBO J 4:1261
King WJ, Greene GL (1984) Nature (London) 307:745
Knowland J, Theulaz I, Wright CVE, Wahli W (1984) Proc Natl Acad Sci USA 81:5777
Matusik RJ, Rosen JM (1978) J Biol Chem 253:2343
Mirkovitch J, Mirault M-E, Laemmli UK (1984) Cell 39:223
Ng WC, Baker BS, Tata JR (1984a) FEBS Lett 178:217
Ng WC, Wolffe AP, Tata JR (1984b) Dev Biol 102:238
O'Malley BW (ed) (1982) Gene Regulation. Academic Press, New York.
O'Malley BW, et al. (1983) In Roy AK, Clark JH (eds) Gene Regulation by Steriod Hormones II. Springer-Verlag, New York
Palmiter RD (1975) Cell 4:189
Payvar F, DeFranco D, Firestone GL, Edgar B, Wrange O, Okret S, Gustafsson J-A, Yamamoto KR (1983) Cell 35:381
Perlman AJ, Wolffe AP, Champion J, Tata JR (1984) Mol Cell Endocrinol 38:151
Renkawitz R, Schutz G, Van der Ahe D, Beato M (1984) Cell 37:503
Robinson ST, Small D, Idzerda R, McKnight GS, Vogelstein B (1983) Nucl Acids Res 11:5113
Ryffel GU, Wahli W (1983) In: Maclean N, Gregory SP, Flavell RA (eds) Eukaryotic Genes. Butterworths, London. p 329
Sato GH, Pardee AB, Sirbasku DA (eds) (1982) Growth of Cells in Hormonally Defined Media. Vol 9, Cold Spring Harbor Conferences on Cell Proliferation.
Schlesinger MJ, Ashburner M, Tissieres A (eds) (1982) Heat Shock. From Bacteria to Man. Cold Spring Harbor Laboratory
Searle PF, Tata JR (1981) Cell 23:741
Shapiro DJ, Brock ML (1985) In Litwack G (ed) Biochemical Actions of Hormones XII. Academic Press, Orlando. p 139
Simmen RCM, Means AR, Clark JH (1984) Endocrinol 115:1197
Tata JR (1982) In O'Malley BW (ed) Gene Expression. Academic Press, New York. p 291
Tata JR, Baker BS (1985) EMBO J 4:3253
Tata JR, James TC, Watson CS, Williams JL, Wolffe AP (1983) In: Porter R, Whelan J (eds) Molecular Biology of Egg Maturation. Pitman Books, London (Ciba Foundation Symp 98). p 96
Tata JR, Smith DF (1979) Recent Prog Horm Res 35:47
Tata JR, Wolffe AP, Perlman AJ, Ng WC (1985) In: Smuckler EA, Clawson GA (eds) Nuclear Envelope Structure and RNA Maturation. Alan R. Liss, New York. p 357
Tenniswood MPR, Searle PF, Wolffe AP, Tata JR (1983) Mol Cell Endocrinol 30:329
Wahli W, Dawid IB, Ryffel GU, Weber R (1981) Science 212-298
Wahli W, Dawid IB, Wyler T, Jaggi RB, Weber R, Ryffel GU (1979) Cell 16:535
Wahli W, Germond J-E, Heggeler BT, May FEB (1982) Proc Natl Acad Sci USA 79:6832
Walker P, Brown-Luedi M, Germond J-E, Wahli W, Meijlink FCPW, van het Schip AD, Roelink H, Gruber M, AB G (1983) EMBO J 2:2271
Walker P, Germond J-E, Brown-Luedi M, Givel F, Wahli W (1984) Nucl Acids Res 12:8611

Westley B, Knowland J (1978) Cell 15:367
Westley B, Knowland J (1979) Biochim Biophys Res Commun 88:1167
Williams JL, Tata JR (1983) Nucl Acids Res 11:1151
Wolffe AP, Glover JF, Martin SC, Tenniswood MPR, Williams JL, Tata JR (1985) Eur J Biochem 146:489
Wolffe AP, Glover JF, Tata JR (1984a) Exp Cell Res 154:581
Wolffe AP, Perlman AJ, Tata JR (1984b) EMBO J 3:2763
Wolffe AP, Tata JR (1983) Eur J Biochem 130:365
Wolffe AP, Tata JR (1984) FEBS Lett 176:8
Wyatt GR (1980) In Locke M, Smith DS (eds) Insect Biology of the Future Academic Press, New York. p 201
Yaniv M (1984) Biol Cell 50:203

Discussion of the Paper Presented by J.R. Tata

ROY: Would you like to comment on the report of extensive cell division in the frog liver after estrogen administration and its implication in the regulation of vitellogenin synthesis. Also, the tissue-specific factor, which you described, can that be a secondary mediator produced by the estrogen?

TATA: I don't know the answer to the second question. As to the first question, the observations of Dr. Lawrence Wangh's are irrelevant to our present studies, in that it is only after several days of estrogen treatment that one observes cell proliferation in the liver. In our studies, we are dealing with acute, very short-term effects. I don't think that the two are related.

O'MALLEY: This interesting effect from the S 100, what is the time-course of that? What is the relative degree of effect on initiation versus elongation in the assay?

TATA: It is difficult to deduce those values from the histograms I presented, but I would say that the values for vitellogenin transcripts initiated are about 10% of the overall rate of elongation when we added competent S 100 to naive male liver nuclei.

O'MALLEY: So you'll make radioactive RNA in assay, for what period of time in vitro?

TATA: About one hour.

O'MALLEY: About one hour, so then it plateaus? Can you add more S 100?

TATA: We haven't done that experiment.

O'MALLEY: How about antiestrogen-treated S 100?

TATA: The trouble is that in frogs we haven't yet found a good antiestrogen. Tamoxifen has an agonist effect.

O'MALLEY: But you have tried some other hormones?

TATA: Yes. We get no effect of any other hormone, except estrogen.

O'MALLEY: Thyroxine, that was a particular interest. Have you tried it?

TATA: Thyroxine has no effect if you don't have estrogen. Estrogen is obligatory.

RINGOLD: I guess I'm a little unclear about the way the experiment is done. You isolate nuclei from naive animals in which there is essentially no vitellogenin transcription ongoing, and you incubate that with S 100 with estrogen-treated cells. You have to preincubate them for 90 min, and then you see some vitellogenin transcription. I'm not sure I understand the distinction between the initiation and

elongation. It seems if you see it, it has to be initiated since there was no initiation before. I'm not sure what you are talking about.
TATA: What I was talking about was that what was measured was a combination of initiation and elongation.
RINGOLD: When do you add the heparin?
TATA: Well, we've added it at various times. In that particular experiment described in my talk it was added right at the beginning.
RINGOLD: But then I don't understand if heparin is blocking initiation, all the transcription had to have been initiated. The second issue is that it is conceivable that you have some vitellogenin RNA in the S 100, and during that preincubation period you allow some potential RNA DNA hybrids to form and are actually elongating and initiating off of those RNAs rather than from authentic transcription and start sites.
TATA: We cannot detect any vitellogenin message in S 100. We have also done the converse experiments in which we've taken nonradioactive DNA, scaled up the whole experiment 10- to 20-fold, and then carried out slot blots using labeled probes. That again confirms that it's not likely to be due to some message that was present in the S 100.
MILGROM: My question was about where it initiated. Have you done any S1 mapping experiments in this? Was the proper promoter used and so on?
TATA: No, not yet.
MILGROM: The second short question was at what ionic strength do you prepare the S 100? In other terms, do you get receptors in there or is it completely depleted?
TATA: That is a very good question because there is some receptor in our S 100 preparations, especially from estrogen-treated animals. But I don't think that receptor is the only factor producing the effects we observed. The ionic strength of S 100 preparations is 0.15.

Discussants: E. MILGROM, B. O'MALLEY, G. RINGOLD, A.K. ROY, and J.R. TATA

Chapter 14

Intra- and Intercellular Aspects of the Hormonal Regulation of α_{2u}-Globulin Gene Expression

A.K. ROY, F.H. SARKAR, C.V.R. MURTY, D. MAJUMDAR, AND W.F. DEMYAN

Introduction

Initial studies of Jensen and Jacobsen (1962), followed by results from many other laboratories, have led to the emergence of the present concept of molecular mechanism of hormone action whereby steroid hormones can ultimately influence gene expression by site-specific receptor-DNA interaction (Payvar et al., 1981). Using various model systems, most of the investigators have come to the conclusion that steroid hormones generally regulate cell functions through pretranslational regulation of target gene transcription. All of these conceptual models are based on the effect of one particular steroid hormone on a specific cell type. However, recent results from several sources have indicated the important, and in some cases critical, role of multicellular interactions in the mediation of hormone action (Kratochwil, 1969; Sakakura et al., 1976; Cunha et al., 1983). Initial studies in our laboratory were directed to explore the molecular mechanism of steroidal regulation of α_{2u}-globulin synthesis in the rat liver. Further studies of the α_{2u}-globulin model led to the appreciation of the multihormonal regulation of α_{2u}-globulin gene expression (Roy, 1979; Roy et al., 1983). In this chapter we highlight recent results from our laboratory which indicate that the tissue architecture and cell-cell interactions can profoundly influence the ultimate hormone responsiveness in the regulation of α_{2u}-globulin gene expression.

About 80% of the cellular mass of the liver is composed of histologically indistinguishable parenchymal cells that are also known as hepatocytes. Hepatocytes are epithelial cells derived from the ectoderm during development. The rest of the liver is composed of Kupffer cells (belonging to the reticuloendothelial system), fibroblasts, and cellular components of blood vessels and bile ducts. Although these cells have diverse developmental origin, it is of interest to note that rudiments of blood vessels are laid down as aggregations of mesenchymal cells.

Both from architectural and hemodynamic standpoints, the mammalian liver can be subdivided into a collection of closely packed polygonal prisms called lobules. Within this lobular structure, incoming blood is received from the branches of the hepatic artery and the portal vein at the periphery of the vertices of the polygon (periportal zone), while the outflowing blood drains into a single central vein located at the center of the polygon (perivenous zone). Such a vascular arrangement creates a biochemical gradient between the periportal and perivenous regions of the hepatic lobule. In this gradient environment, hepatocytes situated around the periportal region are preferentially exposed to blood enriched in O_2, nutrients, and endocrine factors. This hemodynamic arrangement is manifested in the intralobular zonation of synthetic and catabolic activities during feeding and fasting cycles. In fact, enzyme histochemistry and microdissection analysis have clearly demonstrated differential distribution of key metabolic enzymes within the periportal and perivenous zones (Jungermann and Sasse, 1978).

α_{2u}-Globulin and Its Hormonal Regulation

α_{2u}-Globulin is the major urinary protein of the mature male rat (Roy et al., 1966; Roy et al., 1983). This protein is synthesized and secreted by the hepatic parenchymal cells. α_{2u}-Globulin is coded by a multigene family comprising about 30 gene copies per haploid genome (Kurtz, 1981). Primarily because of its low molecular weight, it is rapidly filtered through the kidneys and appears as the principal urinary protein. It is normally present only in the liver of mature male rats, where it appears at puberty (approximately 40 days) and reaches a peak level at about 75–85 days of age. A gradual age-dependent decline begins about 150 days, and ultimately about 800 days there is a total cessation of α_{2u}-globulin synthesis (Fig. 1). α_{2u}-Globulin can be induced in the female rat after ovariectomy followed by androgen treatment. Besides androgen, several other hormones including pituitary growth hormone, insulin, glucocorticoid, and thyroxine synergistically influence the multihormonal regulation of α_{2u}-globulin (Roy, 1973; Roy et al., 1980). In the mature male rat, normal hepatic synthesis of α_{2u}-globulin can be totally inhibited by daily treatment with estrogenic hormones (Roy et al., 1975).

Studies in our laboratory have also identified a 3.5S cytoplasmic androgen-binding (CAB) protein that does not translocate into the nucleus (Roy, 1979). A close correlation between the presence of this CAB protein and the ability of the liver to respond to androgenic induction has been established. For example, both prepubertal and senescent rats that lack the CAB protein also do not synthesize α_{2u}-globulin under androgenic stimulation. Because of its cytoplasmic localization, CAB may have a cytoplasmic regulatory role in mRNA stabilization as proposed for a similar cytoplasmic estrogen-binding protein of the frog liver (Brock and Shapiro, 1983).

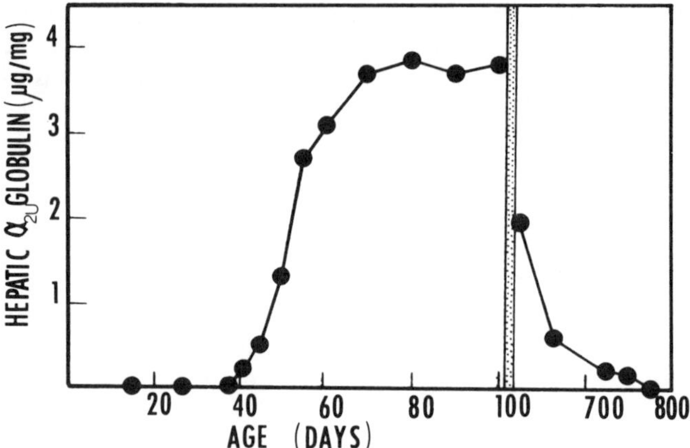

Fig. 1. Age-dependent changes in hepatic α_{2u}-globulin in the male rat. A change in the scale from a unit of 20 days to 100 days is represented at the *dotted bar*. (From Roy et al., (1983) J Biol Chem 258: 10123)

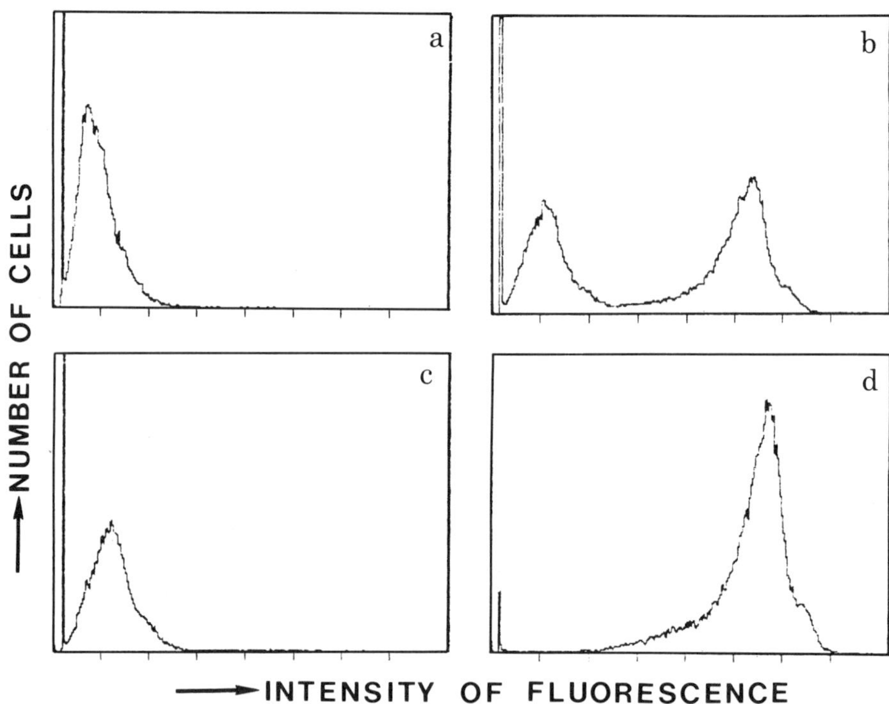

Fig. 2. FACS pattern of the two subpopulations of hepatocytes. **a**: hepatocytes from a prepubertal rat; **b**: hepatocytes from a young-adult male rat; **c**: sorted negative cells from (**b**); **d**: sorted positive from (**b**).

Cellular Specialization in α_{2u}-Globulin Synthesis and Perivenous Location of α_{2u}-Globulin-Producing Hepatocytes

Immunochemical studies of isolated hepatocytes with fluorescence-activated cell sorter (FACS) and immunohistochemical staining of the liver sections have demonstrated two distinct subpopulations of cells, only one of which actively synthesize α_{2u}-globulin (Antalky et al., 1982; Motwani et al., 1984). Figure 2 shows the FACS pattern of hepatocytes derived from prepubertal and postpubertal male rats and separated on the basis of α_{2u}-globulin immunoreactivity. Unlike hepatocytes from prepubertal male rats which only show one nonreactive cell population, the hepatocytes from postpubertal male rats show two cell populations, only one of which specifically binds the antibody. These two cell populations can be collected separately and when reprocessed through the cell sorter retain their distinctive fluorescent pattern. This indicates lack of interconversion due to possible loss of the labeled antibody from the cell surface. Histochemical analysis of the α_{2u}-globulin-producing hepatocytes revealed preferential anatomical localization of these cells around the perivenous area. Figure 3 shows a photomicrograph of the perivenous area of the liver obtained from an androgen-treated female rat and stained for α_{2u}-globulin. After 4

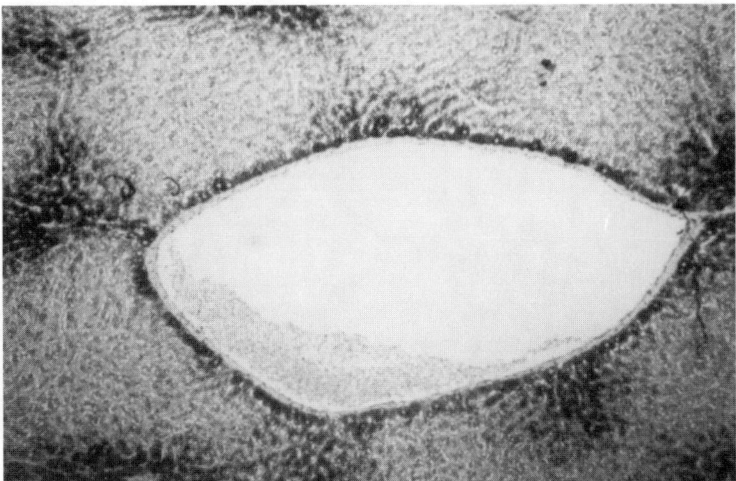

Fig. 3. Immunohistochemical localization of α_{2u}-globulin in the perivenous region of the androgen-treated female rat liver. The liver sample was obtained from sexually mature ovariectomized female rats that received four daily androgen treatments. An intense immunostaining of hepatocytes around the central vein can be seen.

days of androgen treatment, a ring of α_{2u}-globulin-producing hepatocytes around the central vein can be seen. Most of these α_{2u}-globulin-producing cells are found to be in direct contact with the venous wall. Control female rats without any androgen treatment do not show any immunoreactive α_{2u}-globulin-containing hepatocytes.

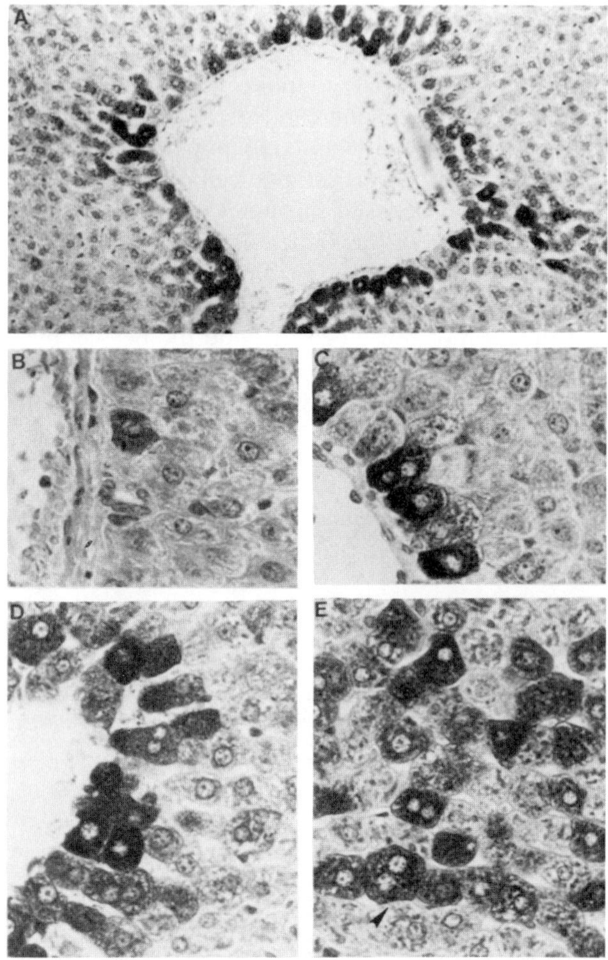

Fig. 4. Appearance of immunoreactive α_{2u}-globulin-producing cell in maturing male rats during puberty. **A**: liver section from a 40-day-old rat showing α_{2u}-globulin-producing cells all around the central vein. **B-D** show progressive conversion of the nonproducing to producing type during maturation (38, 40, and 42 days). Frame **E** (45 day-old) shows side-by-side coexistence of producing and nonproducing cellular cords. The *arrow* points to membrane contact between cells of producing and nonproducing cords.

Perivenous to Periportal Progression of Cell Recruitment During Induction

Initial induction of α_{2u}-globulin in the hepatocytes that are in the proximity of the central vein and propagation of this inductive response toward the periportal region can be visualized better in pictures of the liver sections obtained from maturing male rats (Fig. 4). As shown in Fig. 4, the early stage of the induction process can be divided into three temporal phases: an initial phase when one or few hepatocytes attached to the central vein becomes competent to synthesize α_{2u}-globulin (Fig. 4**B**); a second phase that involves lateral progression of the competency process along the venous wall (Fig. 4**C**); and the third phase involving vertical progression of the competency process along the cords of hepatocytes, away from the central vein (Fig. 4**D**). The vertical cell-to-cell progression of the competency process seems to proceed in an almost all-or-none rather than in a gradient fashion. Furthermore, despite indication of lateral membrane contacts between cells of two adjacent cords (Fig. 4**E**, *arrow*), no evidence for lateral progression of the competency process in phase three was observed.

Lack of Cell Proliferation During Periportal Progression

Because of a recent report indicating massive (40%–70%) cell proliferation in the frog liver after estrogenic stimulation (Spolski et al., 1985), we have considered the possibility that the vertical progression of the inductive response may in fact represent proliferation of hepatocytes attached to the central vein after hormonally mediated cytodifferentiation. However, autoradiographic analysis of the hepatic tissue of the androgen-treated female failed to show any unusual enhancement of the mitotic activity (Fig. 5). Low mitotic activity (1%) in the liver tissue of the adult rat, as observed in this case, is consistent with known data in this field (Greengard et al., 1972). It should also be noted that a few mitotically active hepatocytes which can be identified by autoradiography are mostly located in the periportal rather than perivenous region.

Possible Role of Vasculature Hepatocyte Interaction in the Regulation of α_{2u}-Globulin Synthesis

Despite their locational disadvantage (farthest end of the biochemical gradient), hepatocytes situated at the perivenous site are first to show hormone responsiveness for α_{2u}-globulin synthesis. Furthermore, the first batch of competent cells are always found to be in close physical contact

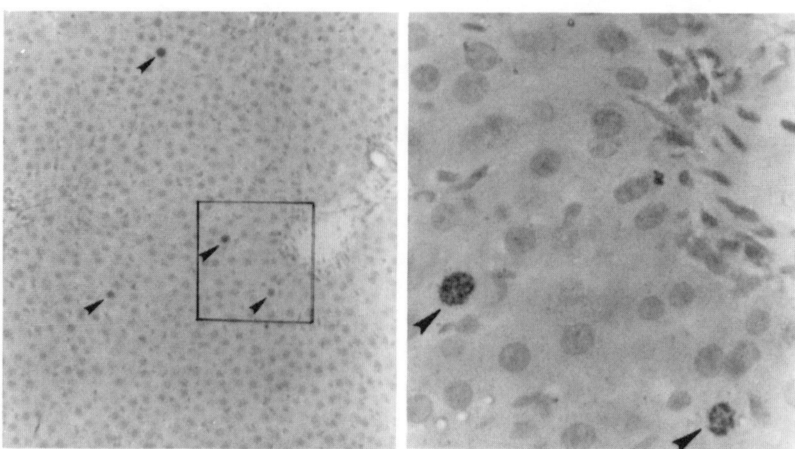

Fig. 5. Autoradiogram of the mitotically active hepatocytes in the ovariectomized female rat liver after 4 days of androgen treatment. The labeled nuclei are marked with *arrow heads*. Right frame shows a magnified picture of the periportal zone (boxed area in the left frame).

with the wall of the central vein. These observations suggest that cell-to-cell interaction and cytodifferentiation rather than preferential accessibility to hormones and substrates within the lobular domain play critical roles in the regulation of α_{2u}-globulin gene expression. Since initial appearance of the competent hepatocytes is found to be associated with the central vein, it is possible that the cellular components of the central vein could be the primary target of the inductive hormone. In addition, the biochemical basis of the apparent all-or-none type of transition of the noncompetent to competent cells in the hepatic cord is also of great interest.

In recent years several important examples of the regulatory role of cell-to-cell interactions in the mediation of hormone action have been reported. In some of these cases steroid hormones have been shown to influence epithelial cell function via its primary action on the underlying stromal elements (Kratochwil, 1969; Sakakura et al., 1976; Cunha et al., 1983). Cunha et al. (1980) have explored the secondary effect of the androgen through tissue recombinants composed of urinary bladder epithelium and urogenital sinus mesenchyme derived from testicular feminized male (Tfm) and normal mice. Results of these studies clearly show that the androgen-mediated cytodifferentiation of the epithelial layer is dependent on the presence of nuclear androgen receptor in the stromal tissue. Presence or absence of androgen receptor in the epithelial layer seems to be inconsequential for the cytodifferentiation of the epithelial layer from bladder to prostate type. Similar conclusions were also drawn earlier by Kratochwil and Schwartz (1976) concerning androgen-induced regression of the mammary epithelium in Tfm $\times$ normal tissue recombinants. These studies suggest that certain cellular component of the mesenchymal element is the primary target for the androgen and that such a primary event induces the synthesis of a paracrine regulatory factor that initiates cytodifferentiation and differential gene expression in the overlying epithelium. It is tantalizing to note that the hepatic blood vessels are derived from the mesechymal tissue and that localization of hormone receptors in the cells of the blood vessels has also been reported (Stumpf and Sar, 1976; Colburn and Buonassisi, 1978; McGill and Sheridan, 1981).

Autoradiographic analysis of the liver sections after ^{3}H-thymidine incorporation indicates that the vertical propagation of the inductive response does not require cell division. We postulate that this step may involve mediation of a paracrine factor. Furthermore, vertical propagation of the competency process seems to be a unidirectional phenomenon that proceeds through the hepatic cords in a perivenous to periportal direction. It is possible that specific junctional complexes may act as a channel for the perivenous to periportal flow of the putative paracrine mediator. If the mediator is less than 1,700 daltons, gap junctions can serve as a channeling mechanism.

Paracrine factors that mediate cell-to-cell communication are known to function through receptor mechanism. In most cases of effector-receptor system, the presence of an excess "spare receptors" is generally the rule rather than the exception. Sequestration of the excess effector molecules by these "spare receptors" within a competent cell may prevent their flow into the next noncompetent cell, long after the competent cell has attained a maximum inductive response. Such a mechanism can account for the apparent all-or-none type of vertical progression of the competency process along the hepatic cord.

A Model for Hormonal Induction Through Hepatocyte-Vasculature Interaction

The critical role of cell-cell communication in cytodifferentiation and organogenesis has long been recognized by developmental biologists. The biochemical nature of such communicative signals, however, has not been clearly established. In recent years studies from several laboratories, especially those of Cunh, Sirbasku and Moscona (Cunha et al., 1983; Sirbasku, 1981; Linser and Moscona, 1982), have indicated that stimulation of specific gene expression and proliferation of the epithelial cells by androgens, estrogens, and glucocorticoid may be indirectly mediated through secondary mediators. These secondary mediators are produced after the primary interaction of the steroid hormone on the underlying nonepithelial cells. Our results are consistent with such a concept. However, it should also be emphasized that although androgens may indirectly influence the perivenous hepatocytes to cause cytodifferentiation and confer competency to produce α_{2u}-globulin, modulation of α_{2u}-globulin synthesis in the competent cells is also directly influenced by the sex hormones. This conclusion is supported by the results of the in vitro perfusion experiments and the immunocytochemical demonstration of nondifferential inhibition of α_{2u}-globulin synthesis in the competent cells after estrogen treatment.

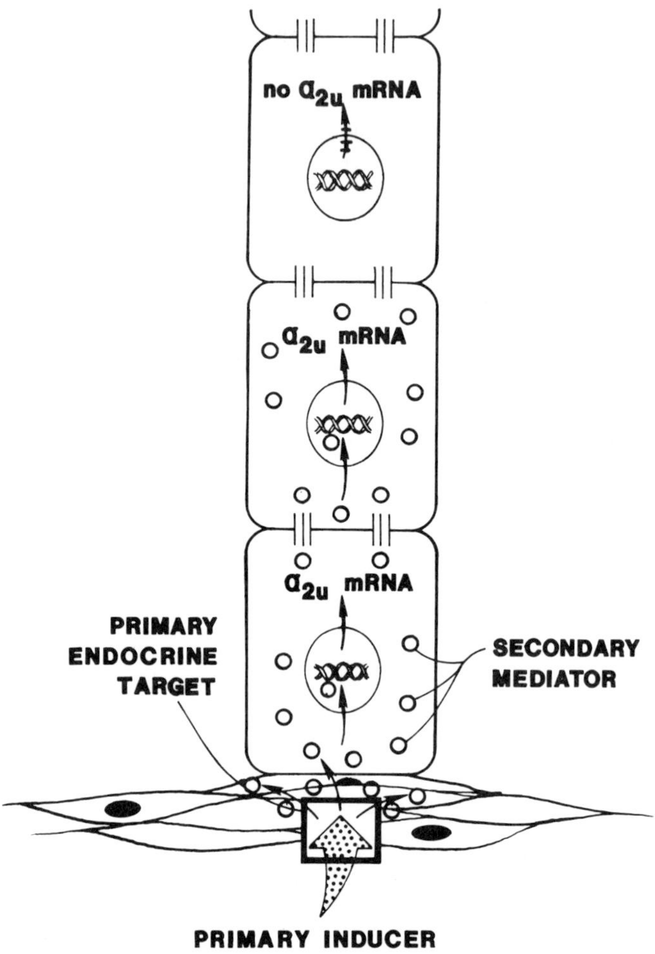

Fig. 6. A model for the hormonal induction of α_{2u}-globulin in the perivenous hepatocytes and the vertical propagation of the inductive influence through a secondary mediator. Cells in the venous wall are depicted as the primary target for the inducing hormone. This primary action produces a secondary mediator that can be channeled within the cells of the same hepatic cord. The secondary mediator is considered to be essential for the coversion of noncompetent cells to competent cells. In addition to the cardinal influence of the secondary mediator, the model also proposes a direct role of the androgen on the α_{2u}-globulin-producing cells.

The CAB protein may be involved in this regulatory component of α_{2u}-globulin synthesis. Despite membrane contact in both lateral and vertical directions, the propagation of competency only proceeds within the component cells of a cord in a venous to portal direction. This may suggest the role of specific junctional complexes rather than simple membrane contact as a prerequisite for the transmission of the inductive signal. As mentioned earlier, if the signalling process is mediated via a small molecule (1,700 daltons or less), gap junctions may provide a channel for such a mediator. Vertical transmission of competency in an all-or-none fashion may indicate a well-known endocrinological phenomenon, i.e., presence of vast excess of "spare receptors" for the inductive signal. In every case examined, the target tissue contains an amount of receptor that is greater than the number required for evoking the maximum cellular response. If the same mechanism is operative in the transfer of competency, spare receptors in the newly recruited competent cell will continue to sequester the effector molecules and prevent their flow into the adjacent cell even after it has become totally competent to express α_{2u}-globulin gene. The effector molecules can only overflow into the next cell after the spare receptors are saturated. A model based on these considerations is presented in Fig. 6. The basic tenets of this model can be subjected to experimental verification.

References

Antalky T, Lynch KR, Nakhasi HL, Feigelson P (1982) Am J Anat 165: 211–224
Brock ML, Shapiro DJ (1983) Cell 34: 207–224
Colburn P, Buonassisi V (1978) Science 201: 817–819
Cunha GR, Chung LWK, Shannon JM, Reese BA (1980) Biol Reprod 22: 19–42
Cunha GR, Chung LWK, Shannon JM, Taguchi O, Hirohiko F (1983) Recent Prog Horm Res 39: 559–598
Greengard O, Federman M, Knox WE (1972) J Cell Biol 52: 261–272
Jensen EV, Jacobsen HJ (1962) Recent Prog Horm Res 18: 387–414
Jungermann K, Sasse D (1978) Trends Biochem Sci 3: 198–200
Kratochwil K (1969) Dev Biol 20: 46–71
Kratochwil K, Schwartz P (1976) Proc Natl Acad Sci USA 73: 4041–4044
Kurtz DT (1981) J Mol Appl Gen 1: 29–38
Linser P, Moscona AA (1982) In: Moscona AA, Monroy A (eds) Current Topics in Developmental Biology Academic Press, New York. pp 115–188
McGill HC, Sheridan PJ (1981) Circ Res 48: 238–244
Motwani NM, Caron D, Demyan WF, Chatterjee B, Hunter S, Poulik MD, Roy AK (1984) J Biol Chem 259: 3653–3657
Payvar F, Wrange O, Carlsedt-Duke J, Okret S, Gustafsson J, Yamamoto K (1981) Proc Natl Acad Sci USA 78: 6628–6632
Roy AK (1973) J Endocrinol 56: 295–301
Roy AK (1979) Biochem Act Horm 6: 481–517
Roy AK, Chatterjee B, Demyan WF, Milin BS, Motwani NM, Nath TS, Schiop MJ (1983). Recent Prog Horm Res 39: 425–461

Roy AK, Chatterjee B, Prasad MSK, Unakar NJ (1980) J Biol Chem 255: 11614–11618
Roy AK, McMinn D, Biswas NM (1975) Endocrinology 97: 1501–1508
Roy AK, Nath TS, Motwani NM, Chatterjee B (1983) J Biol Chem 258: 10123
Roy AK, Neuhaus OW, Harmison CR (1966) Biochim Biophys Acta 127: 72–81
Sakakura T, Nishizuka Y, Dawe EJ (1976) Science 194: 1439–1441
Sirbasku DA (1981) Branbury Report 8: 425–443
Spolski RJ, Schneider W, Wangh LJ (1985) Dev Biol 108: 332–340
Stumpf WE, Sar M (1976) In: JR Pasqualini (ed) Receptors and Mechanism of Action of Steroid Hormones, Part 1. Marcel-Dekker, New York. pp 41–84

Discussion of Paper Presented by A.K. Roy

O'MALLEY: Just a couple of quick questions, and then one comment. Are there any other examples of this type of heterogeneity in the liver that you know of?

ROY: Yes. Results based on microdissection and morphometric analysis indicate that enzymes involved in certain types of metabolic functions such as oxidative energy metabolism and β-oxidation are preferentially enriched in the periportal hepatocytes while those involved in glycolysis, lipogenesis, and detoxification are present in higher amounts in the perivenous hepatocytes. However, none of these examples shows all-or-none type of distribution.

O'MALLEY: The second point is that it seems this communication (intercellular communication for cell recruitment) must take place through the gap junction.

ROY: This is a distinct possibility.

O'MALLEY: In addition there must be a polarity to these cells.

ROY: Well, the polarity could be explained on the basis of concentration gradient. If the androgen is required to synthesize a secondary mediator, you automatically generate polarity based on the position of the cell that contains the nuclear androgen receptor.

O'MALLEY: I think this is interesting. People working with early development of the fertilized egg realized the polarity of cells, and the cell-cell communication, and I think in the differentiated state we don't usually see that. Certainly gap junctions are one of the first things set up in early development and if you interfere with them, you interfere with development of the tissue and introduce genetic defects, and they are set up before membranes are totally formed. This type of communication must exist many more times than we generally realize.

SCHRADER: Did I understand correctly that you think that the androgen target is actually a cell in the vasculature? Is that right?

ROY: Yes, we think the primary action of the androgen may be on one of the components of the central vein.

SCHRADER: If that is true, how do you explain the induction in the perfused liver?

ROY: As I explained in my talk, there is a multi-tier control for the regulation of α_{2u}-globulin gene expression. One of the critical events during hormonal induction is the acquisition of cellular competency based on vasculature hepatocyte interaction. There seems to be other levels of control besides the acquisition of cellular competency where the androgen can influence α_{2u}-globulin synthesis. Such regulatory influence can be exerted on the competent hepatocytes, which are known to contain CAB protein.

UNAKAR: In the hepatic lobule, you showed positive hepatic chords coexisting side by side with negative hepatic chords. Quite often the hepatic chords branched and the interesting thing that I saw was that the branches of the positive chords were also positive. This may have implications concerning the junctions that exist between the same chord.

HAQUE: Being a clinician, I feel like a fish out of water, but there seems to be considerable clinical implications about the cellular heterogeneity in the liver. In a clinical situation, especially with diabetes, we see anomalous insulin response. Could that be due to changes in the distribution of insulin receptors within the cell population?

ROY: What is known about the distribution of the insulin receptor goes back to Bert's [O'Malley] question. The insulin receptor in the periportal and the perivenous area fluctuates differentially, depending on the time of the meal. Thus, one finds not only heterogeneity with regard to the distribution of the gene activity in the hierarchy of the hepatic structure, but also certain dynamic changes taking place within the cells themselves.

GUSTAFSSON: I would like to hear more about how you visualize the interplay between growth hormone and androgen. Is it possible for you to induce α_{2u}-globulin just by androgen in a hypophysectomized rat?

ROY: The hypophysectomized rat, as you know, is deficient in multiple hormones. It is deficient in growth hormone, in glucocorticoid, in thyroxine, and in many other things, so androgen alone won't be enough.

GUSTAFSSON: You have to have other hormones as well. If you work with primary hepatocyte cultures, can you induce the protein by androgen or do you need to add a mix of hormones?

ROY: Possibly because of the critical role of multicellular interactions in the regulation of α_{2u}-globulin synthesis, we have not been able to show androgenic induction of this protein in pure hepatocyte culture system.

GUSTAFSSON: Where do you think the primary target cell is for growth hormone? Is it also on the mesenchymal cells?

ROY: At present I do not have enough data to give you the right answer.

GUMUCIO: I think it is really important when you look at any aspect of this kind of heterogeneity in liver to realize that Rappaport defined the acinus as being a group of liver cells that surrounds the terminal portal venule, not the subterminal portal venule, from which the blood enters and then moves across the lobule to the terminal hepatic vein. Thus you have a gradient of blood supply. If you look, as you did, at the very large terminal hepatic veins or the very large portal veins, you are not sampling the functional unit of the liver parenchyma. The terminal hepatic veins are separated from the hepatocytes by only a thin layer of collagen and occasionally a thin cell layer, but there is no contact between the cell layer and the hepatocyte. Then you have to assume that something is going on in the very last terminal cells that are exposed to a lowest concentration of substrates and hormones. These cells have to be highly sensitive to androgen stimulus and then, as you mentioned, the secondary mediator has to be passed backward along the cell chord. I would also like to comment that this is a familiar situation with respect to toxic injury to the liver.

BARDIN: When you did your fluroescent cell sorting did you measure the mRNA level of the nonfluorescent cells?

ROY: We did not estimate the mRNA level; however, we did in situ hybridization on liver sections and the results follow exactly what we see in immunocytochemistry.

LIAO: I just want to comment on the androgen receptor content in the liver. Before it was very difficult to demonstrate the hepatic androgen receptor because people used dihydrotestosterone. More recently people have been using not only R-1881 but also dimethyl-19-nortestosterone, which we also used. With these ligands, in the liver you can find 10%–15% of androgen receptor compared with prostate. There are a lot of androgen receptors in the liver. On the average you can calculate that there are a couple of thousands of androgen-receptor molecules per cell.

ROY: Although two laboratories (Lucier; Ota) have claimed nuclear translocation of the cytoplasmic androgen receptor of rat liver, in our hand the CAB protein, which we discovered in early seventies, still remains in the cytoplasm. We have consistently failed to show its nuclear translocation. Thus we feel that it must be playing some cytoplasmic regulatory role.

Discussants: W. BARDIN, D. GUMUCIO, J.A. GUSTAFSSON, N. HAQUE, S. LIAO, B. O'MALLEY, A.K. ROY, W. SCHRADER, and N. UNAKAR

Chapter 15

Hormonal Regulation of Sexually Differentiated Isozymes of Cytochrome P-450 in Rat Liver

C. MacGeoch, E.T. Morgan, and J.Å. Gustafsson

Introduction

Cytochrome P-450 proteins represent a group of enzymes (molecular weight 45,000–60,000) that are present in all eukaryotes and in some prokaryotic organisms. These heme-containing enzymes are a major component of the defenses that protect living organisms against the toxic chemicals in their environment. The P-450 enzymes add oxygen to a wide range of compounds. They are needed for the synthesis of endogenous biologically active agents, including steroid hormones and prostaglandins, but also work on foreign chemicals, helping in their detoxification and converting them to a form that can be more readily excreted from the body. Paradoxically, however, some of the foreign chemicals, which are initially not capable of causing cancer, are converted to active carcinogens by P-450 catalyzed reactions.

Because many of the P-450 substrates are important drugs, compounds involved in steroid hormone synthesis, potential toxins, or carcinogens, a great deal of effort has gone into the purification and characterization of cytochrome P-450 in recent years. Two early discoveries contributed greatly to an understanding of the biochemical properties of cytochrome P-450: Omura and Sato (1962) demonstrated that it was a cytochrome of the *b* type, and Estabrook and associates (Estabrook, et al., 1963) provided evidence that it was a component of the mixed function oxidase system leading to the hydroxylation of steroids and drugs.

Preliminary attempts to solubilize cytochrome P-450 were unsuccessful owing to instability of the enzyme in the presence of a variety of solubilizing agents—such as detergents, lipases, proteases, organic solvents, and high-salt concentrations—which convert cytochrome P-450 to an inactive, spectrally distinct form denoted as cytochrome P-420. Ichikawa and Yamano (1967) first reported that polyols stabilize detergent-treated cytochrome P-450. Shortly thereafter, Lu and Coon (1968) reported the solubilization, resolution, and reconstitution of a microsomal hydroxylase system, which consisted of cytochrome P-450, nicotinamide adenine dinucleotide phosphate, reduced (NADPH) cytochrome P-450 reductase, and a lipid factor. Rabbit liver microsomes were treated with deoxycholate in the presence of glycerol and resolved into the three components by diethylaminoethyl (DEAE)-cellulose column chromatography. When these three fractions were recombined, laurate hydroxylase activity could be reconstituted. Subsequently, the mixed-function oxidase activities of such reconstituted systems were demonstrated using steroids, drugs, and alkanes as substrates (Lu, et al., 1976; 1978).

Multiplicity of Cytochrome P-450

The existence of more than one liver microsomal drug-metabolizing enzyme in different animal species was first postulated more than 25 years ago. An early study by Conney et al. (1959) suggested not only that the drug-metabolizing enzymes of various species differ but that the liver of a single species contains several enzymes that catalyze the same reaction. They found that the administration of benzo(a)pyrene to rats markedly increased the microsomal metabolism of substrates such as benzo*[a]*pyrene, acetanilid, and zoxazolamine, yet decreased the metabolism of meperidine and benadryl and had no effect on the metabolism of chlorpromazine.

In the mouse, liver microsomes from animals treated with polycyclic aromatic hydrocarbons such as 3-methylcholanthrene were shown to differ from the liver microsomes from untreated animals or phenobarbital(PB)-treated animals with respect to their CO-difference spectrum, substrate specificity, and genetic control of cytochrome P-450 induction (Nebert and Gielen, 1972). Thus, pretreatment of animals with various inducers results in differential effects on the metabolism of a variety of drugs, carcinogens, and endogenous compounds (Mannering, 1971). The degree of induction is dependent on the sex, species, strain, and inducer as well as on the particular substrate and metabolic pathway being studied. The use of different inducers to manipulate the biochemical and biophysical properties of the microsomal hydroxylation system played a major role in establishing the existence of multiple forms of cytochrome P-450. However, definite proof for the existence of multiple forms had to await the isolation and purification of different forms of cytochrome P-450. In 1974 two groups independently reported the successful purification of hepatic microsomal P-450 of PB-treated rabbits to a gel-electrophoretically homogeneous state (Imai and Sato, 1974; Van der Hoeven et al., 1974). Since then, cytochrome P-450s have been purified from a variety of induced and noninduced species and characterized, supporting the concept of multiple forms of cytochrome P-450. Table 1 lists various forms of cytochrome P-450 that have been purified from rat liver microsomes. This is by no means a comprehensive list, as the precise number of forms of cytochromes P-450 is un-

Table 1. Various Cytochrome P-450 Isozymes Purified From Rat Liver Microsomes

Isozyme	Inducer[a]	Corresponding forms purified by other groups[a]
P-450_a	Aroclor	P-450 PB-3, P-450 UT-F
P-450_b	PB	P-450 PB-4, P-450 PB-B
P-450_c	3-MC/Isosafrole	βNF-B
P-450_d	3-MC/Isosafrole	P-450/ISF-G
P-450_e	PB	P-450 PB-5, P-450 PB-D
P-450_f	Noninduced	
P-450_g	Noninduced	
P-450_h	GH	P-450 16α, P-450 2c, P-450 UT-A, P-450 RLM_5, P-450-male
P-450_i	GH	P-450 15β, P-450 2d, P-450-female
P-450_j	Isoniazid	LM_{3a}
P-450_p/P-450_{TAO}	PCN, DEX, TAO, BP	LM_{3c}

[a]Purified cytochromes P-450 from the laboratories of Levin et al., P-450_a - $P450_j$; Guzelian et al., $P450_p$; Waxman et al., PB-3, PB-4, PB-5, P-450 2c, P-450 2d; Guengerich et al., UT-F, UT-A, PB-B, βNF-B, ISF-G; Gustafsson et al., P-450 16α, P-450 15β; Schenkman et al., RLM_5; Kamataki et al., P-450-male, P-450-female; Coon et al., LM_{3a}, LM_{3c}.
Note: PB = phenobartital; 3-MC = methylcholanthrene; GH = growth hormone; PCN = pregnenolene-16α-carbonitrile; DEX = dexamethasone; TAO = triacetyloleandomycin; BP = benzo[(*a*)]pyrene; RLM = rat liver microsomal; LM = liver microsomal (from rabbit); βNF = β-naphthoflavone; UT = untreated; ISF = isosafrole.

known and not every laboratory is represented, but it is included to illustrate the complexity of the cytochrome P-450 system itself and the nomenclature in current use. Reference will be made throughout this chapter to some of the isozymes listed.

In general, the following criteria have been used to demonstrate that various purified cytochromes P-450 are distinct proteins:

1. Mobility on sodium dodecyl sulfate polyacrylamide gel electrophoresis (SDS-PAGE) has proved to be an extremely useful tool in the characterization of cytochrome P-450 since it can be used not only to evaluate the purity of an enzyme preparation but also to determine under denaturing conditions the subunit molecular weight of different preparations. Since SDS-gel electrophoresis can normally resolve proteins whose subunit molecular weights differ by as little as 1,000 daltons, it has been used as a convenient and sensitive technique for distinguishing multiple forms of cytochrome P-450.
2. The various purified cytochrome P-450 preparations possess different but overlapping specificities. The difference in turnover numbers among forms is dependent on the substrate. For example, various liver cytochrome P-450 isozymes catalyze the *N*-demethylation of benzphetamine or the hydroxylation of benzo*[a]*pyrene at rates that can differ by a factor of 100 (Guengerich, 1977; Johnson and Mueller-Eberhard, 1977; Ryan et al., 1979). In contrast, the rates at which a variety of

different forms of cytochrome P-450 hydroxylate aniline are quite similar (Ryan et al., 1979; Fujita and Mannering, 1973; Lu et al., 1972). Many cytochrome P-450 isozymes also exhibit positional selectivity and stereoselectivity, e.g., in the metabolism of such compounds as benzo*[a]*pyrene (Deutsch et al., 1979), testosterone (Haugen et al., 1975), and warfarin (Fasco et al., 1978).

3. A number of immunochemical experiments have also provided evidence for the multiplicity of cytochrome P-450. Immunochemical approaches have been instrumental in proving the nonidentity of many of these isozymes as well as common structural features in other isozymes (Guengerich et al., 1982; Reik et al., 1982). Specific antibodies have also provided the basis for assays to determine the levels of individual isozymes in microsomal samples in the presence of a multitude of both unrelated and related proteins (Guengerich et al., 1982; Thomas et al., 1983). However, the heterogeneity of polyclonal antibodies presents certain problems as to their use as specific probes of cytochrome P-450 diversity. Polyclonal antibodies are directed against many different epitopes on large proteins such as cytochrome P-450, and an antibody directed against any given epitope is usually characterized by a heterogeneous family of immunoglobins with diverse binding affinities. Thus, antibody cross-reactivity is common among cytochrome P-450 isozymes.
4. Another valuable criterion for evaluating structural similarities or dissimilarities of isolated cytochrome P-450 species is a comparison of the peptide fragments of different enzyme preparations after limited proteolysis. Peptide maps obtained from major forms of rat and rabbit cytochrome P-450 are very different (Guengerich, 1978; Johnson et al., 1979), suggesting that the primary structure of these cytochrome P-450 species is distinct. However, the ultimate proof of a different primary structure is a comparison of the total amino acid sequence of each isozyme. Along with results from peptide mapping, distinct end-terminal sequences provide further support for the multiplicity of cytochromes P-450 and suggest that these cytochromes P-450 are separate gene products rather than posttranslational modifications of a common precursor.

Despite the isolation of many cytochrome P-450 isozymes from rat liver, cytochromes P-450 have not been studied so extensively in humans. However, forms have been purified to apparent homogeneity. One form, P-450_{DB}, has relatively high catalytic activity toward the drugs debrisoquine, sparteine, encainide, bufuralol, and propranolol. It appears to be the enzyme involved in the polymorphic distribution of oxidative activities toward these substrates in humans (Distlerath and Buengerich, 1984). Approximately 10% of whites are poor metabolizers of the antihypertensive drug debrisoquine. One clinical consequence of this deficiency is that poor metabolizers are more sensitive to the hypotensive effects of debrisoquine and can also be especially sensitive to the therapeutic or toxic effects of several other drugs. These biochemical studies provide a basis for better understanding the mechanisms that underlie genetic polymorphisms involving P-450 cytochromes in humans.

Extrahepatic Forms of Cytochrome P-450

Microsomes or mitochondria prepared from tissues other than liver, such as adrenal and kidney, have substrate specificities that differ considerably from that of liver microsomes, suggesting that the cytochrome P-450 isozymes in various tissues may be different. In the adrenal cortex, the biogenesis of steroid hormones from cholesterol is catalyzed by four distinct forms of cytochrome P-450 that have been purified and characterized (Waterman and Simpson, 1985). Cytochromes P-450$_{scc}$ and P-450$_{11\beta}$ are mitochondrial enzymes whereas cytochromes P-450$_{C21}$ and P-450$_{17\alpha}$ are located in the microsomal fraction. The microsomal hydroxylase electron transport system only consists of cytochrome P-450 and a flavoprotein, NADPH cytochrome P-450 reductase, whereas the mitochondrial hydroxylase system consists of cytochrome P-450, a flavoprotein, NADPH adrenodoxin reductase, and an iron-sulfur protein, adrenodoxin. The 21-hydroxylation pathway of progesterone metabolism is largely localized in the adrenal cortex of most species and is not usually found in liver. However, rabbit hepatic cytochrome P-450 I is unusual in that it catalyzes the 21-hydroxylation of progesterone (Dieter et al., 1982) but it does not seem as if P-450 I is closely related to the adrenal 21-hydroxylase since monoclonal antibodies to P-450 I do not detect the presence of a related protein in adrenal microsomes (Johnson et al., 1985).

Two distinct forms of cytochrome P-450 have also been purified from rabbit lung microsomes (Wolf et al., 1979). These two forms, P-450$_{I}$ and P-450$_{II}$, have been shown to be distinct and unrelated based on structural, immunochemical, spectral, , and catalytic properties. However, P-450$_{I}$ has been concluded to be the same enzyme as P-450$_{PB}$, the major rabbit hepatic form induced by phenobarbital, i.e., LM$_2$ (Serabjit-Singh et al., 1979).

Cytochrome P-450 has been purified from bacteria, e.g., *Pseudomonas putida* (Tyson et al., 1972) and *Bacillus megaterium* (Berg et al., 1975), where it is localized in the soluble cytoplasmic fraction. In *P. putida,* the cytochrome P-450 system is very specific for the hydroxylation of camphor (Tyson et al., 1972) and no other compounds can be metabolized by this enzyme system. Undoubtedly, cytochrome P-450 is very widely distributed among various forms of life, from primitive bacteria to highly developed mammals, and participates in diversified metabolic reactions as the oxygen-activating component of monooxygenase systems.

Cytochrome P-450 Protein and Gene Structure

In order to answer the question of the number of cytochrome P-450 isozymes, the extent of their structural similarities, and their regulation, the application of recombinant DNA techniques has been initiated in several laboratories. In this approach one isolates cDNA and genomic clones for specific P-450 isozymes and determines the number of related genes in the rat genome and the extent of homology of the different genes and their encoded polypeptides. To date, there is experimental evidence that the P-450 superfamily consists of at least five gene families: (1) PB-inducible, (2) 2,3,7,8-tetrachlorodibenzo-*p*-dioxin(TCDD)-inducible (3) pregnenolone-16α-carbonitrile(PCN)-inducible, (4) P-450_{scc} gene(s) responsible for cholesterol side chain cleavage in mitochondria, and (5) the gene(s) encoding steroid C-21 hydroxylation.

There has been much discussion about how the multiplicity of P-450 enzymes might have arisen. Each P-450 isozyme appears to have its own complete gene, and it is the prevailing opinion that it is gene conversion, i.e., the introduction of new DNA segments into genes, that has generated the diversity. There has also been a great deal of discussion about whether the various gene families evolved from the same ancestral gene or from separate, unrelated ancestral genes. The data so far are contradictory, and sequences of genes from more species and P-450 families may help to resolve the issue.

Regulation of Cytochrome P-450 Gene Expression

The induction of specific P-450 isozymes by inducing agents such as PB, 3-methylcholanthrene (3-MC), and PCN provides a convenient model system for studying P-450 gene regulation. The underlying mechanism is particularly intriguing since a wide variety of compounds, some structurally unrelated, can induce the production of the same isozyme (Thomas et al., 1981). In the case of PB, 3-MC, and PCN, the increase in the level of the corresponding isozymes results from an enhanced rate of synthesis that is a consequence of the accumulation of the respective mRNAs (Adesnik et al., 1981; Tukey et al., 1982; Hardwick et al., 1983). For PB it has been shown that this results from the rapid transcriptional activation of the corresponding genes (Hardwick et al., 1983; Atchison and Abesnik, 1983). On the basis of extensive genetic and biochemical studies on the mechanism of induction of mouse aryl hydrocarbon hydroxylase (P_1-450) by 3-MC or TCDD, strong evidence has been obtained that the inducer interacts with a cytosolic receptor (the Ah or TCDD receptor), which translocates to the nucleus and then presumably enhances specific gene transcription (Tukey et al., 1982; Nebert et al., 1981). Recently, a possible receptor-binding site with enhancer characteristics was found in the 5′ flanking region of the mouse P_1-450 gene (Jones et al., 1985).

On the basis of studies in primary cultures of adult rat hepatocytes of the effects of glucocorticoids and antiglucocorticoids, e.g., PCN and other steroid hormones, on de novo synthesis of P-450_p, it was proposed that steroid inducers act through a unique stereospecific "receptor-like" mechanism that can be clearly distinguished from that of the classical glucocorticoid receptor (Schuetz et al., 1984; Schuetz and Guzelian, 1984). It has also been shown that P-450_p is induced by the nonsteroid macrolide antibiotic triacetyloleandomycin (TAO) (Wrighton et al., 1985). TAO has been proved to be the most efficacious single inducer of rat liver microsomal cytochrome P-450 yet reported (Guengerich et al., 1982; Heuman et al., 1982). It has been proposed that TAO may induce P-450_p by altering the metabolism of an endogenous steroid inducer or by covalently binding to the P-450 protein and in some way slowing its degradation rate.

Evidence has been provided that adrenocorticotrophic hormone (ACTH) stimulates the mitochondrial P-450_{scc} and P-$450_{11\beta}$-hydroxylase and the microsomal P-$450_{17\alpha}$ hydroxylase and P-450_{c21}-hydroxylase genes via a cell surface receptor and that the induction is mediated via cyclic adenosine 5′-monophosphate (cAMP) (Waterman and Simpson, 1985; Kramer et al., 1984). The stimulation of the P-450 11β gene by ACTH has recently been shown to be cycloheximide sensitive, implying a requirement for synthesis of a transcriptional factor (John et al., 1985). It has also been suggested that posttranscriptional regulation (e.g., stabilization of the P-450_{scc} mRNA) may be important in the mechanism of certain types of P-450 induction.

Sexual Differentiation of Hepatic Steroid Metabolism

The hepatic metabolism of several drugs and xenobiotics, e.g., ethylmorphine (Castro and Gillette, 1967), aniline (El Defrawy et al., 1974), 7-hydroxycoumarin (Kamataki et al., 1980), lidocaine, and imipramine (Skett et al., 1980) in the rat, have been shown to be sex-differentiated. Sex differences in liver metabolism of corticosteroids (Yates et al., 1958), estrogens (Conney et al., 1965), and androgens (Yates et al., 1958; Conney et al., 1965; Einarsson et al., 1973) have also been well documented. Generally, males exhibit higher oxidative metabolic activities than females. Many other liver proteins and enzymes such as rat α_2-urinary globulins, mouse major urinary protein, prolactin receptors, monoamine oxidase, and aldehyde oxidase show considerable sexual dimorphism (Roy and Chatterjee, 1983). Hepatic steroid and drug metabolism in humans has also been shown to be sex dependent (MacLeod et al., 1979), although the differences do not seem to be as marked as in the rat. Thus, it is of both scientific and clinical interest to ascertain the mechanism of control of the sex differences noted.

Steroid hormones are extensively metabolized in the liver. The main metabolic reactions are (1) oxidoreduction by 5α- and 5α-ring A-reductases and by various hydroxysteroid oxidoreductases; (2) hydroxylation by the cytochrome P-450 monooxygenase system; and (3) conjugation of free steroids with either sulphuric acid or glucuronic acid.

The capacity of mammalian liver for steroid hormone hydroxylation is often dependent on sex, suggesting that at least some of the liver microsomal P-450 isozymes active in steroid hydroxylation might be expressed in a sex-dependent fashion. Of particular interest in this respect are two sex-specific P-450-dependent steroid hydroxylase activities known to be present in rat liver microsomes. One is a predominantly female activity, 15β-hydroxylase, active on steroid sulfates (e.g., 5α-androstane-3α, 17β-diol 3,17-disulfate) (Gustafsson and Ingelman-Sundberg, 1974; 1975) and the second is a male-specific steroid 16α-hydroxylase active on various androgens, including testosterone, androstenedione, and dehydroepiandrosterone (Einarsson et al., 1973; Pasleau et al., 1981).

Of all the cytochrome P-450-dependent activities of rat liver microsomes that show sex differences, the steroid 16α-hydroxylase has been the most thoroughly characterized in terms of its hormonal regulation. The physiological role of the steroid 16α-hydroxylase is not known with certainty, but it probably contributes significantly to the oxidative metabolism of androgens in adult male rat liver and may help to regulate circulating androgen levels by catalyzing formation of less active and more readily eliminated hydroxylated derivatives. It is, however, possible that it may also serve to biosynthesize specifically hydroxylated steroids, which have important biological functions that are presently undefined.

The female specific 15β-hydroxylase system, also under regulation by hormonal factors, is unique for a number of other reasons. As mentioned, almost all previously described cytochrome P-450-catalyzed reactions have been found to be more efficient in male than in female rat liver. The P-450-dependent steroid sulfate 15β-hydroxylase activity is therefore unusual in that it is found to be present at high levels in female microsomes and undetectable in both males and immature females (Gustafsson and Ingelman-Sundberg, 1974; 1975). In contrast to most hepatic cytochrome P-450 enzymes that participate in the transformation of lipophilic compounds to more polar products, the microsomal 15β-hydroxylase system prefers water-soluble, highly polar compounds such as steroid sulfates. It has been shown that the 15β-hydroxylase system metabolizes a variety of steroid sulfates, including conjugates of deoxycorticosterone, but only those having a sulfate group at positions 17β or 21 can act as substrates.

In the human fetus and in adult rats, 50%–90% of all plasma and liver steroids are sulfate conjugated. It is evident that steroid sulfates may be regarded as potentially important substrates for enzyme systems in the liver and that 15β-hydroxylation is a major physiological pathway, since 15β-hydroxylated sulfate conjugates are the major metabolites of corticosterone in the feces, bile, and urine of female rats. Steroid sulfates in female rats are excreted almost entirely as 15β-hydroxylated products, whereas such metabolites are essentially absent from excreta of male rats (Carlstedt-Duke et al., 1975; Eriksson et al., 1971).

The demonstration that administration of PB, 3-MC, and 16α-cyano-pregnenolone, drugs known to induce the amount of liver microsomal cytochrome P-450 (Conney et al., 1973), did not influence or actually decreased the specific activity of the sulfate-specific hydroxylase system (Gustafsson and Ingelman-Sundberg, 1974), further indicated the presence of a novel type of hydroxylase system in female liver microsomes.

Gonadal Hormonal Regulation of Hepatic Steroid Metabolism

Although the sex differences in hepatic steroid metabolism are seen only in adult animals, the sexual differentiation is predetermined or "imprinted" in the neonatal period when testicular androgen imprints the ability to express a male pattern of metabolism later in life (Einarsson et al., 1973; Denef and DeMoor, 1972; Chung and Chao, 1980; Mode et al., 1981). Neonatal castration of male animals leads to a completely feminized pattern of hepatic steroid metabolism in the adult animal, but treatment of neonatally castrated male rats with testosterone in the adult period only partially restores the male type of liver steroid metabolism (Gustafsson and Stenberg, 1974a; 1974b). The effect of ovarian estrogens in hepatic steroid metabolism seems to be of less importance since ovariectomy of female rats of varying ages does not lead to any changes in liver metabolism (Gustafsson and Stenberg, 1974b; Gustafsson et al., 1975).

At least three classes of P-450-dependent steroid hormone hydroxylases have been distinguished on the basis of their sex specificity and responsiveness to gonadal hormones (Einarsson et al., 1973; Gustafsson and Stenberg, 1974b). Enzymes in one group including androstenedione 16α-hydroxylase and androstanediol disulfate 15β-hydroxylase are "imprinted" or programmed for their induction or lack of induction, respectively, at puberty by exposure to androgen in the neonatal male rat. Enzymes in a second group, best exemplified by a steroid 6β-hydroxylase that is more active in male than in female rat liver, are reversibly inducible by androgens in adult rat liver. Included in the third group is a steroid 7α-hydroxylase that is present at moderately higher levels in female as compared with male rat liver. This enzyme appears to be regulated primarily by nongonadal factors.

Pituitary Control of Hepatic Steroid Metabolism

Denef (1974) and our own laboratory (Gustafsson and Stenberg, 1974a) independently presented data that indicated that the pituitary gland was essential for the maintenance of the female-type steroid metabolism. Hypophysectomy of male and female animals abolished the sex differences.

The necessity of the pituitary gland gained further support when it was shown that implantation of a pituitary gland under the kidney capsule in hypophysectomized animals resulted in a feminized pattern of liver steroid metabolism, irrespective of the sex of the donor animal (Denef, 1974; Gustafsson and Stenberg, 1976). It was suggested that the neonatal surge of androgens in the male animal imprinted the brain by activating an inhibitory center suppressing the release of a "feminizing factor" responsible for the female type of hepatic steroid metabolism (Gustafsson et al., 1975). This hypothesis gained support when it was shown that destruction of the hypothalamus in male animals caused a complete feminization of the liver steroid metabolism whereas the effects of this lesion in female rats were minor (Gustafsson et al., 1976; 1978).

Nature of the "Feminizing Factor"

The ectopic pituitary gland has been shown to secrete prolactin (PRL) and growth hormone (GH) and only small amounts of other anterior pituitary hormones (Eneroth et al., 1977; Lam et al., 1976). Injection of the dopaminergic agonist 2-Br-α-ergocryptine, an inhibitor of PRL secretion (Corrodi et al., 1973; Fuxe et al., 1974), into normal male and female rats led to a decrease of the serum level of PRL without abolishing the sex difference in hepatic steroid metabolism (Skett et al., 1978). This argued against PRL being identical to the feminizing factor. However, when human growth hormone (hGH) was administered continuously to normal male rats, the hepatic metabolism of 4-androstene-3, 17-dione was completely feminized (Mode et al., 1981). The pituitary hormone responsible for the regulation of sex-dependent liver enzyme levels has now been shown to be identical to GH (Mode et al., 1983).

However, initial experiments did not show any correlation between the regulatory effect of GH and its levels in the serum since the mean level of GH was about the same in both sexes. This was later explained when a sex difference in the secretory pattern of GH was demonstrated (Eden, 1979; Terry et al., 1977). In the male there are regular surges every 3–4 h with low, often undetectable, levels between the peaks, whereas in the female the peak pattern is more irregular, with lower peak heights and higher levels between the peaks. The possibility that the female secretory pattern of GH is the feminizing principle of GH action was investigated in a series of experiments (Mode et al., 1981; 1982). The same daily dose of hGH was administered to hypophysectomized rats by continuous infusion or by subcutaneous injections at 3-, 6-, or 12-h intervals. The results indicated that continuous or frequent administration, imitating a female plasma GH pattern, caused a feminization of the liver steroid metabolism, whereas intermittent GH injections, imitating a male plasma GH pattern, did not. The secretory pattern of GH is therefore essential for its effects on hepatic steroid metabolism. The effects of androgens and estrogens

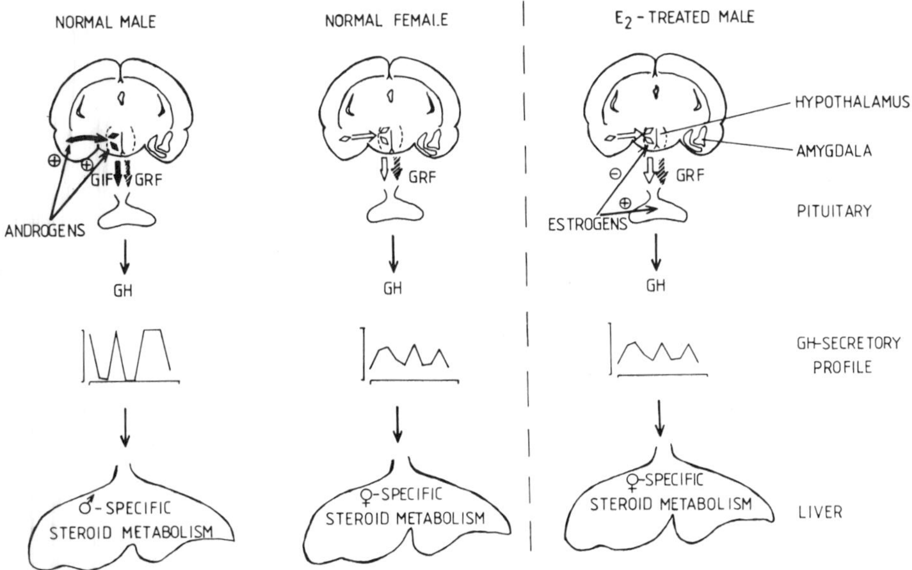

Fig. 1. Hypothetical scheme for the regulation of steroid metabolism in rat liver. *Open arrows* indicate less prominent influences. GIF = growth-hormone-inhibiting hormone or somatostatin; GRF = growth-hormone-releasing factor.

with regard to hepatic steroid metabolism may be mediated via their central regulation of the secretory pattern of GH (Fig. 1). The regulatory role of GH secretion pattern has been shown for other different liver proteins. The liver receptors for both GH and PRL are sex dependent, being more abundant in females than males. PRL receptors in rat liver are positively regulated by GH (Norstedt et al., 1981). In hypophysectomized rats, PRL receptor concentrations fall and can be restored by subsequent administration of GH but not sex steroids. For GH to be effective in elevating receptor concentrations it must be supplied continuously (Norstedt, 1982).

There is some controversy regarding the regulation of GH receptors by growth hormone. Some reports state that in male rats hypophysectomy leads to an increase in liver GH-binding sites and that GH administration reverses this increase (Messina et al., 1982; Maes et al., 1983; Picard et al., 1984), suggesting that GH down-regulates its own receptor. Other reports indicate that GH receptors are increased or up-regulated by GH (Baxter et al., 1982). Induction of the mouse major urinary protein (MUP), predominant in males, has been shown to require pulsatile occupancy of GH receptors, which is achieved naturally in males or by injection of GH (Norstedt and Palmiter, 1984). MUPs are related to α_{2u}-globulins of rats, and recent results indicate that GH regulates the synthesis of α_{2u}-globulin by increasing the amount of mRNA (Roy and Chatterjee, 1983).

Objective of the Present Study

We wished to purify the steroid 16α-hydroxylase and steroid sulfate 15β-hydroxylase from liver and prepare antibodies against them for the following reasons:

1. Since many metabolic reactions of drugs and steroids have been shown to be catalyzed by one or more isozymes of cytochrome P-450 (Lu and West, 1980), we wished to investigate whether the known difference in steroid 16α- and 15β-hydroxylase activities in male and female rat liver was due to the existence of sex-specific forms of cytochrome P-450. Furthermore, at the time of initiation of this study, although isozymes of P-450 had been purified and characterized from a variety of sources, efforts had been concentrated on the major inducible P-450 species, and very few uninduced and/or sex-specific species had been studied.
2. Assuming existence of the sex-specific isozymes and with the production of antibodies raised against the individual isozymes, we wished to investigate whether the regulation of their activities in microsomes by the hypothalamo-pituitary axis occurred by an allosteric effect on the enzymes or by an effect on the levels of the specific apoproteins.
3. In particular, we wished to test the hypothesis that the episodic plasma pattern of GH levels is the ultimate determining factor in the expression of the sexually differentiated P-450 15β and P-450 16α.
4. With the availability of both male- and female-specific isozymes, an opportunity would be provided to determine, at the molecular level, the mechanism of positive and negative regulation of cytochrome P-450 isozyme expression by the same hormone. The possible contribution of other hormones to the regulation could also be investigated.

Results and Discussion

Purification of Male and Female P-450 Isozymes

The procedure developed for the purification was based in part on the methods used by Cheng and Schenkman (1982) for the purification of two constitutive forms of cytochrome P-450 from male rat liver microsomes, namely RLM_3 and RLM_5. RLM_5 was shown to be a major constitutive isozyme in the male, possessing a high steroid 16α-hydroxylating activity. However, no evidence was presented that this protein was sexually dimorphic.

A partial purification procedure developed earlier in our laboratory for P-450 15β had resulted in a 50% pure fraction, possessing high 15β-hydroxylase activity and containing only one protein in the cytochrome P-450 region (MW 50,000) and a single contaminating protein, (MW 70,000) (C. MacGeoch, J. Halpert, J-Å. Gustafsson, unpublished observations, 1982). This information enabled us to easily identify the P-450 15β protein band on sodium dodecyl sulfate-polyacrylamide gel electrophoresis (SDS-PAGE) when the method of Cheng and Schenkman was applied to female rat liver. (The procedure removed the 70,000 MW contaminating protein by lauric acid chromatography).

Solubilization of female rat liver microsomes with Emulgen 913 was followed by column chromatography on lauric acid-AH-Sepharose and CM-Sepharose. By stepwise elution with 5-90 m*M* sodium phosphate on CM-Sepharose a major P-450 peak was eluted at 60 m*M*. This 60 m*M* fraction showed only two protein bands when analyzed by SDS-PAGE. Purification of the two female forms was effected on a DE52 column or a DEAE-Sepharose column where form DEa eluted in the unbound fraction and form P-450 15β eluted in the salt gradient. Purified P-450 DEa had a molecular weight of 52,500 and a specific content of 16.7 nmol P-450 per milligram of protein. Purified P-450 15β had a molecular weight of 50,000 and a specific content of 17 nmol/mg of protein. A protein-staining band corresponding to P-450 15β was never observed in the CM-Sepharose 60 m*M* fraction from untreated male rats.

When the same purification procedure was applied to male rat liver, the major peak of male P-450 eluted from the CM-Sepharose column at 45 m*M* sodium phosphate in contrast to the female CM-Sepharose column profile. Highly purified P-450 16α was obtained by final chromatography on DE52 and contained 16.7 nmol of P-450 per milligram of protein and had a minimum molecular weight of 52,000. P-450 16α could not be purified from female liver using the same procedure.

Catalytic Activities

Purified P-450 15β was capable of catalyzing the hydroxylation of 5α-androstane-3α, 17β-diol 3,17-disulfate in the 15β position with a turnover number of 2.6 nmol/min/nmol P-450, approximately five times higher than that observed in microsomes. P-450 15β did not metabolize testosterone or androstenedione to any significant extent. Although catalytic similarities were observed between P-450 15β and P-450 DEa with regard to, e.g., aniline, benzphetamine, and aminopyrine; 5α-androstane-3α, 17β-diol 3,17-disulfate was poorly metabolized by P-450 DEa.

The purified P-450 16α had high turnover numbers for 16α-hydroxylation with testosterone and androstenedione (4.5 and 7.5, respectively) as substrates and for 2α-hydroxylation of testosterone (3,4). P-450 16α was also an efficient catalyst of the *N*-demethylation of aminopyrine, ethylmorphine, and benzphetamine. Because isozymes P-450 16α and P-450 DEa showed very similar chromatographic and electrophoretic behavior, the possibility existed that these were identical, nonsexually differentiated isozymes. However, forms P-450 16α and P-450 DEa had distinct substrate preferences for drugs and steroids. Form P-450 DEa also metabolized aminopyrine, ethylmorphine, and benzphetamine but to a much lesser extent than P-450 16α and was ineffective in catalyzing the oxidation of testosterone and androstenedione at the 16α position.

Experiments involving mixing the 16α and DEa proteins in the catalytic assays excluded the possibility that form DEa lacks 16α-hydroxylase activity because of a dissociable, endogenous inhibitor.

Immunological Evidence for Sexual Differentiation of P-450 15β and P-450 16α

Antibodies against the individual highly purified cytochromes P-450 15β and P-450 16α were raised in rabbits. When analyzed by Western blot immunoassay, antibodies to both isozymes were observed to cross-react with each other and with isozyme DEa. The antibodies were therefore immunoabsorbed with a total P-450 fraction from male or female liver coupled to Affi gel 10. In this way, specific anti-P-450 15β and anti-P-450 16α, respectively, were obtained, which were subsequently used for immunological detection and quantitation of proteins using the Western blot procedure.

When blots of male and female liver microsomes from untreated rats were incubated with immunoabsorbed anti-P-450 15β, only one band could be detected in liver microsomes from female rats having the same apparent molecular weight as that of the purified P-450 15β. The immunoreactive protein band was essentially absent in male liver microsomes. The content of 15β protein thus measured immunologically could account for 50% of the total spectrally measurable P-450, although variation was observed in the absolute levels of P-450 15β (and also P-450 16α) in microsomes as measured by the Western blot immunoassay.

From the immunological evidence as well as from catalytic properties, isozyme 15β is proposed to be the major isozyme responsible for the predominance of 15β-hydroxylated steroid sulfate metabolites in excreta of female rats and their absence in males.

Similarly, when male and female microsomes were incubated with immunoabsorbed anti-P-450 16α, only male microsomes showed a single major immunodetectable protein, corresponding to P-450 16α. P-450 16α may comprise about 40% of the total spectrally measurable P-450 in the male. The conclusion that P-450 16α is the sexually differentiated 16α-hydroxylase in rat liver microsomes was further supported when it was shown that anti-P-450 16α specifically inhibited the 16α-hydroxylase activity of male rat liver microsomes but had no effect on the much lower testosterone 16α-hydroxylating activity of female rat liver microsomes. The 2α-hydroxylating activity of male microsomes was also inhibited by anti-P-450 16α to the same extent. Estradiol 2α-hydroxylase is also a well-characterized male-specific activity and is a known activity of RLM_5 (Lu et al., 1968). Interestingly, Jellinck et al. (1985) recently showed that normal and recombinant hGH administered by constant infusion feminizes catechol estrogen formation by rat liver microsomes.

It is not certain whether form DEa is sex dependent but it is believed to be a female-specific isozyme of cytochrome P-450, since no protein corresponding to form DEa was seen in the equivalent column fractions during purification of male cytochrome P-450 or in Western blots of male microsomes incubated with antibodies that cross-reacted with purified cytochrome P-450 DEa.

While our work was in progress, several other laboratories reported the purification of various cytochrome P-450 isozymes from untreated animals. Purification, catalytic, and spectral properties have shown that P-450 16α is identical to Cheng and Schenkman's RLM_5 (1982) and to Waxman's P-450 2c (1984), which corresponds to the male-specific and developmentally induced steroid 16α-hydroxylase of rat liver. The male-specific P-450 ("P-450 male") purified by Kamataki (1983), form UT-A of Guengerich et al. (1982), and $P\text{-}450_h$ of Ryan et al. (1984) are speculated to be identical to RLM_5 and P-450 16α. Marked sequence homology exists between the NH_2-terminal sequences of cytochrome $P\text{-}450_h$ and P-450 RLM_5 (Haniu et al., 1984). $P\text{-}450_h$ is purified in only trace amounts from female rats (Ryan et al., 1984) and P-450 male has been shown to occur in levels about 30 times higher in the male (Kamatki et al., 1983). Form UT-A also appears to be sexually differentiated.

Biochemical characterization has indicated that the female-specific P-450 15β probably corresponds to the forms "P-450 female" isolated by Kamataki (1983), $P\text{-}450_i$ (Ryan et al) (1984), and P-450 2d purified by Waxman (1984). Immunoquantitation results have shown that "P-450 female" is a major component of hepatic microsomes from adult female rats and is not detectable in hepatic microsomes from adult male (Maeda et al., 1984). Of nine isozymes purified by Ryan et al. (1984), only $P\text{-}450_i$ catalyzed the metabolism of 5α-androstane-3α, 17β-diol 3,17-disulfate. $P\text{-}450_i$ had very low catalytic activity toward testosterone and was not purified from male rat liver in agreement with our data.

Harada and Negishi (1984; 1984) purified and characterized sexually differentiated isozymes of cytochrome P-450 from an inbred strain of mice. These authors demonstrated that the microsomal testosterone 16α-hydroxylase activity (expressed predominantly in adult male 129/J mice) and testosterone 15α-hydroxylase activity (expressed predominantly in adult female 129/J mice) can be explained by the existence of highly specific cytochrome P-450s for 16α- and 15α-hydroxylation activities in male and female mice, respectively.

Hormonal Regulation of the Sexually Differentiated P-450 15β and P-450 16α

Since the corresponding steroid hydroxylating activities have been shown to be sexually differentiated and regulated by the pituitary, we wished to study the effects of gonadectomy, hypophysectomy, and treatment with pituitary and gonadal hormones on P-450 16α and P-450 15β apoprotein levels in rat liver to determine if the hormones regulated the levels of the specific P-450 apoproteins. This was done using the antibodies raised against the individual cytochromes together with the Western blot technique.

The development of P-450 15β and P-450 16α expression in liver microsomes of male and female rats from the ages of 6 days to 8 months was investigated. P-450 15β and P-450 16α levels were low in both male and female animals until the age of puberty. At 35 days, just after onset of puberty and up to 8 months, the 15β level in the female and the 16α level in the male became fully developed. The development of P-450 16α and P-450 15β expression in rat liver coincided with the development of the corresponding microsomal steroid hydroxylase activities and of the sexually dimorphic pattern of GH secretion, which becomes apparent at 25–30 days.

Age-related changes in pharmacological responses have been accounted for in part by the decrease in the hepatic capacity to metabolize foreign compounds (Schmucker, 1979). Fujita et al. (1982) showed that a marked age-related decrease in the drug-metabolizing capacity is observable in male but not in female rats. In a recent study, Kamataki et al. (1985) showed that liver microsomes from old male rats (older than 25 months) contain only P-450 female instead of P-450 male in liver microsomes. These authors propose that the population of forms of cytochrome P-450 in liver microsomes is altered at least twice in the life of male rats, since they had previously reported that P-450 female appeared in male rats during a certain early period after birth (around 25 days of age) before P-450 male appeared (Maeda et al., 1984).

The effects of castration on P-450 15β and P-450 16α were also investigated. Castration of male rats in the neonatal period led to an increase in the levels of P-450 15β and a decrease in the levels of P-450 16α in the adult animals. However, castration of male rats postpubertally had no effect on the specific microsomal content of either isozyme (Fig. 2). These results confirm that the sexual differentiation of expression of the two apoproteins is predetermined in the neonatal period by androgenic imprinting. This imprinting appears to determine the pattern of hypophyseal GH secretion since it has been shown that neonatal castration of male animals eliminates the GH troughs characteristic of male rats.

Injecting the potent synthetic androgen, methyltrienolone (125 μg subcutaneously) once daily for 14 days into female rats led to a masculinization of both the P-450 15β and P-450 16α levels. Conversely, treatment of adult

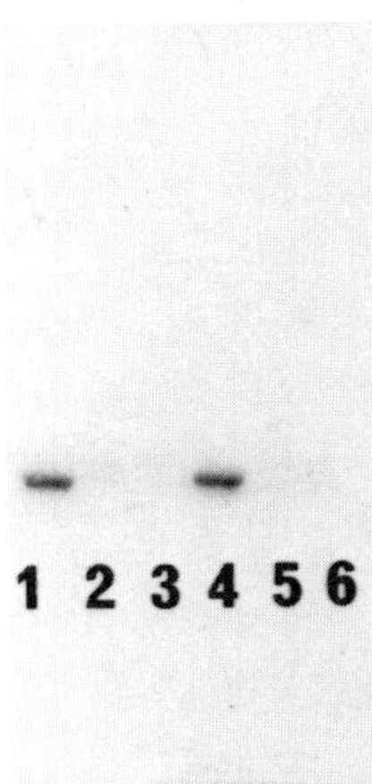

Fig. 2. Neonatal imprinting of P-450 15β expression in rat liver: autoradiogram of an immunoblot incubated with antibodies against P-450 15β. The protein samples were electrophoresed on 7.5% polyacrylamide gels and blotted on nitrocellulose filters. Each well contained 5 μg microsomal protein. Lane 1, neonatally castrated male rat; lane 2, sham-operated neonatal male; lane 3, intact male; lane 4, intact female; lane 5, male castrated as adult; lane 6, sham-operated adult male.

male rats with a single intramuscular injection of 125 μg estradiol valerate led to a complete feminization of both apoprotein levels. Hypophysectomized rats of both sexes have been shown to be insensitive to sex steroids with regard to activities, indicating that androgens and estrogens affect P-450 15β and P-450 16α via the hypothalamus. The effect of estrogens may be mediated via somatostatin and GH secretion, as it has been shown that estrogens change the distribution frequency of plasma GH levels. (Mode et al., 1982).

Hypophysectomy of normal male and normal female animals abolished the sex difference so that hypophysectomized rats of either sex had P-450 16α levels intermediate to those seen in intact rats of each sex and P-450 15β similar to normal male animals. Administration of hGH at a rate of 5 μg/h for 7 days, mimicking the female pattern of blood GH levels, led to a feminization of the levels of both the apoproteins in intact male animals and hypophysectomized rats of both sexes, although feminization of the hypophysectomized animals was never complete (Fig. 3).

Since hGH can bind to both GH and PRL receptors in rat liver (Postel-Vinay, 1976), we tested the specific somatogenic hormone rat growth hormone (rGH) for effects on P-450 15β and P-450 16α levels. The rGH infusion in male rats caused a feminization of both P-450 15β and P-450 16α levels, although this hormone was less potent than hGH. Ovine prolactin (oPRL) infusion had no effect on either apoprotein level. This suggests that P-450 15β and P-450 16α are at least mainly regulated via the GH

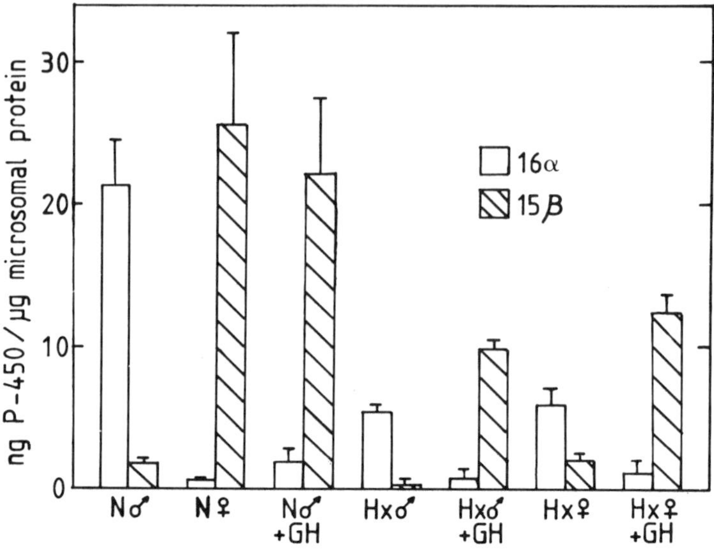

Fig. 3. Effects of hypophysectomy (Hx) and administration (osmotic minipumps) of hGH on hepatic levels of P-450 15β and P-450 16α, assayed by immunoblotting.

receptor. Injecting rGH at the same dose (240 μg/d) intermittently into intact males every 12 h had no effect on either the P-450 16α level or the level of P-450 15β (Fig. 4). Administration of hGH (50 μg/d) to hypophysectomized rats by 12 h, subcutaneous injections caused an increase in their P-450 16α levels, i.e., causing a masculinization of the levels and reversal of the effect of hypophysectomy. These results emphasize the importance of continuous levels of GH in the blood for a feminization and of intermittent GH pulses for a masculinization of the P-450 15β and P-450 16α levels.

To summarize, in all cases it was found that the levels of the two isozymes were oppositely affected by the various treatments. It was also shown that the effects of sexual development, androgens, estrogens, and GH on the microsomal content of the P-450 15β and P-450 16α apoproteins, in every case correlated with the known or measured effects on the microsomal 15β and 16α hydroxylase activities.

Thus, the pituitary gland and the gonadal hormones control hepatic metabolism by modulation of levels of specific P-450 isozymes in rat liver. These results provide direct evidence for the regulation of expression of hepatic P-450 by a peptide hormone and further strengthen the evidence that the cytochrome P-450s isolated are those responsible for the activities observed in microsomes.

It is possible that the sexually differentiated isozymes, P-450 16α, P-450 15β, and (possibly) P-450 DEa, may be products of a previously un-

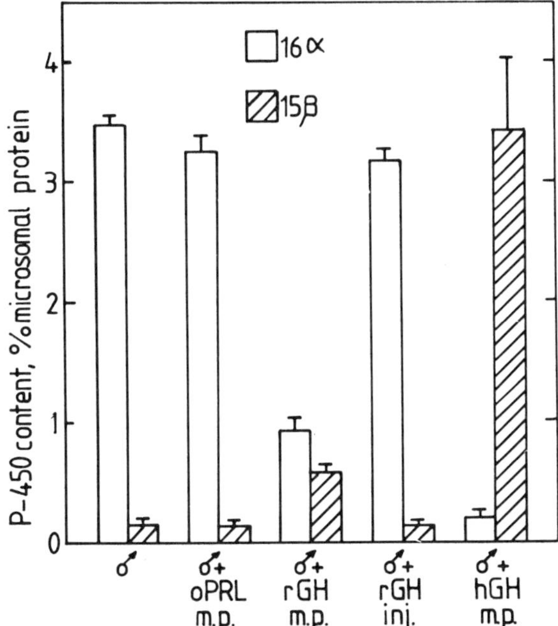

Fig. 4. Effects of oPRL, rGH, and hGH, administered by way of osmotic minipumps (m.p.) or intermittent injections (inj., see text), on hepatic levels of P-450 15β and P-450 16α, assayed by immunoblotting.

identified distinct family of genes. This is supported by the high degree of immunochemical relatedness, since the polyclonal antibodies to P-450 16α cross-react with forms 15β and DEa, and there is reciprocal cross-reactivity of the anti-15β serum. Furthermore, there are distinct similarities in the peptide maps for the P-450 16α and P-450 DEa isozymes. Similarities between the peptide patterns of the sex-specific cytochromes P-450 2c and P-450 2d have been observed (Waxman, 1984), suggesting significant homologies in their primary structure. Bandiera et al. (1985) recently showed that $P\text{-}450_g$ and $P\text{-}450_h$, which are present in untreated male and not female rats, and cytochrome $P\text{-}450_i$, occurring only in female rats, possess a high degree of immunochemical relatedness, supporting our observations.

Comparisons of cytochrome P-450 proteins have revealed that isozymes with high structural homology as defined by cDNA sequence data and immunochemical studies are often regulated by the same inducer. For example, cytochromes $P\text{-}450_c$ and $P\text{-}450_d$ inducible by 3-MC have been shown to be immunochemically related by the use of monoclonal antibodies (Thomas et al., 1984). The structural basis for this has now been determined from the complete amino acid sequences of these proteins. $P\text{-}450_c$ and $P\text{-}450_d$ possess greater than 65% homology (Kawajiri et al.,

1984; Yabusaki et al., 1984). Isozymes P-450_b and P-450_e, inducible by PB, have been shown to be immunochemically identical. This is not surprising since these two isozymes are known to possess 97% total sequence homology (Fujii-Kuriyama et al., 1982; Yuan et al., 1983). In addition, P-450_b has been shown to have 77% sequence homology with PB-inducible cytochrome P-450 LM_2 from rabbit liver microsomes (Kawajiri et al., 1984) but only 14%–25% sequence homology with cytochromes P-450_{scc}, P-450_c, and P-450_d (Kawajiri et al., 1984; Yabusaki et al., 1984; Morohashi et al., 1984). Antibodies to P-450_b cross react with P-450 LM_2 but not with the other cytochromes.

The best known function of GH is to stimulate somatic growth. However, GH also produces effects on cell proliferation and glucose and fatty acid metabolism in various target tissues (Isaksson et al., 1985). It is now generally accepted that binding of GH to its receptor in the plasma membrane is the primary event in the action of GH on target cells. However, knowledge is meager concerning the cellular events following GH interaction with its hepatic receptor. An inhibition of adenylate cyclase with a concomitant decrease in intracellular cAMP might be an important early cellular event in the course of GH action, but it is not known whether or how this change in nucleotide metabolism relates to the various expressed effects of the hormone.

Considerable evidence has accumulated to suggest that the effect of GH on different growth processes is mediated by GH-dependent plasma factors—insulinlike growth factors (IGFs)—that are produced in the liver in response to GH (Isaksson et al., 1985). The main basis for this theory is the fact that most investigators have failed to demonstrate stimulatory effects of GH in vitro on cartilage metabolism, whereas IGFs, when added in vitro, stimulate a number of processes that are associated with both cell multiplication and growth. Although somatomedins undoubtedly stimulate a number of cellular processes in vitro, their postulated mediator role for GH action in vivo is still controversial.

The possible mechanisms whereby the pattern of blood GH levels concomitantly causes the up-regulation of one cytochrome P-450 isozyme and the down-regulation of another are intriguing. The regulation of P-450 15β and P-450 16α by GH is likely to involve an effect on gene transcription, since in all cases studied so far, other inducible hepatic or extrahepatic cytochrome P-450 isozymes are regulated at this level. Development of cDNA probes for these enzymes will be essential to allow detection and quantitation of the relevant mRNAs in cellular extracts and as products of in vitro nuclear transcription. In addition, such cDNAs may also be used to isolate the genomic genes for the P-450 isozymes from a genomic DNA library. Using the polyclonal antibodies to P-450 15β to screen a female rat liver cDNA library cloned in the *Pst* I site of pBR322, we isolated a cDNA of 1,200 base pairs that hybridize to a single-size class of mRNA from female rat liver, but not from male liver. This cDNA is presently being characterized by DNA sequencing and hybridization selection using polyclonal and monoclonal anti-P-450 15β antibodies to precipitate in vitro translation products from mRNA selected by hybridization to the immobilized cDNA probe.

General Summary

The main results of this investigation can be summarized as follows:

1. A sex-differentiated form of cytochrome P-450 was purified to homogeneity from liver microsomes of untreated female rats. The final preparation was active in the hydroxylation of 5α-[^{3}H] androstane-3α, 17β-diol 3,17-disulfate in the 15β position with a turnover number of 2.6 nmol/min/nmol P-450. The purified preparation, termed P-450 15β, contained 17 nmol of P-450 per milligram of protein and had a minimum molecular weight of 50,000. Immunological and chromatographic evidence showed that it was absent in male rats. Thus, convincing evidence is provided that the sexually differentiated steroid sulfate 15β-hydroxylase of female rat liver is due to the presence of a specific isozyme of cytochrome P-450 in female liver (MacGeoch et al., 1984).

   The same purification procedure for female rat liver microsomes also yielded another purified cytochrome P-450, characterized by a minimum molecular weight of 52,500, termed P-450 DEa, which was inefficient in the 15β-hydroxylation of 5α-[^{3}H]androstane-3α, 17β-diol 3,17-disulfate. This isozyme may also be sexually differentiated, as it was not found in the corresponding fraction from male rat liver. Further work is needed, however, to elucidate this point.
2. A sexually differentiated form of cytochrome P-450 was purified to homogeneity from untreated male rat liver. The purified isozyme was highly active in the 16α-hydroxylation of androstenedione and testosterone. It also hydroxylated testosterone at the 2α position and metabolized a number of drug substrates at a high rate. The final preparation, termed P-450 16α, had a minimum molecular weight of 52,000 and contained 16 nmol of P-450 per milligram of protein. The isozyme P-450 16α was not present in female liver. The data indicate that P-450 16α is the isozyme responsible for the sexually differentiated 16α-hydroxylase activity of male rat liver microsomes (Morgan et al., 1985). This conclusion is supported by our observation of a strong inhibition of 16α-hydroxylase activity by anti-P-450 16α in microsomes from male rats but not those from female.
3. Using rabbit antibodies to cytochromes P-450 15β and P-450 16α in a Western blot immunoassay, it was not only proved that the expression of these isozymes in rat liver is sex specific, but also that they are regulated by the hypothalamo-pituitary axis in precisely the same way

as described for sexually differentiated steroid-metabolizing enzyme activities. The effects of gonadectomy, hypophysectomy, and treatment with pituitary and gonadal hormones on P-450 15β and P-450 16α apoprotein levels in rat liver were studied. In all cases it was found that the levels of the two isozymes were oppositely affected by the various treatments. Thus, estrogen treatment of males lowered the levels of the P-450 16α and increased the levels of the P-450 15β, whereas androgen treatment of female rats had the reverse effect. Hypophysectomy abolished the sex difference (Morgan et al., 1985; MacGeoch et al., 1985).

4. GH was shown to be a major regulatory factor for the expression of the two sexually differentiated isozymes. Continuous infusion of hGH completely feminized the levels of P-450 16α and P-450 15β in the livers of male rats. Thus, P-450 15β levels increased while P-450 16α levels decreased. If the same daily dose of hGH is administered by injection at 12-h intervals to male rats, no effect was seen. The somatogenic property of hGH seems to be responsible for the feminizing effect since oPRL infusion had no effect on the levels of either isozyme (Morgan et al., 1985; MacGeoch et al., 1985).

Acknowledgment. This study was supported by a grant from the Swedish Medical Research Council (No 13X-2819).

References

Adesnik M, Bar-Nun S, Maschio F, Zunich M, Lippman A, Bard E (1981) J Biol Chem 256:10340–10345

Atchison M, Adesnik M (1983) J Biol Chem 258:11285–11295

Bandiera S, Ryan DE, Levin W, Thomas PE (1985) Arch Biochem Biophys 240:478–482

Baxter RC, Zaltsman Z, Turtle JR (1982) Endocrinology 111:1020–1022

Berg A, Gustafsson J-Å, Ingelman-Sundberg M, Carlström K (1975) J Biol Chem 251:2831–2838

Carlstedt-Duke J, Gustafsson J-Å, Gustafsson SA (1975) Biochemistry 14:639–648

Castro JA, Gillette JR (1967) Biochem Biophys Res Commun 28:426–430

Cheng K-C, Schenkman JB (1982) J Biol Chem 257:2378–2385

Chung LWK, Chao H (1980) Mol Pharmacol 18:543–549

Conney AH, Gillette JR, Inscoe JK, Trams ER, Posner HS (1959) Induced synthesis of liver microsomal enzymes which metabolize foreign compounds. Science 130:1478–1479

Conney AH, Schneidman K, Jacobson M, Kuntzman Z (1965) Ann NY Acad Sci 123:98–109

Conney AH, Lu AYH, Levin W, Somogyi A, West S, Jacobson M, Ryan D, Kuntzman R (1973) Clin Pharmacol Ther 14:727–741

Corrodi M, Fuxe K, Hökfelt T, Lidbrink P, Ungerstedt U (1973) J Pharm Pharmacol 25:409–411
Denef C, DeMoor P (1972) Endocrinology 91:374–384
Denef C (1974) Endocrinology 94:1577–1582
Deutsch J, Vatsis KP, Coon MJ, Leutz JC, Gelboin HV (1979) Mol Pharmacol 16:1011–1018
Dieter HH, Muller-Eberhard U, Johnson EF (1982) Biochem Biophys Res Commun 105:515–520
Distlerath LM, Guengerich FP (1984) Proc Natl Acad Sci USA 81:7348–7352
Eden S (1979) Endocrinology 105:555–560
Einarsson K, Gustafsson J-Å, Stenberg Å (1973) J Biol Chem 248:4987–4997
El Defrawy El Masry S, Mannering GJ (1974) Drug Metab Dispos 2:279–284
Eneroth P, Gustafsson J-Å, Skett P, Stenberg Å (1977) Mol Cell Endocrinol 7:167–175
Eriksson H, Gustafsson J-Å, Sjövall J (1971) Eur J Biochem 19:433–441
Estabrook RW, Cooper DY, Rosenthal O (1963) Biochem Z 338:741–755
Fasco MJ, Vatsis KP, Kaminsky LS, Coon MJ (1978) J Biol Chem 253:7813–7820
Fujii-Kuriyama Y, Mizukami Y, Kawajiri K, Sogawa K, Muramatsu M (1982) Proc Natl Acad Sci USA 79:2793–2797
Fujita T, Mannering GI (1973) J Biol Chem 248:8150–8156
Fujita T, Uesugi T, Kitagawa H, Suzuki T, Kitani K (1982) In: Kitani K (ed) Liver and Aging. Elsevier Biomedical Press, Amsterdam, pp 55–74
Fuxe K, Corrodi H, Hökfelt T, Lidbrink P, Ungerstedt U (1974) Med Biol 52:121–132
Guengerich FP (1977) J Biol Chem 252:3970–3979
Guengerich FP (1978) Biochem Biophys Res Commun 82:820–827
Guengerich FP, Dannan GA, Wright ST, Martin MV, Kaminsky LS (1982) Biochemistry 21:6019–6030
Gustafsson J-Å, Stenberg Å (1974a) Endocrinology 95:891–896
Gustafsson J-Å, Stenberg Å (1974b) J Biol Chem 249:711–718
Gustafsson J-Å, Ingelman-Sundberg M (1974) J Biol Chem 249:1940–1945
Gustafsson J-Å, Ingelman-Sundberg M (1975) J Biol Chem 250:3451–3458
Gustafsson J-Å, Ingelman-Sundberg M, Stenberg Å (1975) J Steroid Biochem 6:543–649
Gustafsson J-Å, Stenberg Å (1976) Proc Natl Acad Sci USA 73:1462–1465
Gustafsson J-Å, Ingelman-Sundberg M, Stenberg Å, Hökfelt T (1976) Endocrinology 98:922–926
Gustafsson J-Å, Eneroth P, Hökfelt T, Skett P (1978) Endocrinology 103:141–151
Haniu M, Ryan DE, Iida S, Leiber CS, Levin W, Shively JE (1984) Arch Biochem Biophys 235:304–311
Harada N, Negishi M (1984) J Biol Chem 259:1265–1271
Harada N, Negishi M (1984) J Biol Chem 259:1265–1271
Hardwick JP, Gonzalez FJ, Kasper CB (1983) J Biol Chem 258:8081–8085
Haugen DA, Van der Hoeven TA, Coon MJ (1975) J Biol Chem 250:3567–3570
Heuman DM, Gallagher EJ, Barwick JL, Elshourhagy NA, Guzelian PS (1982) Mol Pharmacol 21:753–760
Ichikawa Y, Yamano T (1967) Biochim Biophys Acta 131:490–497
Imai Y, Sato R (1974) Biochem Biophys Res Commun 60:8–14
Isaksson OGP, Eden S, Jansson JO (1985) Ann Rev Physiol 47:483–499

Jellinck PH, Quail JA, Crowley CA (1985) Endocrinology 117:2274–2279
John ME, John MC, Simpson ER, Waterman MR (1985) J Biol Chem 260:5760–5767
Johnson EF, Muller-Eberhard U (1977) Biochem Biophys Res Commun 76:644–651
Johnson EF, Zounes MC, Muller-Eberhard U (1979) Arch Biochem Biophys 192:282–289
Johnson EF, Finlayson MJP, Raucy JL (1985) In: (Boobis AR, Caldwell J, DeMatteis F, Elcombe CR eds) Proceedings of 6th International Symposium on Microsomes and Drug Oxidations. Taylor and Francis, London, pp 3–12
Jones PBC, Galeazzi DR, Fisher JM, Whitlock JP (1985) Science 227:1499–1502
Kamataki T, Ando M, Yamazoe Y, Ishii K, Kato R (1980) Biochem Pharmacol 29:1015–1022
Kamataki T, Maeda K, Yamazoe Y, Nagai T, Kato R (1983) Arch Biochem Biophys 225:758–770
Kamataki T, Maeda K, Shimada M, Kitani K, Nagai T, Kato R (1985) J Pharmacol Exp Ther 233:222–228
Kawajiri K, Gotoh O, Sogawa K, Takashira Y, Muramatsu M, Fujii-Kuriyama Y (1984) Proc Natl Acad Sci USA 81:1649–1653
Kramer RE, Rainey WE, Funkstein B, Dee A, Simpson ER, Waterman MR (1984) J Biol Chem 259:707–713
Lam PCO, Morishige WK, Rotchild I (1976) Proc Soc Exp Biol Med 152:615–617
Lu AYH, Coon MJ (1968) J Biol Chem 243:1331–1332
Lu AYH, Jacobson M, Levin W, West SB, Kuntzman R (1972) Arch Biochem Biophys 153:294–297
Lu AYH (1976) Fed Proc 35:2460–2463
Lu AYH, West SB (1978) Pharmacol Ther A 2:337–358
Lu AYH, West SB (1980) Pharmacol Rev 31:277–295
MacGeoch C, Morgan ET, Halpert J, Gustafsson J-Å (1984) J Biol Chem 259:15433–15439
MacGeoch C, Morgan ET, Gustafsson J-Å (1985) Endocrinology 117:2085–2092
MacLeod SM, Giles HG, Bengert B, Liu FF, Sellers EM (1979) J Clin Pharmacol 19:15–19
Maeda K, Kamataki T, Nagai T, Kato R (1984) Biochem Pharmacol 33:509–512
Maes M, De Hertogh R, Watrin-Granger P, Keteslegers JM (1983) Endocrinology 113:1325–1332
Mannering GJ (1971) Metabolism 20:228–245
Messina JL, Kostyo JL, Eden S (1982) Changes in bovine GH binding to male rat liver membranes induced by GH deficiency or GH administration. Program of the 64th Annual Meeting of the Endocrine Society, San Francisco, CA, p 99 (Abstract).
Mode A, Norstedt G, Eneroth P, Hökfelt T, Gustafsson J-Å (1981) In: (Fuxe K, Gustafsson J-Å, Wetterberg L eds) Steoid Hormone Regulation of the Brain Permagon Press, New York, pp 61–70
Mode A, Norstedt G, Simic B, Eneroth P, Gustafsson J-Å (1981) Endocrinology 108:2103–2108
Mode A, Gustafsson J-Å, Jansson JO, Eden S, Isaksson O (1982) Endocrinology 111:1692–1696
Mode A, Norstedt G, Eneroth P, Gustafsson J-Å (1983) Endocrinology 113:1250–1260

Morgan ET, MacGeoch C, Gustafsson J-Å (1985) Mol Pharmacol 27:471–479
Morgan ET, MacGeoch C, Gustafsson J-Å (1985) J Biol Chem 260:11895–11898
Morohashi K, Fujii-Kuriyama Y, Okada Y, Sogawa K, Hirose T, Inayama S, Omura T (1984) Proc Natl Acad Sci USA 81:4647–4651
Nebert DW, Gielen JE (1972) Fed Proc 31:1315–1325
Nebert DW, Eisen HJ, Negishi M, Lang MA, Hjelmeland LM (1981) Ann Rev Pharmacol Toxicol 21:431–461
Norstedt G, Mode A, Eneroth P, Gustafsson J-Å (1981) Endocrinology 108:1855–1866
Norstedt G (1982) Endocrinology 110:2107–2112
Norstedt G, Palmiter R (1984) Cell 36:805–812
Omura T, Sato R (1962) A new cytochrome in liver microsomes. J Biol Chem 237:pc 1375
Pasleau F, Kolodzici C, Kremers P, Gielen JE (1981) Eur J Biochem 120:213–220
Picard F, Postel-Vinay MC (1984) Endocrinology 114:1328–1333
Postel-Vinay MC (1976) FEBS Lett 69:137–142
Reik L, Levin W, Ryan DE, Thomas PE (1982) J Biol Chem 257:3950–3957
Roy AK, Chatterjee B (1983) Ann Rev Physiol 45:37–50
Ryan DE, Thomas PE, Korzeniowski D, Levin W (1979) J Biol Chem 254:1365–1374
Ryan DE, Iida S, Wood AW, Thomas PE, Lieber CS, Levin W (1984) J Biol Chem 259:1239–1250
Ryan DE, Dixon R, Evans RH, Ramanathan L, Thomas PE, Wood AW, Levin W (1984) Arch Biochem Biophys 233:636–642
Schmucker DL (1979) Pharmacol Rev 30:445–456
Schuetz EG, Wrighton SA, Barwick JL, Guzelian PS (1984) J Biol Chem 259:1999–2006
Schuetz EG, Guzelian PS (1984) J Biol Chem 259:2007–2012
Serabjit-Singh CJ, Wolf CR, Philpot RM (1979) J Biol Chem 254:9901–9907
Skett P, Eneroth P, Gustafsson J-Å (1978) Biochem Pharmacol 27:1713–1716
Skett P, Mode A, Rafter J, Sahlin L, Gustafsson J-Å (1980) Biochem Pharmacol 29:2759–2762
Terry L, Saunders A, Aadat J, Willoughby J, Brazeau P, Martin J (1977) Clin Endocrinol 6:19s–28s
Thomas PE, Reik LM, Ryan DE, Levin W (1981) J Biol Chem 256:1044–1052
Thomas PE, Reik LM, Ryan DE, Levin W (1983) J Biol Chem 258:4590–4598
Thomas PE, Reik LM, Ryan DE, Levin W (1984) J Biol Chem 259:3890–3899
Tukey RH, Hannah RR, Negishi M, Nebert DW, Eisen HJ (1982) Cell 31:275–284
Tyson CA, Lipscomb JD, Gunsalus IC (1972) J Biol Chem 247:5777–5783
Van der Hoeven TA, Haugen DA, Coon MJ (1974) Biochem Biophys Res Commun 60:569–575
Waterman MR, Simpson ER (1985) Mol Cell Endocrinol 39:81–89
Waxman DJ (1984) J Biol Chem 259:15481–15490
Wolf CR, Smith BR, Ball LM, Serabjit-Singh CJ, Bend JR, Philpot RM (1979) J Biol Chem 254:3658–3663
Wrighton SA, Maurel P, Schuetz EG, Watkins PB, Young B, Guzelian PS (1985) Biochemistry 24:2171–2178
Yabusaki Y, Shimizu M, Murakami H, Nakamura K, Oeda K, Ohkawa H (1984) Nucleic Acids Res 12:2929–2938

Yates FE, Herbst AL, Urquhart J (1958) Endocrinology 63:887–902
Yuan PM, Ryan DE, Levin W, Shively JE (1983) Proc Natl Acad Sci USA 80:1169–1173

Discussion of Paper Presented by J.Å. Gustafsson

MUELLER: Have you ever performed any experiments where you attempt to mimic this hormonal effect in cell culture?
GUSTAFSSON: This is, of course, the obvious thing to do. We have been working with hepatocytes for several years and they are extremely tricky to work with. We have performed experiments where we have observed effects on the 5α-reductase/16α-hydroxylase ratio following addition of growth hormone to hepatocytes in culture, but the effects are relatively limited, unfortunately. We have also observed effects on prolactin receptors following administration of growth hormone. Prolactin receptors are growth-hormone controlled, and occur at higher concentrations in female as compared with male rats. We are currently attempting to optimize our conditions for studying hormonal control of hepatocytes in culture.
MUELLER: On the other hand, in the intact rat, have you had any experiments where you attempt to look at growth hormone receptors in the two different states of the liver?
GUSTAFSSON: Yes, male and female rats do not differ significantly with regard to hepatic growth hormone receptor concentrations.
ROY: Explanation of the sexual dimorphic pattern of hepatic gene expression based on growth hormone alone seems to be incomplete. As I showed you this morning, androgen does have a direct affect on α_{2u}-globulin synthesis in perfused rat liver. Direct effects of estrogen on the vitellogenin gene expression in the amphibian and avian liver are also well known.
GUSTAFSSON: I agree with you. I think the α_{2u}-globulin story is slightly different from the P-450 15β/16α story. It is true that when reinducing the α_{2u}-globulin in the hypophysectomized rat, it is not enough to give growth hormone alone, you also have to administer a hormone cocktail consisting of thyroxine, glucocorticoid, and androgen, whereas in the case of restoring the P-450 15β/16α ratio, it is enough only to give growth hormone.
TATA: My comment is related to the question raised by Gerry Mueller. About 3 years ago, in order to optimize the effect of estrogen on male hepatocytes, we studied the metabolism of various estrogens by male and female hepatocytes. In primary cell culture, you do get a partial feminization with estrogen of the pattern of metabolism of different estrogens. In fact, it was very important that in order to optimize the induction of vitellogenin genes in male cells, we needed to deliver estrogen in a certain pattern. Instead of a mini pump what I had was a very kind graduate student in a sleeping bag and he added pulses of estrogen every 2 h in order to show that the gene was operating at a maximum level in response to estrogen. But it is also true that you can produce a partial feminization with just estrogen added to male hepatocytes.
SCHRADER: The down-regulation of growth hormone receptors in the liver has actually been studied and the time it takes for those liver cells to recover growth hormone receptors has, I believe, been published. I can't think at the moment who did it, so it might be interesting to see whether the timing of your pulses is

related to the timing at which the growth hormone receptors are down-regulated. I think you already said that at you had not looked at the growth hormone receptors in response to Gerry Mueller's question. My question is, do you know whether it is the same region of the growth hormone polypeptide that is required for this activity as is required for growth hormone activity? Have you ever used defined fragments of growth hormone in either injections or infusions?

GUSTAFSSON: Regarding the first comment, growth hormone seems to up-regulate its own receptor. With regard to the second question, yes we have tested the 20K fragment of growth hormone and it is both somatogenic and has a feminizing effect. Indeed, we have studied several lactogens and somatogens from various species, and we have found that there is always a good, positive correlation between somatogenic and feminizing effects.

CLARK: That was very nicely presented. Let me make sure this is not the way you tell males from females in Sweden, is it? You know Mother Nature usually says that the fetus will grow up to be a female unless it gets an androgen signal, which really is a signal that says well, this time around, don't make a female. Its essentially turning on the masculine situation, keeping the animal from being a female. Your hypothesis seems to be the reverse of that, i.e., androgen instructing the brain now to instruct the liver to be male, maybe it isn't the reverse of that.

GUSTAFSSON: Neonatal imprinting of the brain probably includes effects of androgens on the somatostatin/GRF systems resulting in a male secretory pattern of growth hormone leading to a masculine type of liver steroid metabolism. With regard to your first comment, the reason why we have such beautiful women in Sweden is that they have an osmotic minipump containing GH operated at birth.

BAXTER: Talking about women in Sweden, to what extent do you think this is all true in humans?

GUSTAFSSON: Well that of course is a very interesting question. Indeed, there exist sex differences in the growth hormone secretory pattern in humans. There may actually be other examples where the effects of sex steroids are mediated by growth hormone. For instance, it is possible that anabolic steroids act indirectly on skeletal muscle by affecting the growth hormone secretory pattern.

BAXTER: What about the 15 hydroxylations, do they occur in man?

GUSTAFSSON: 15β-hydroxylation of the type characteristic of the rat does not seem to occur in the human. On the other hand, the human fetus has the capacity to 15-hydroxylate steroids.

Discussants: J. BAXTER, J. CLARK, J.Å. GUSTAFSSON, G. MUELLER, A.K. ROY, W. SCHRADER, and J.R. TATA

Chapter 16

Interaction of Thyroid Hormone and Carbohydrates on Hepatic Gene Expression

C.N. MARIASH, W.B. KINLAW, H.L. SCHWARTZ,
H. FREAKE, AND J.H. OPPENHEIMER

Introduction

Although 14 years have elapsed since the initial description of the triiodothyronine (T3)-nuciear receptor (Oppenheimer et al., 1972), the precise mechanisms by which T3 regulates hepatic gene expression remain unknown. Nevertheless, developments in cell and molecular biology have led to significant advancements in our understanding of hormonal regulation of hepatic gene expression. Studies on the regulation of hepatic lipogenic enzymes have provided a particularly fruitful model for studying the physiology of thyroid hormone action at a molecular level. Since these enzymes are also regulated by alterations in carbohydrate feeding (Fitch and Chaikoff, 1960; Tepperman and Tepperman, 1964; Gibson et al., 1972), they provide the opportunity to study the interaction of multiple hormones on the regulation of a common group of genes.

In this chapter we provide evidence from our laboratory indicating that thyroid hormones and a product of carbohydrate metabolism interact synergistically to regulate the level of the lipogenic enzymes and their respective mRNAs. Such studies have been aided by the development of two-dimensional gel electrophoresis (O'Farrell, 1975), in vitro translational assays (Pelham and Jackson, 1975), and computerized videodensitometry (Mariash et al., 1982). The development of cDNA technology has allowed us to examine the multifactorial regulation of one mRNA, mRNA-S14, which represents an ideal model for the hepatic lipogenic enzymes.

Initiation of Thyroid Hormone Action

The evidence supporting the T3-nuclear receptor as the site of initiation of thyroid hormone action has been the subject of a number of recent reviews (Oppenheimer, 1979; Samuels, 1983). The receptor appears to be an acidic protein of Mr-55,000, found only in the nucleus of the cell in both occupied and unoccupied states, which binds T3 with single-site kinetics. Based on multiple analogue studies, it appears that hormone action is initiated by any analogue that binds to the receptor. At present there are no known antagonists to thyroid hormone. This finding stands in contrast to that for the steroid receptors, where relative analogue binding is not necessarily reflected in relative potency of response.

On the other hand, recent data have shed some light on the differential regulation of tissue responses to thyroid hormone analogues. It has been known for a number of years that the dextro-enantiomer of thyroid hormone (D-T3) has about 15% (Cuthbertson et al 1960; Boyd and Oliver 1960) of the activity exhibited by the levo-enantiomer (L-T3). However, the two enantiomers bind to the T3-nuclear receptor with nearly the same affinity and binding capacity (Fig. 1) (Schwartz et al., 1983), whether whole nuclear assays or salt-extracted receptor assays are used. The discrepancy between isolated receptor binding and in vivo activity is explained if one measures the in vivo occupancy of the nuclear receptor by D-T3 or L-T3. For any given molar concentration of in vivo administered T3, the integrated concentration of D-T3 on the nuclear receptor is about 15% of that of L-T3 (Schwartz et al., 1983) (Fig. 2).

There appeared to be two possible mechanisms for the diminished in vivo binding of D-T3 to the T3-nuclear receptor. The first possibility, enhanced metabolism of D-T3, did not appear to be the cause. This left the possibility that diminished transport of D-T3 into the nucleus accounted for the relative differences in in vivo binding. This result was surprising because, until recently (Rao et al., 1976; Krenning et al., 1981; Rao et al., 1981), most evidence indicated that there was little to no active transport of T_3 in hepatic tissue. That is, movement of T3 from the plasma into the cell and subsequently into the nucleus occurred by diffusional processes only (Lein and Dowben, 1961; Hillier, 1969; Surks et al., 1975).

To test this hypothesis, measurements of free hormone concentrations were made in three compartments: rat plasma, hepatic cytosol, and hepatic nuclei (Oppenheimer and Schwartz, 1985). Plasma-free hormone measurements were made by standard equilibrium dialysis techniques. Cytosolic-free hormonal concentrations were also determined by equilibrium dialysis and tracer T3 kinetics after correction for appropriate cytosolic dilution and extrahepatic contamination of tissue homogenates. The free-hormone concentration in the nucleus was calculated on the basis of the law of mass action, given the known in vitro equilibrium association constant. That is, if the in vivo occupancy of the nuclear receptor is known, simple substitution into the following equation will yield the intranuclear free hormone concentration.

$$K_a = \frac{[T3N]}{[M - T3N][T3]}$$

where K_a is the equilibrium association constant, [T3N] is the concen tration of bound nuclear receptors, M is the total concentration of nuclear

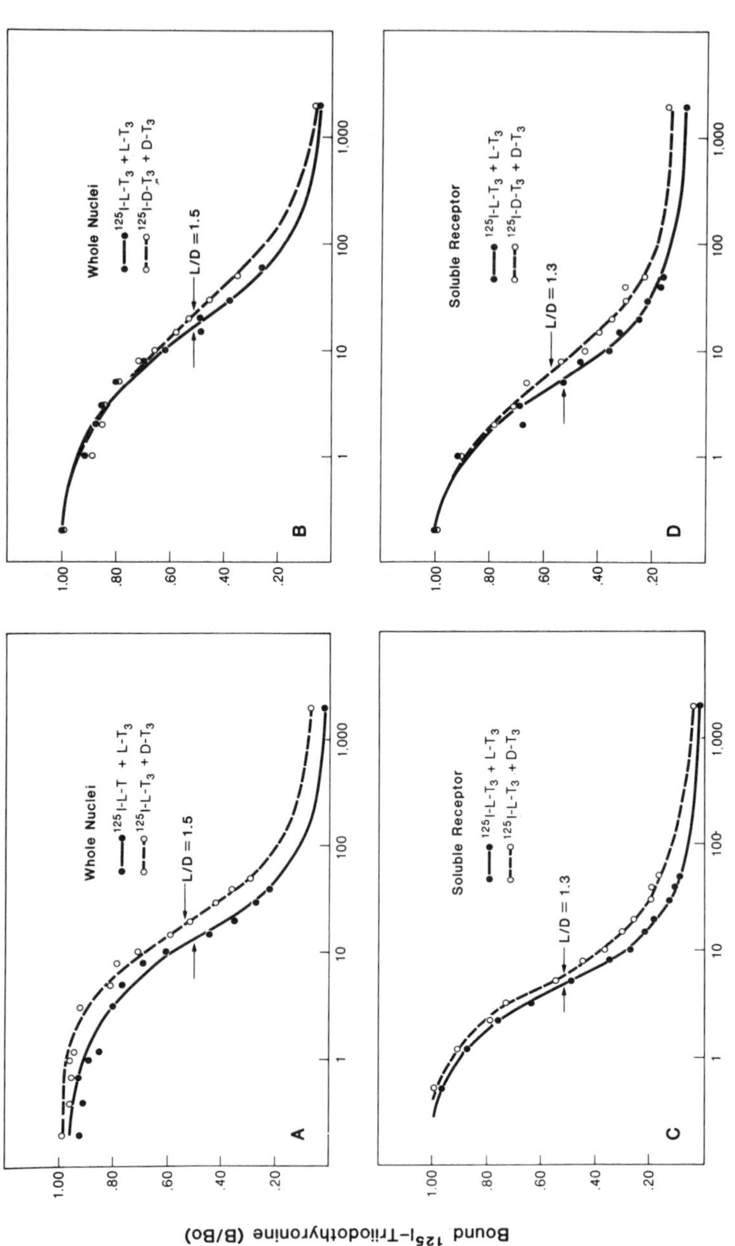

Fig. 1. Relative affinity of binding of $[^{125}I]$L- and D-T3 to the T3-nuclear receptor. Panels **A** and **C** represent displacement of $[^{125}I]$L-T3 by unlabeled L-T3 or D-T3. Panels **B** and **D** represent displacement of $[^{125}I]$L-T3 by unlabeled L-T3 or $[^{125}I]$D-T3 by unlabeled D-T3. *Arrows* indicate the ED_{50} for each curve. (From Schwartz et al., 1983, © The Endocrine Society)

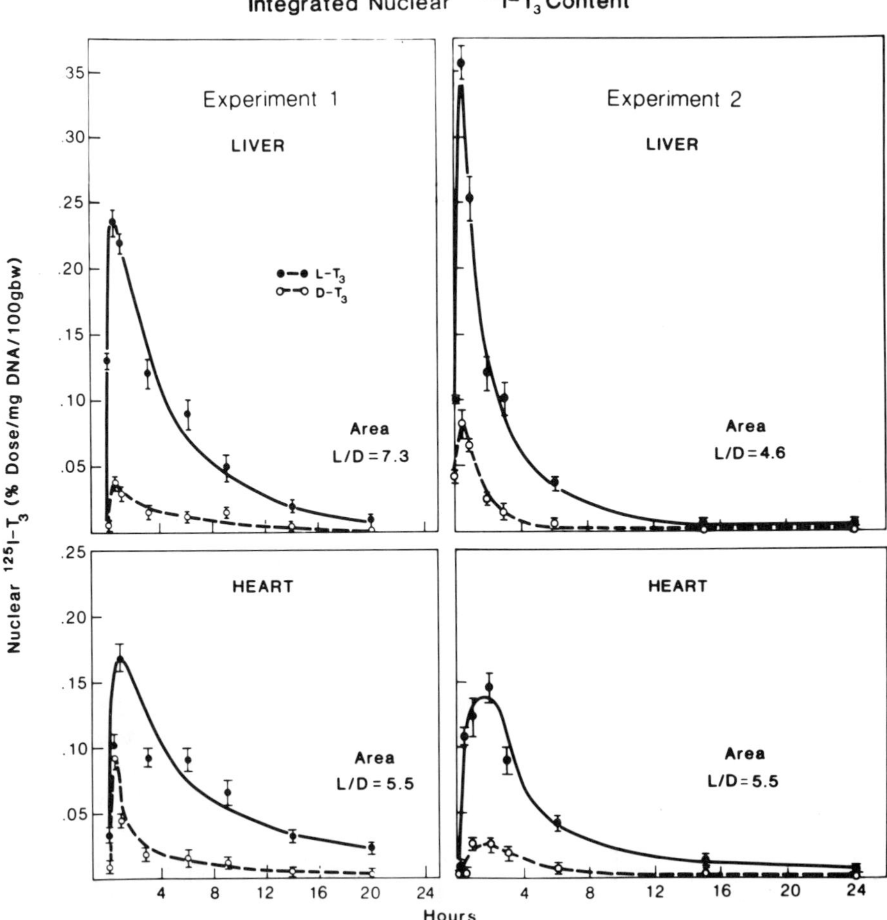

Fig. 2. Nuclear concentration of [^{125}I]L-T3 or [^{125}I]D-T3 as a function of time. Tracer doses of the labeled T3 were injected intravenously at 0 time and animals were killed at the indicated times. Nuclei were prepared from the liver and heart and the concentration of receptor expressed as a percentage of the injected dose. The data from two separate experiments are shown. The area under the curve was calcuated by extrapolation of the terminal slope to t = ∞. (From Schwartz et al., 1983, © The Endocrine Society)

receptors, and T3 is the concentration of nuclear free T3. Since the K_a, M, and T3N are known (Oppenheimer et al., 1974), the nuclear free T_3 concentration can readily be calculated. When free-hormone measurements were made in plasma and in hepatic cytosol, and compared with the calculated free-hormone concentration in the nucleus of the cell, a substantial gradient was found between the cytosol and the nucleus for L-T3 compared with D-T3 (Table 1). The nuclear/cytosolic gradient stands

Table 1. Free-Hormone Concentration Gradients in Various Tissues

Tissue	Fc/Fp L	Fc/Fp D	Fn/Fc L	Fn/Fc D
Liver	2.8	21.6	58.2	3.7
Kidney	1.2	63.3	55.9	1.5
Heart	1.5	2.1	80.6	24.9

Note: Fc = free T3 in cytosolic, Fp = free T3 in plasma, Fn = free T3 in nucleus. L and D refer to L-T3 and D-T3, respectively.
Source: Data reproduced from Oppenheimer and Schwartz, 1985.

in contrast to the well-known ability of the liver and kidney cytosol to concentrate the D-enantiomer from the plasma. These data suggest that there is an "active" transport system that is both stereospecific and tissue specific for the movement of T3 into the nuclear compartment.

This transport system was investigated in two different cell systems, the acutely isolated hepatocyte and the GH1 pituitary tumor cell line. Although there was a significant gradient of free hormone from cytosol to nucleus, the magnitude of this gradient was considerably less than exists in the liver in situ (Mooradian et al., 1985). We found that the free-hormone gradient from cytosol to nucleus was one seventh that found in the intact liver. Furthermore, there appears to be a partial loss of the stereospecific nature of the transport. The reason for the diminished nuclear uptake was not determined in these studies, but did not appear to be due to a loss of cellular energy or transient damage to plasma membranes by the isolation process, because similar findings were obtained in 24-h cultured cells. Because the acutely isolated cells exhibit less nuclear transport, we were hampered in pursuing further the mechanism of the stereospecific transport of thyroid hormones into the nucleus of liver cells.

In a manner analogous to the isolated hepatocyte, the GH cell line also displayed a diminished nuclear/cytosolic free-hormone gradient compared with the intact liver (Freake et al., 1986). Although based on examination of only two cell types, these data raise the possibility that removal of cells from a normal environment is associated with a loss of the nuclear transport function.

Our studies indicate that the movement of T3 from plasma into cytosol, and ultimately into the nucleus, is highly stereospecific. Further, this process appears to have the characteristics of an active transport mechanism since large concentration gradients are established between the plasma compartment and the cell nucleus, with the greatest step-up occurring between the cytosol and the nucleus. The mechanisms and further physiologic significance of the T3-nuclear transport system remain to be determined. It is interesting to speculate on the possibility that such a transport system may also be present for other hormones.

Interaction of T_3 with Dietary Factors

Many of the studies that were used to identify the site of action of T3 involved measurements of the hepatic enzyme, malic enzyme. When the enzyme decarboxylates malic acid, nicotinamide adenine dinucleotide phosphate (NADP) is reduced to nicotinamide adenine dinucleotide phosphate, reduced (NADPH), placing malic enzyme in the class of enzymes involved in the regulation of long-chain fatty acid synthesis. One of the characteristics of this class of enzymes is the coordinate regulation by dietary factors (Gibson et al., 1972). Thus, starvation is associated with a decrease in enzyme activity and mass, and carbohydrate feeding is associated with an increase in both enzyme activity and mass. Furthermore, Gibson and colleagues established that for several of these enzymes, the dietary alterations led to changes in enzyme synthesis rather than alterations in enzyme turnover.

To gain further insight into the mechanism of action of thyroid hormone, we examined the interaction between thyroidal status and carbohydrate status (Mariash et al., 1980) on four of the lipogenic enzymes (Fig. 3). For these studies, rats were rendered maximally hypothyroid by the combined treatment of surgical thyroidectomy followed by ^{131}I ablation. After such treatment, the level of the four lipogenic enzymes studied were significantly less than the value in euthyroid controls. Moreover, placement of the animals on a high carbohydrate diet produced only minimal increases in the level of these lipogenic enzymes. Thus, hypothyroidism markedly inhibited the normal response to carbohydrate feeding.

Not only does the thyroidal status influence the response to carbohydrate feeding, but the dietary status significantly influences the response characteristics to T3. As depicted by the closed circles in Fig. 3, there is a shift in the dose of T3 which yields 50% of the maximal response (ED_{50}). Carbohydrate feeding lowered the ED_{50} for T3 by nearly one order of magnitude for all four enzymes studied.

Furthermore, the rate of response to T3 is significantly influenced by the diet. When placed on a high carbohydrate diet, small doses of T3 led to large changes in the level of the lipogenic enzymes. This interaction is best depicted for glucose-6-phosphate dehydrogenase. While on a regular chow diet, small doses of T3 produced minimally detectable increases in this enzyme. On the other hand, feeding a high carbohydrate diet to the rats increased the initial rate of response to T3 by 40-fold. Thus, carbohydrate feeding markedly sensitized the liver to T3 administration. It is important to point out that in our studies carbohydrate feeding did not alter T3 metabolism, T3 nuclear receptor number, or relative affinity of T3 for its receptor.

Another example of the interaction between T3 and carbohydrates can be gleaned from examining the response of malic enzyme to T3 in the developing rat (Forciea et al., 1981). This model has been shown to represent a mild form of carbohydrate resistance, similar to Type II diabetes

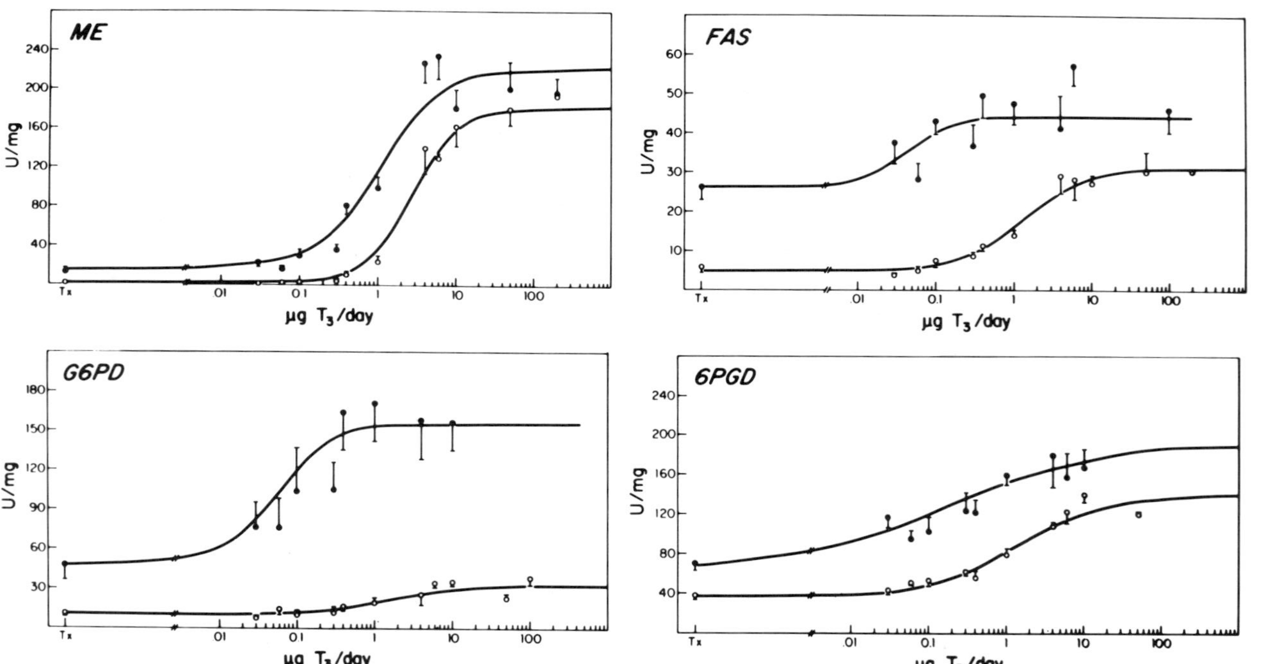

Fig. 3. Dose-response relationship of lipogenic enzymes to T3 on regular and high carbohydrate diets. Rats were given the indicated doses of T3 for 7 days and maintained on either a regular chow diet *(open circles)* or a high carbohydrate diet *(closed circles)*. ME = malic enzyme, FAS = fatty acid synthetase, 6PGD = 6-phosphogluconate dehydrogenase, G6PD = glucose-6-phosphate dehydrogenase. (Reproduced from Mariash et al., The Journal of Clinical Investigation, 1980, by copyright permission of the American Society for Clinical Investigation)

mellitus in man (Bracho-Romero and Reaven, 1977). In this model we found that aging was associated with a progressive decrease in the response to T3 (Table 2). However, the fall in response to T3 was correlated with the fall in baseline malic enzyme activity. Indeed, as shown in Table 2, T3 administration was associated with an approximately 10-fold increase in malic enzyme activity over the baseline at all ages examined. This finding is consistent with the hypothesis that T3 acts as a multiplier of a primary carbohydrate signal. Since the aging rat expresses a relative degree of carbohydrate resistance, the fall in basal malic enzyme activity is due to his carbohydrate resistance. As previously noted in the carbohydrate studies, there were no changes in the level of T3 or in any of the measured parameters associated with the T3 receptor in the aging rat.

Because alterations in diet and T3 administration can lead to changes in a large number of hormones and metabolites, it became necessary to establish an isolated hepatocyte culture system to determine the nature of the factors responsible for the interaction of T3 with carbohydrates. For these studies, we used the technique of culturing the hepatocytes on collagen gels supported by a nylon mesh as described by Sirica et al., (1979). These culture conditions permit the cells to remain viable and responsive to hormonal manipulations for at least one week.

Our initial studies indicated that the dietary carbohydrate factor, which leads to the induction of malic enzyme, was related to the plasma glucose concentration (Mariash et al., 1981). We found that by simply increasing the concentration of glucose in the medium within the physiologic range of portal glucose, a progressive increase in the level of malic enzyme was observed (Fig. 4). Since T3 was absent from the medium, and concentration levels of insulin and dexamethasone were constant, we concluded that the ability of carbohydrate to increase malic enzyme is related to the amount of glucose presented to the hepatic cell and not to alterations in insulin or other fuel related hormones.

We also used the hepatocyte culture to confirm the in vivo finding that

Table 2. Effect of Age and Diet on ME Response to T3

	Regular diet			High CHO diet		
Age (mo)	Basal	+T3	Ratio	Basal	+T3	Ratio
1.0	669	6284	9.4	6232	9051	1.5
1.5	482	3866	8.0	4432	7059	1.6
6.0	215	3179	14.8	2427	3811	1.6
12.0	262	3375	12.9	782	2947	3.8
18.0	251	2577	10.3	1053	2695	2.6

Note: Malic enzyme (ME) is expressed in units/mg DNA. CHO represents a 60% sucrose fat-free diet. Rats were given T3 (15 μg/100-g body weight) for 7 days prior to killing.

Source: Adapted from Forciea et al., The Journal of Clinical Investigation, 1981, by copyright permission of the American Society for Clinical Investigation.

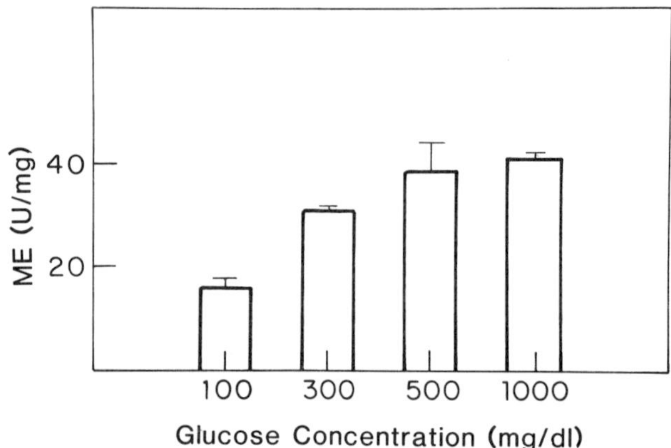

Fig. 4. Induction of malic enzyme by glucose. Hepatocytes were maintained for 5 days in culture with medium supplemented with 10% T3-free calf-serum. Initial concentrations of glucose are indicated. The medium was changed every 48 h. Enzyme activity is expressed in U/mg protein. (Reproduced from Mariash et al., The Journal of Clinical Investigation, 1981, by copyright permission of the American Society for Clinical Investigation)

T3 acts to multiply a primary carbohydrate signal. In the presence of less than maximally effective concentrations of thyroid hormone, T3 led to a constant 1.5-fold increase in the level of malic enzyme over the basal level at all glucose concentrations tested (Table 3). Furthermore, we found in these studies that T3 did not alter the net amount of glucose utilized, nor did the glucose concentration affect the kinetics of T3 metabolism or binding to the receptor. Therefore, the interaction between T3 and carbohydrates appears to occur directly within the liver, does not require changes in an extrahepatic hormone or metabolite, and is not a consequence of a thyroid hormone-induced alteration in the net metabolism of glucose.

We next turned our attention to localizing the intracellular site of the carbohydrate signal responsible for induction of malic enzyme. We made use of the paradigm that insulin stimulates glucose utilization through enhanced glycolysis, and glucagon inhibits glucose utilization by stimulating gluconeogensis. Although glucagon and insulin regulate cellular processes other than glycolysis, it appeared that the major actions of the two hormones on glycolysis should be useful initially in localizing the intracellular carbohydrate regulatory signal. We postulated that if a specific glycolytic intermediate were responsible for the carbohydrate induction of malic enzyme, then addition of this intermediate to the medium in the absence of insulin should lead to induction of malic enzyme. Furthermore, addition of any product in the glycolytic pathway distal to this intermediate should be ineffective either in the presence or absence of insulin in the culture. On the other hand, since glucagon leads to enhanced gluconeogenesis from

Table 3. Effect of T3 on Malic Enzyme in Hepatocyte Cultures[a]

Glucose concentration (m*M*)	Malic enzyme (U/mg protein)		
	0.2 n*M* T3	10 n*M* T3	Fold response
5.5	21	34	1.62
16.5	33	51	1.55
27.5	47	64	1.36
55.0	49	79	1.61

[a]Hepatocytes were cultured for 5 days in 10% calf serum either without T3 supplementation (endogenous T3 concentration is 0.2 n*M*) or supplemented with T3 to a concentration of 10 n*M*. This concentration was calculated to occupy 50% of the T3-nuclear receptor during the culture period.
Source: Adapted from Mariash et al., The Journal of Clinical Investigation, 1981, by copyright permission of the American Society for Clinical Investigation.

the products of glycolysis, addition of such a distal ineffective intermediate might stimulate the formation of the glycolytic intermediate responsible for malic enzyme induction if glucagon is present.

When insulin was present in the medium, we found only those sugars that enter the glycolytic pathway distal to glucose-6-phosphate induced malic enzyme (Mariash and Oppenheimer, 1984). Thus, sugars that do not enter the cell such as sucrose, and sugars that enter the cell but are not readily metabolized such as 3-O-methylglucose and 2-deoxyglucose, did not induce malic enzyme. Furthermore, any glycolytic metabolite up to, and including, pyruvate also led to an increase in malic enzyme. Interestingly, acetate, which readily enters the hepatocyte and is capable of undergoing further metabolic alterations (Dietschy and McGarry, 1974; Beynen et al., 1979), did not produce any significant increase in malic enzyme. On the basis of these data it appeared possible that the generation of pyruvate from glycolysis led to the induction of malic enzyme. If this were true, then addition of pyruvate in the presence of glucagon should stimulate the formation of malic enzyme.

However, we found that in the presence of glucagon, or in the absence of insulin, none of the glycolytic compounds we added was capable of leading to malic enzyme induction. If pyruvate were the inducer of malic enzyme, then it should have led to an increase in the enzyme activity both in the presence of glucagon and in the absence of insulin. Similarly, if any of the phosphorylated glycolytic intermediates, such as phosphoenolpyruvate or glycerophosphate, were the proximate inducer of malic enzyme, then addition of pyruvate to cultures that contained glucagon should lead to enhanced malic enzyme activity. As discussed above, this hypothesis is based on the assumption that glucagon stimulates the formation of these compounds from pyruvate. Since not one of the compounds added was able to overcome the block induced by either glucagon or the absence of insulin, we concluded that the signal responsible for malic enzyme induction was derived somewhere between the formation of pyruvate and the production of acetate. Moreover, this signal was sensitive to the presence of insulin and inhibited by glucagon. The metabolic pathway that fulfills these requirements is the insulin-sensitive mitochondrial enzyme pyruvate dehydrogenase (Denton and Hughes, 1978).

We elected to test this hypothesis further by using the chlorinated acetic acid derivative, dichloracetic acid (DCA). DCA activates the pyruvate dehydrogenase enzyme complex (PDH) by inhibiting the enzyme which phosphorylates PDH (Crabb et al., 1981). The phosphorylated enzyme complex is relatively inactive, and the dephosphorylated state is much more active. DCA can activate PDH even if insulin is absent. We found that DCA supplementation of normal rat chow led to a significant increase in the level of malic enzyme (Mariash and Schwartz, 1986), confirming earlier studies (Anderson et al., 1975). More importantly, we found that with the addition of DCA to hepatocyte culture media that did not contain any insulin, or in the presence of maximally effective amounts of glucagon, we were able to reverse the effect of insulin deficiency or glucagon excess (Fig. 5). Although it is possible that DCA has alternative mechanisms of action, these data are most compatible with our hypothesis that metabolism of pyruvate generates the signal responsible for malic enzyme induction. It is interesting to speculate that this signal may be related to the generation

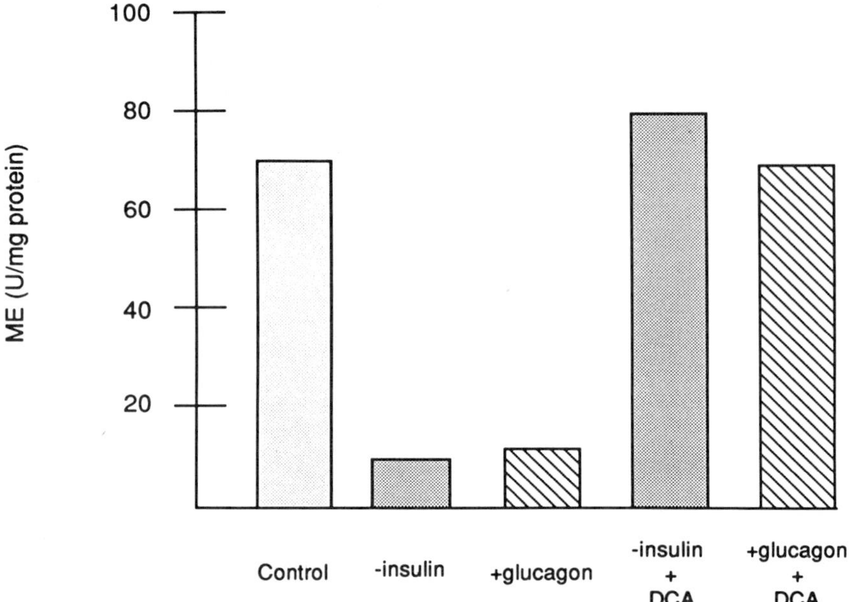

Fig. 5. Dichloroacetic acid (DCA) reversal of insulin deficiency or glucagon excess. Control cultures contained 100 mU/ml insulin. When indicated, glucagon was present at a concentration of 100 n*M*. All cultures contained 27.5 m*M* glucose. DCA was added at a concentration of 10 m*M*. (Adapted from Mariash and Oppenheimer, 1984)

of high-energy compounds such as adenosine 5′-triphosphate (ATP) or NADPH, which ultimately are the inducers of malic enzyme mRNA and perhaps the other carbohydrate responsive mRNAs.

Evaluation of mRNA Activity Profiles by Two-Dimensional Gel Electrophoresis

To gain further understanding of the mechanism of thyroid hormone action and its interaction with carbohydrates and other dietary factors, we made use of the techniques of in vitro translational assays (Pelham and Jackson, 1975) and two-dimensional gel electrophoresis (O'Farrell, 1975). In previous studies we used the rabbit reticulocyte lysate translational assay to measure the relative amount of the hepatic mRNA for malic enzyme (Towle et al., 1980). These studies demonstrated that the previously reported T3 and carbohydrate-induced increase in malic enzyme activity is proportional to the level of translatable mRNA for malic enzymes, similar findings have been observed by Goodridge and colleagues in birds (Siddiqui et al., 1981). Furthermore, as previously discussed, we and others have established that both T3 and carbohydrate feeding coordinately increase the activities, mass, and rate of synthesis of several of the lipogenic enzymes. Therefore, it became important to define the extent of genomic changes that occurred following application of either of these stimuli. Such a question can be approached by an analysis of the pattern of mRNAs translated in vitro by 2-D gel electrophoresis. After in vivo physiological manipulations, hepatic RNA is extracted and translated in the rabbit reticulocyte assay using ^{35}S-methionine to label the synthesized proteins, which are subsequently separated by 2-D gel electrophoresis. After drying, the gel is exposed to x-ray film to produce an autoradiogram. In these studies the assumption is made that the intensity of each of the exposed spots is directly related to the relative amount of the specific mRNA added to the translational assay. With only rare exceptions, this assumption has remained valid (Shull and Theil, 1982; Sonenshein and Brawerman, 1976; Asselbergs et al., 1980; McGuire et al., 1985).

We initially used this technique to show that alterations of thyroidal status lead to changes in approximately 19 of the 250 visible spots (Fig. 6) (Seelig et al., 1981). Both increases and decreases were noted in the transitions from hypothyroid to euthyroid and euthyroid to hyperthyroid states. Thus, nearly 10% of the highly abundant mRNAs are altered by thyroid hormone.

Further studies indicated that a large overlap exists between the mRNAs altered by thyroid hormone and those influenced by carbohydrate feeding (Liaw et al., 1983; Carr et al., 1984). Moreover, we found that nearly one third of all thyroid hormone responsive mRNAs were altered indirectly by a thyroid hormone-induced increase in pituitary growth hormone secretion.

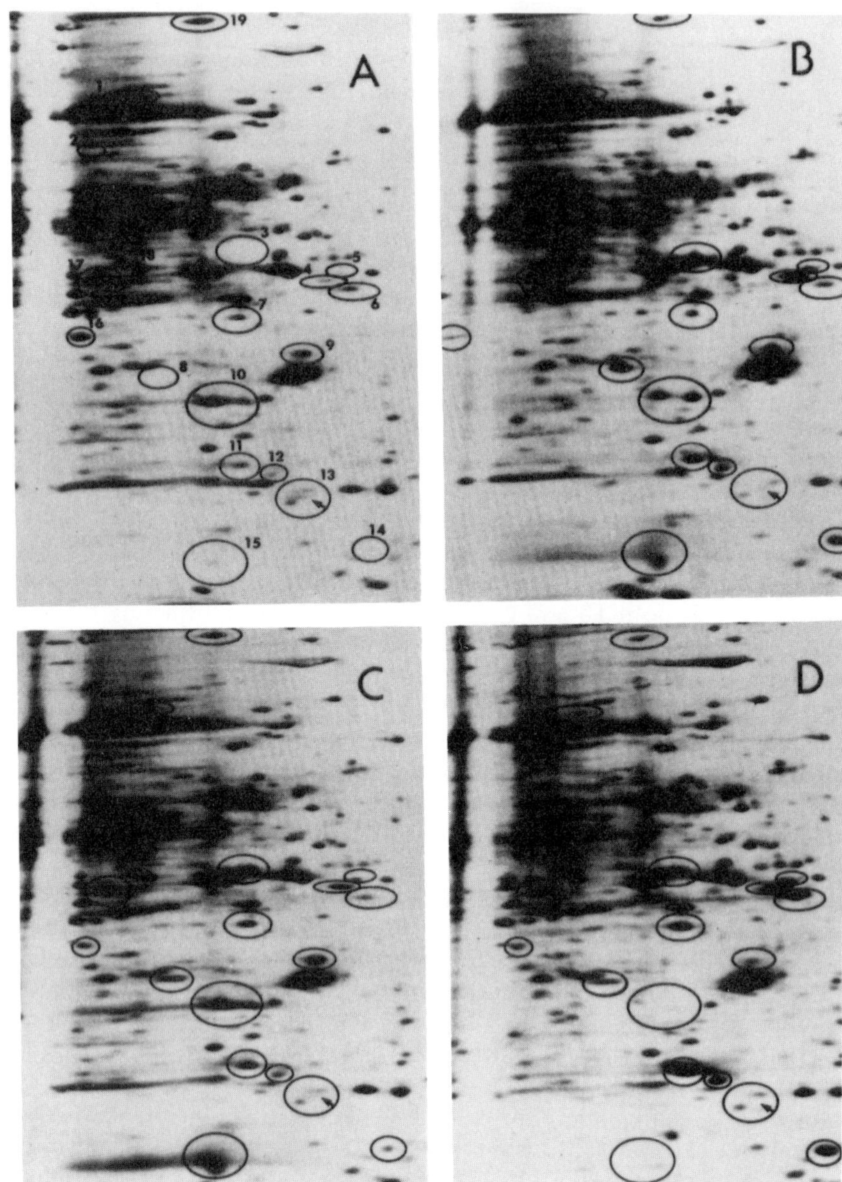

Fig. 6. Two-dimensional fluorogram of [^{35}S]methionine-labeled in vitro translated products of hepatic poly(A+)RNA obtained from thyroidectomized rats (**A**), euthyroid rats (**B**), thyroidectomized rats treated with 0.3 μg/100 g T3 for 7 days (**C**), and thyroidectomized rats treated with 15 μg/100g T3 for 7 days (**D**). The isoelectric focusing range was from pH 7.9 (left) to 4.5 (right). The sodium dodecyl sulfate (SDS) mobility was from 100,000 daltons (top) to 12,000 daltons (bottom).The *circled spots* are those that were found to be altered by changes in thyroidal status. (From Seelig et al., 1981)

The mRNA changes induced both by growth hormone and carbohydrate feeding indicated that many of the responses that were initially thought to be a specific response to thyroid hormone were actually influenced by several stimuli. It became apparent that a more detailed analysis that included quantitative measurements of the changing mRNAs should yield more information about the relationships between many of the pathophysiologic states we have studied. Therefore, we developed the technique of analyzing the autoradiograms by computerized videodensitometry (Mariash et al., 1982). With this development we were able to quantitate all the spots present on a given autoradiogram in under 1 h.

The ability to quantitate all the spots on the 2-D gels gives us the opportunity to analyze a large mass of data. This task is most efficiently handled by the statistical techniques of multivariate analysis (Carr et al., 1984). These statistical analyses are available at most university computer centers in either the SPSS or BMDP package. Each gel consists of an individual case, and each case consists of up to 250 variables (one for each spot, and one to represent the physiological state of the animal). We found that a number of the spots responded similarly in all states examined. The clustering of these spots is analogous to the previously reported coordinate regulation of the hepatic lipogenic enzymes to starvation and carbohydrate feeding.

Of the 250 spots available for analysis, measurements of only 10 were required to statistically describe the gels. We analyzed 50 gels with at least two gels from each of 10 different states, including hypothyroid, euthyroid, and hyperthyroid animals, starved and carbohydrate fed, and diabetic rats. Multivariate analysis allowed transformation of the spot data into a new set of orthoganol variables, canonical variables. We found that three canonical variables separated all the states examined. Each state could then be represented in a three-dimensional space (Fig. 7), each dimension representing one of the canonical variables.

Several interesting relationships are apparent from an examination of Fig. 7. The relatedness of each state can be estimated by the proximity to the other states. For example, the hyperthyroid state and the high carbohydrate states are quite close together in this three-dimensional graph. Likewise, the starved and diabetic state are also close to each other. Such findings are not surprising since they conform to known physiologic relationships. Thus, these relationships provide further validation of this statistical modeling technique. More importantly, we have been able to use quantitative analysis of in vitro translated products separated by 2-D gel electrophoresis to classify the pattern of changes in specific mRNAs. The statistical analysis of the 2-D gels shows that the mRNA profiles of every physiologic condition examined is distinct and allows numeric comparison and assignment of the different states. These conclusions could not have been reached if one were able to quantitate only a few specific mRNAs.

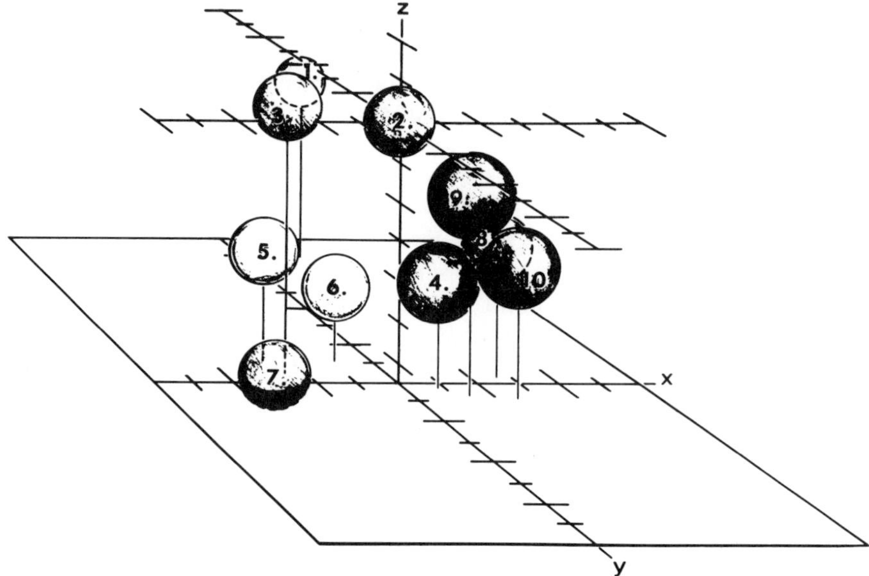

Fig. 7. Three-dimensional representation of 10 physiologic states in terms of the first three canonical variables. The axes were normalized to the mean of the euthyroid state (0,0,0). State identification is: 1-thyroidectomized; 2-euthyroid; 3-fasting; 4-fasting + T3; 5-mild diabetes; 6-mild diabetes + insulin; 7-sever diabetes; 8-high carbohydrate fed; 9-euthyroid + T3; 10-hyperthyroid. (From Carr et al., 1984)

Spot 14 as a Model of the Lipogenic Enzymes

We have also used the technique of 2-D gel electrophoresis of in vitro translated products to characterize the time course of mRNA responses to thyroid hormone administration. As described earlier, we found that several of the hepatic mRNAs respond to T3 indirectly, via a T3-induced increase in pituitary growth hormone secretion. This was suggested from the hepatic mRNA responses to growth hormone administration, as well as the lag time of response after an injection of T3 to hypothyroid rats (Liaw et al., 1983). In the course of these experiments we also found an extremely rapidly responsive mRNA, arbitrarily labeled mRNA-S14. This mRNA was induced 15- to 20-fold by T3 administration, and by 4 h reached nearly 80% of its maximum level (Seelig et al., 1982). Thus, spot 14 appeared to be an excellent candidate for a direct and primary response to T3.

Further studies of the physiology of spot 14 were made possible by the cloning of the mRNA for this protein in the laboratory of Dr. Howard C. Towle (Narayan et al., 1984; Liaw and Towle, 1984). They found that the gene for this mRNA exists as a single copy, consists of two exons separated by one intron, that the protein is coded for entirely in the 5′-exon, and that the 3′-exon has two polyadenilation signals giving rise to a 1325 and a 1525 nucleotide mRNA.

The availability of the cDNA clone for spot 14 has allowed us to apply an mRNA-cDNA "dot blot" assay to measure the relative amount of mRNA-S14 under various physiologic conditions. The dot blot assay demonstrated that both T3 and carbohydrate feeding lead to marked increases in the relative content of mRNA-S14 (Fig. 8) (Jump et al., 1984), thus confirming the earlier results obtained by translational assays. More importantly, the increased sensitivity of this assay allowed us to demonstrate that this mRNA responded to T3 in less than 20 min (Jump et al., 1984), while the heterogeneous nuclear RNA precursor responded within 10 min (Narayan et al., 1984). This mRNA is the most rapidly responding hepatic mRNA to thyroid hormone described to date.

Although computer searches of protein and nucleic acid data banks have not yielded any significant homology to known proteins or mRNAs, based

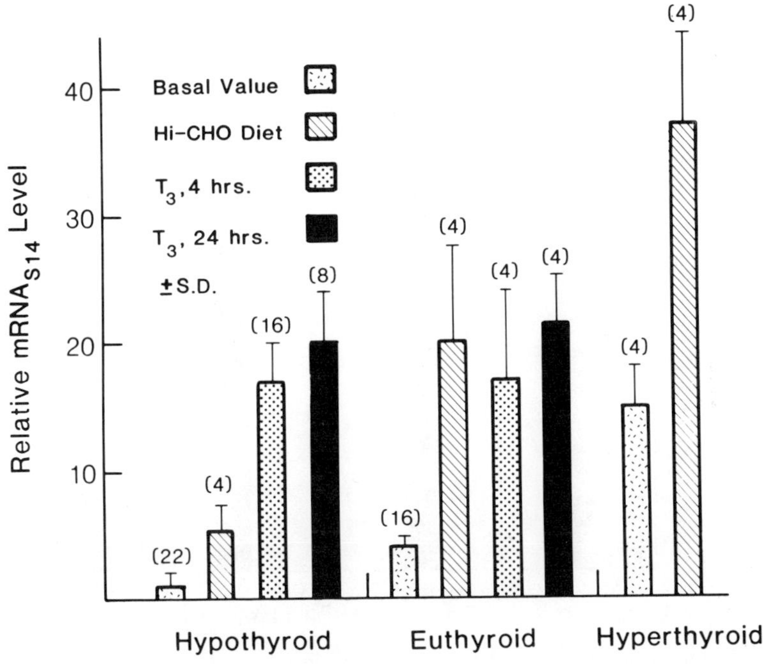

Fig. 8. Relative expression of hepatic mRNA-S14. Hypothyroid, euthyroid, and hyperthyroid (15 μg/100 g body weight for 10 days) were fed either normal chow or high carbohydrate diet for 10 days. The relative abundance of mRNA-S14 was measured by dot-blot assay using either total or poly(A+)RNA and normalized to the thyroidectomized value. Each bar represents the mean ± SD. (From Jump et al., 1984)

on physiologic data we believe this mRNA has some function related to fatty acid synthesis, storage, metabolism, or secretion. Using poly(A+)RNA in the dot-blot assay we found that those tissues actively engaged in lipogenesis for storage and secretion had abundant amounts of this mRNA, whereas nonlipogenic tissue such as kidney, lung, and heart had relatively little mRNA-S14 present (Jump and Oppenheimer, 1985). Furthermore, only the lipogenic tissues responded to T3 with an increase in mRNA-S14 (Fig. 9). The coordinate regulation of mRNA-S14 by diet and hormones, as well as the tissue localization of mRNA-S14, indicates that the protein coded by mRNA-S14 has some function related to lipid production, storage, metabolism, or release.

To obtain a better understanding of the mechanism of regulation of mRNA-S14, we felt it important to determine the kinetics of response of this mRNA to carbohydrate feeding. The intragastric administration of 60% sucrose to euthyroid rats led to a rapid increase in the level of hepatic mRNA-S14. As found following T3 administration to hypothyroid rats, sucrose feeding to euthyroid rats led to a significant increase in mRNA-

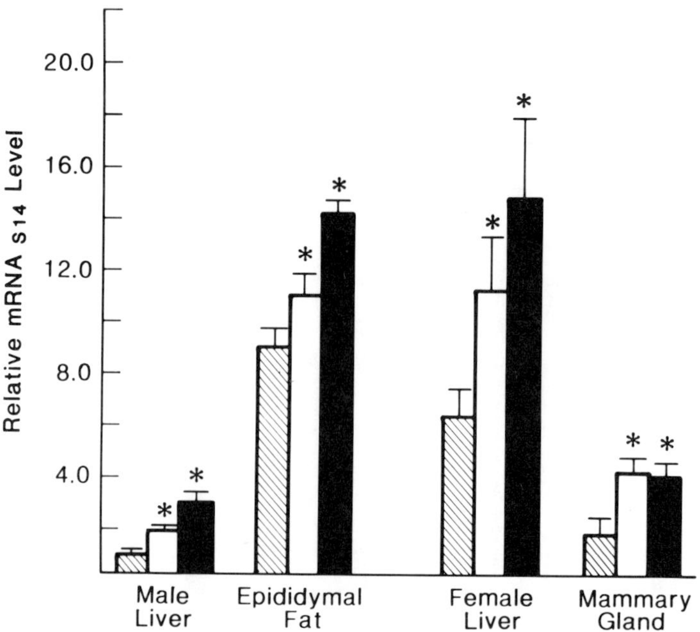

Fig. 9. T3 regulation of mRNA-S14 in liver, epididymal fat, and mammary gland. Male euthyroid rats and lactating female rats were injected with T3 (200 μg/100g body weight) and killed at either 4 h after T3 *(clear bars)* or 24 h after T3 *(black bars)*. The relative abundance of mRNA-S14 was quantitated from total RNA and normalized to the euthyroid male liver value. (From Jump and Oppenheimer, © 1985, The Endocrine Society)

S14 within 30 min. By 4 h, a four- to fivefold increase in the relative level of this mRNA was obtained.

Interestingly, the response of mRNA-S14 in hypothyroid rats to the acute administration of sucrose was markedly diminished. Indeed, in several experiments we were unable to document any statistically significant increase after 4 h. The diminished response to sucrose in hypothyroid rats is analogous to the diminished response of malic enzyme to carbohydrate feeding described earlier. Thus, mRNA-S14 displays a similar synergistic interaction between thyroid hormone and carbohydrates as previously demonstrated for the other lipogenic enzymes.

To determine whether the synergism between T3 and carbohydrates required the intervention of other hormones or metabolites, we examined the response of mRNA-S14 to T3 and glucose in hepatocyte cultures. In the absence of T3 and at low glucose concentrations, the level of mRNA-S14 is only 4.6 ± 1.2% of the level found in the liver of the intact euthyroid control rat (internal standard). Supplementation of the medium with maximally effective concentrations of T3 or glucose yields levels of mRNA-S14 of 11.2% and 11.0% of the euthyroid control value, respectively. However, if the hepatocyte medium is supplemented with both T3 and high glucose, the level of mRNA-S14 reaches 134.8% of the euthyroid control. Thus, the synergism between T3 and carbohydrates occurs directly within the hepatocyte and does not require the alteration in any other hormone or metabolite. However, both T3 and glucose supplementation of hepatocyte media lead to changes in many specific mRNAs (Topliss et al., 1983). Thus, it is possible that the interaction between T3 and carbohydrates does not occur directly and immediately on the production of mRNA-S14, but requires the synthesis of some precursor proteins or mRNAs.

Further evidence supporting the direct and immediate interaction of T3 and carbohydrates on the induction of mRNA-S14 comes from an examination of the time course of response to these two stimuli in vivo. If the interaction between these two stimuli required the synthesis of other proteins, such as carbohydrate-metabolizing enzymes, one would anticipate a lag time of many hours to days before evidence of the interaction could be demonstrated. On the other hand, if the interaction were immediate and did not require the induction of new proteins, there should be no delay in the synergistic interaction between T3 and carbohydrates. To differentiate between these two possibilities, we administered either 400 ng T3 intravenously, or 1 cc of 60% sucrose by gavage, or both, to overnight starved hypothyroid rats. The basal mRNA-S14 level was 3.5% of the internal standard. This value increased to 12.9% and 11.4% of the internal standard after 4 h of either T3 or sucrose. However, if T3 was given with sucrose, the level of mRNA-S14 after 4 h was 74.5% of the internal control, a value indistinguishable from that obtained in euthyroid rats given sucrose.

Thus, the interaction between T3 and carbohydrate does not require alterations in extrahepatic hormones or metabolites, occurs directly within the liver, and appears to be an immediate response. These findings indicate that T3 interacts synergistically (multiplies) with a carbohydrate-generated signal to increase the level of mRNA-S14. The mRNA for spot 14 is an excellent model for the study of the regulation of the lipogenic enzymes by diet and hormones, since its response to multiple physiologic stimuli in vivo and in culture is similar to that observed for malic enzyme and the other lipogenic enzymes. Moreover, this mRNA has the advantage of responding exceedingly rapidly to these stimuli and reaches near steady state levels after just 4 h, presumably because of a relatively short $t_{1/2}$ of the mRNA.

Recent studies have provided further insight into the regulation of mRNA-S14. It had been noted earlier (Jump et al., 1984) that mRNA-S14 underwent diurnal changes. The peak level occurred near 8 PM, and the trough was present at 8 AM. Since rats are nocturnal eaters, and mRNA-S14 is responsive to dietary factors, we proposed that the increase in mRNA-S14 that occurred at night was due to increased eating activity at this time. However, we found that the diurnal changes were entrained primarily to the photoperiod and not to the eating cycle of the rat (Kinlaw et al., 1986). By providing food only in the AM or PM to two groups of rats, they found that the diurnal rhythm persisted, and the peak levels occurred at night whether the rats were fed only in the day or only in the night.

Diurnal factors were also found to have a role in the response of mRNA-S14 to hormonal manipulation. Although we had previously demonstrated that mRNA-S14 levels fell after glucagon administration in hepatocyte cultures, we found that glucagon caused a decrease in mRNA-S14 only when given at night. When a similar dose was administered at 8 AM, there was no further decline in the level of this mRNA. On the other hand, only when given in the morning did T3 lead to an increase in mRNA-S14. If T3 were given at 7 PM, no further increase in the level of this mRNA was noted. This observation accounts for the unexpected fall in the level of mRNA-S14 observed in our earlier time-course studies (Jump et al., 1984). Since all rats were killed between 8 AM and noon, those animals which received their injections of T3 at midnight (8 h prior to killing) had much less of a response than either the 4-h or 16-h injected rats. More importantly, these studies demonstrate that the response of specific mRNAs to hormonal and metabolic stimuli can be modulated by diurnal factors. An understanding of the mechanism responsible for the diurnal regulation of mRNA-S14 should provide further insight into the molecular mechanisms of T3 and carbohydrate regulation of gene expression.

An interesting aspect of mRNA-S14 regulation relates to the molecular mechanisms that lead to enhanced quantities of the mature mRNA. We found that addition of either glucagon or Actinomycin-D to hepatocyte cultures led to similar monoexponential decay rates of mRNA-S14 both in the presence and absence of T3 (Mariash et al., 1984). Therefore, T3 induction of mature mRNA-S14 is due to an increase in the rate of synthesis of this mRNA. On the other hand, nuclear "run-on" assays, as reported by Narayan and Towle (1985), have shown that enhanced transcriptional rates do not account for the major increase in the level of either the nuclear precursor or the mature form of mRNA-S14. They have postulated that T3 increases the formation of mature mRNA-S14 by a selective stabilization of the nuclear precursor for this mRNA. We have obtained further evidence in support of this hypothesis in recent studies in the neonatal rat (Jump et al., 1986). While the content of mRNA-S14 was 1/200th the level of 60-day old rats, the level of gene transcription, as measured by nuclear "run-on" assays, was fully 72% of the 60-day control rats. Thus, not only does an alteration in thyroidal status lead to a discrepency between transcription rates and levels of mature mRNA, but the ontogeny of this mRNA is also associated with major differences between mRNA levels and gene transcription. The mechanism of stabilization of nuclear heterogeneous RNA remains a challenging aspect of hormonal regulation of gene expression.

Concluding Remarks

In this chapter we have emphasized the evidence that thyroid hormone interacts with carbohydrates to regulate the level of several lipogenic enzymes and their respective mRNAs. The mechanism of thyroid hormone regulation of gene expression begins after T3 is transported into the nucleus of the target tissue against an energy gradient and binds to its specific receptor.

The ability of T3 to alter the level of the hepatic lipogenic enzymes and their mRNA is modulated by the availability of readily metabolized carbohydrate. These two stimuli interact synergistically to increase the activity, mass, rate of synthesis, and relative amount of the specific mRNAs. Thus, in the absence of T3, carbohydrate administration is relatively ineffective, whereas carbohydrate feeding markedly sensitizes the hepatic response to T3. T3 acts to multiply a carbohydrate signal that is derived from the oxidation of pyruvate in the mitochondrion.

We have used 2-D gels of in vitro translated products directed by hepatic mRNAs to assess the scope of responses to various hormonal and metabolic signals. Analysis of a large number of mRNAs showed that many of the same mRNAs respond to multiple different stimuli. On the other hand, the overall pattern of mRNA response is unique for every physiologic state we have investigated. This technique allows one to readily classify any physiologic intervention by the pattern of mRNA responses and determine the relatedness of the states by statistical criteria.

The 2-D gel analysis also allowed us to discover a new mRNA, mRNA-S14, which responds rapidly to T3 administration. The generation of a cDNA to this mRNA provided the tool required to gain a better understanding of the molecular mechanisms of T3 action and the interaction of T3 with carbohydrates. We have found this mRNA to be an excellent model for the study of the interaction of thyroid hormone with carbohydrates and other hormonal and metabolic stimuli. Our initial studies on mRNA-S14 demonstrate the complexity by which multiple factors regulate hepatic gene expression. Further studies on mRNA-S14 should yield a greater understanding of the molecular mechanism of thyroid hormone regulation of gene expression.

Acknowledgments. The research for this chapter was supported by National Institutes of Health Grants AM32885 (CNM), AM19812(JHO), Clinical Investigator Award AM01277 (WBK), and Training Grant AM07203 (JHO).

The authors wish to thank the many collaborators and colleagues who have contributed to these studies, especially Drs. Howard C. Towle, Donald B. Jump, Steven Seelig, and Fran E. Carr. The technical assistance of Robert Gunville, Ana Martinez-Tapp, Mary Ellen Domeir, and Deborah Iden is greatfully appreciated. We also wish to thank Patrice Schaus and Kate Steinmeyer for administrative and secretarial support.

References

Anderson JW, Karounos D, Yoneyama T (1975) Proc Soc Exp Biol Med 149:814

Asselbergs FAM, Meulenberg E, van Venrodij WJ, Bloemendal H (1980) Eur J Biochem 109:159

Beynen AC, Vaartjes WJ, Geelen MJH (1979) Diabetes 28:828

Boyd GS, Oliver MF (1960) J Endocrinol 21:25

Bracho-Romero E, Reaven GM (1977) J Am Geriatr Soc 25:299

Carr FE, Bingham C, Oppenheimer JH, Kistner C, Mariash CN (1984) Proc Natl Acad Sci USA 81:974

Crabb DW, Yount EA, Harris RA (1981) Metabolism 30:1024

Cuthbertson WFJ, Elcoate PV, Ireland DM, Mills DCB, Shearley P (1960) J Endocrinol 21:45

Denton RM, Hughes WA (1978) Int J Biochem 9:545

Dietschy JM, McGarry JD (1974) J Biol Chem 249:52

Fitch W, Chaikoff I (1960) J Biol Chem 235:554

Forciea MA, Schwartz HL, Towle HC, Mariash CN, Kaiser FE, Oppenheimer JH (1981) J Clin Invest 67:1739

Freake HC, Mooradian AD, Schwartz HL, Oppenheimer JH (1986) Mol Cell Endocrinol 44:25

Gibson D, Lyons R, Scott D, Muto Y (1972) Adv Enzyme Regul 10:187

Glock G, McLean P (1955) Biochem J 61:390

Hillier AP (1969) J Physiol (Lond) 203:419

Jump DB, Oppenheimer JH (1985) Endocrinology 117:2259

Jump DB, Narayan P, Towle HC, Oppenheimer JH (1984) J Biol Chem 259:2789

Jump DB, Tao TY, Towle HC, Oppenheimer JH (1986) Endocrinology 118:1892

Kinlaw WB, Towle HC, Tao TY, Jump DB, Schwartz HL, Mariash CN, Oppenheimer JH (1986) Proceedings 9th International Thyroid Congress (in press)
Krenning EP, Docter R, Bernard B, Visser T, Henneman G (1981) Biochem Biophys Acta 676:314
Lein A, Dowben RM (1961) Am J Physiol 200:1029
Liaw C, Seelig S, Mariash CN, Oppenheimer JH, Towle HC (1983) Biochemistry 22:213
Liaw CW, Towle HC (1984) J Biol Chem 259:7253
Mariash CN, Oppenheimer JH (1984) Metabolism 33:545
Mariash CN, Schwartz HL (1986) Metabolism 35:452
Mariash CN, Jump DB, Oppenheimer JH (1984) Biochem Biophys Res Commun 123:1122
Mariash CN, Kaiser FE, Schwartz HL, Towle HC, Oppenheimer JH (1980) J Clin Invest 65:1126
Mariash CN, McSwigan CR, Towle HC, Schwartz HL, Oppenheimer JH (1981) J Clin Invest 68:1485
Mariash CN, Seelig S, Oppenheimer JH (1982) Anal Biochem 121:388
McGuire DM, Chan L, Smith LC, Towle HC, Dempsey ME (1985) J Biol Chem 260:5435
Mooradian AD, Schwartz HL, Mariash CN, Oppenheimer JH (1985) Endocrinology 117:2449
Narayan P, Towle HC (1985) Mol Cell Biol 5:2642
Narayan P, Liaw CW, Towle HC (1984) Proc Natl Acad Sci USA 81:4687
O'Farrell PH (1975) J Biol Chem 250:4007
Oppenheimer JH (1979) Science 203:971
Oppenheimer JH, Schwartz HL (1985) J Clin Invest 75:147
Oppenheimer JH, Koerner D, Schwartz HL, Surks MI (1972) J Clin Endocrinol Metab 35:330
Oppenheimer JH, Schwartz HL, Koerner D, Surks MI (1974) J Clin Invest 53:768
Pelham HRB, Jackson RJ (1975) Eur J Biochem 67:247
Rao GS, Eckel J, Rao ML, Breuer H (1976) Biochem Biophys Res Commun 73:98
Rao GS, Rao ML, Thilman A, Quendau HD (1981) Biochem J 198:457
Samuels HH (1983) In: Oppenheimer JH, Samuels HH (eds) Molecular Basis of Thyroid Hormone Action. Academic Press, New York. p. 36
Schwartz HL, Trence D, Oppenheimer JH, Jiang NS, Jump DB (1983) Endocrinology 113:1236
Seelig S, Jump DB, Towle HC, Liaw C, Mariash CN, Schwartz HL, Oppenheimer JH (1982) Endocrinology 110:671
Seelig S, Liaw C, Towle HC, Oppenheimer JH (1981) Proc Natl Acad Sci USA 78:4733
Shull GE, Theil EC (1982) J Biol Chem 257:14187
Siddiqui UA, Goldflam T, Goodridge AG (1981) J Biol Chem 256:4544
Sirica AE, Richards W, Tsukada Y, Sattler CA, Pitot HC (1979) Proc Natl Acad Sci USA 76:283
Sonenshein GE, Brawerman G (1976) Biochemistry 15:5501
Surks MI, Koerner DH, Oppenheimer JH (1975) J Clin Invest 55:50
Tepperman H, Tepperman J (1964) Am J Physiol 206:357
Topliss DJ, Mariash CN, Seelig S, Carr FE, Oppenheimer JH (1983) Endocrinology 112:1868
Towle HC, Mariash CN, Oppenheimer JH (1980) Biochemistry 19:579

Discussion of the Paper Presented by C. Mariash

ROY: You didn't show the effect of dichloroacetic acid on the expression of spot 14 gene?

MARIASH: Yes, I didn't show that. We have given DCA both to the intact animal and to hepatocytes in culture, followed with an examination by two-dimensional gel electrophoresis of a wide spectrum of mRNAs. We found, as one would predict based on the studies I've shown, that DCA administration stimulates virtually all the mRNA changes found if one feeds carbohydrate to the rat, or enhances the culture medium glucose concentration. DCA administration leads to enhanced pyruvate metabolism, which generates the same postulated intermediate as carbohydrate feeding, ultimately inducing spot 14 mRNA as well as all the other carbohydrate-responsive mRNAs. We have not looked at the time course of response to DCA.

RINGOLD: In the experiment that you mentioned on the hypothyroid versus euthyroid rat, you said that the response to sucrose was markedly diminished. In fact, it seems like the basal level is what was diminished. In fact, the responsiveness to sucrose remained the same if one looked at the fold induction. Is my understanding correct?

MARIASH: Yes, you are indeed correct. That slide originally had 4.5-fold written above both hypothyroid and euthyroid responses to sucrose. Thus, in that experiment the fold response in the two thyroidal states was similar.

RINGOLD: So what do you really think is going on?

MARIASH: I think it depends on how one views the mechanism of induction. If you think that a specific factor is necessary to induce the mRNA, then one can say there is a diminished production of that factor in hypothyroidism. With less of this factor present, the absolute increment in spot 14 mRNA will be diminished in hypothyroidism. On the other hand, one can postulate that both T3 and carbohydrate act by directly multiplying the basal level of expression of this gene. When one gives maximally effective amounts of carbohydrate, the basal level of gene expression is multiplied by some unknown mechanism. If one gives T3, again the basal level is multiplied by another unknown mechanism. Both act by multiplying the basal expression of the gene. Together they should multiply each other. At this point, I don't think we have sufficient data to support one hypothesis over the other.

RINGOLD: The second point I just made is that your analysis of spot 14 mRNA content in various tissues suggests that it is involved in fatty acid synthesis, metabolism, or storage seems reasonable. But one thing also to realize is that other tissues that were completely negative or very low, such as lung, which produces tremendous amounts of lipid surfactant, and cardiac muscle, which utilizes fatty acids directly as a source of energy, express many of the same things. Therefore, it may be involved in something other than fat biosynthesis.

MARIASH: Yes, that is possible. However, the fatty acids produced by lung, as you know, are mostly unsaturated fatty acids, whereas the tissues that express spot 14 mRNA are involved in the synthesis of saturated fatty acids. Nevertheless, until we really know the function of the protein, any inference as to the role of mRNA-S14 remains speculative.

BAXTER: Your conclusion is that the free hormone gradient, or the specific uptake, is primarily at the nuclear membrane. How do you place the surface sites characterized most extensively by Ira Pastan's group into this context?

MARIASH: We are not sure that the findings in the GH cell model are relevant to the intact liver. Nevertheless, Dr. Cheng has recently reported at the International Thyroid Meetings that there appears to be a direct transfer of T3 from the plasma membrane into the nucleus without entering the cytoplasm. Therefore, it is entirely possible that in the GH cell or GC cell model the plasma membrane has specific carriers and specific transporters that allow T3 to get from the outside into the nucleus against this gradient. We have also observed a nuclear-free hormone gradient in the GC cell line. On the other hand, we have tried extensively to show a plasma membrane transport process in the intact liver. With the exception of potassium cyanide, we have been unable to abolish the gradient of free hormone from the plasma compartment into the nucleus. Therefore, we are not sure whether a similar phenomenon occurs in the liver, as shown by Cheng and Pastan in the pituitary cell model. It is possible that it does exist, and we are using inappropriate technology to measure intracellular T3.

SARKAR: Have you any information as to whether the stimulation of mRNA S14 that you are getting is a result of increased stability of the mRNA or synthesis at the nuclei? I'm asking this because in an analogous system, the effect of T3 on tubulin induction in brain, where we can get an equally rapid response, we see that approximately half of the effect of the hormone is due to an increased stability of the protein. Do you have any comment on this?

MARIASH: I didn't present any of the data on mRNA transcription, most of which has been worked out by Dr. Howard Towle. His laboratory has shown that the nuclear precursor level of mRNA-S14 rises before the level of mature cytoplasmic mRNA-S14 rises. On the other hand, using the nuclear "run-on" assay, he has demonstrated that the transcription rate for spot 14 mRNA does not change much at all. He finds relatively high transcription rates in the hypothyroid state and hyperthyroid state. These data have been used to postulate that T3 administration leads to a stabilization of the nuclear precursor form of the mRNA. This is consistent with our data demonstrating that thyroid hormone does not change the stability of the mature form of mRNA-S14.

Discussants: J. BAXTER, C. MARIASH, G. RINGOLD, A.K. ROY, and P.K. SARKAR

Index